Springer Series in **Materials Science** 27

Edited by U. Gonser

Springer
Berlin
Heidelberg
New York
Barcelona
Budapest
Hong Kong
London
Milan
Paris
Singapore
Tokyo

Springer Series in *Materials Science*

Advisors: M. S. Dresselhaus · H. Kamimura · K. A. Müller
Editors: U. Gonser · R. M. Osgood Jr. · H. Sakaki
Managing Editor: H. K. V. Lotsch

Volumes 1–25 are listed at the end of the book.

Francisco E. Fujita (Ed.)

Physics of
New Materials

Second, Updated Edition

With Contributions by
R. W. Cahn, F. E. Fujita, H. Fujita, U. Gonser,
J. Kamamori, K. Motizuki, R. W. Siegel, N. Sumida,
K. Suzuki, N. Suzuki

With 200 Figures

 Springer

Professor Francisco E. Fujita

Higashiyama 7.12.701
Ashiya 659, Japan

Series Editors:
Prof. Dr. U. Gonser

Fachbereich 12.1, Gebäude 22/6
Werkstoffwissenschaften
Universität des Saarlandes
D-66041 Saarbrücken, Germany

Prof. R. M. Osgood Jr.

Microelectronics Science Laboratory
Department of Electrical Engineering
Columbia University
Seeley W. Mudd Building
New York, NY 10027, USA

Prof. H. Sakaki

Institute of Industrial Science
University of Tokyo
7-22-1 Roppongi, Minato-ku
Tokyo 106, Japan

Managing Editor:
Dr.-Ing. Helmut K. V. Lotsch

Springer-Verlag, Tiergartenstrasse 17
D-69121 Heidelberg, Germany

Library of Congress Cataloging-in-Publication Data.

Physics of new materials / F. E. Fujita, ed.; with contributions by R. W. Cahn... [et al.]. – 2nd, updated ed. p. cm. – (Springer series in materials science; v. 27) Includes bibliographical referfences. ISBN-13: 978-3-642-46864-3 1. Solid state physics. 2. Materials. 3. Electronic structure. 4. Phase transformations (Statistical physics) I. Fujita, F. E. (Francisco Eiichi) II. Cahn, R.W. (Robert W.), 1924– . III. Series. QC176.P483 1998 620.1'699–dc21 98-26048 CIP

ISSN 0933-033X
ISBN-13: 978-3-642-46864-3   e-ISBN-13: 978-3-642-46862-9
DOI: 10.1007/978-3-642-46862-9

Typesetting: K+V Fotosatz, Beerfelden
Cover concept: eStudio Calamar Steinen
Cover producition: *design & production* GmbH, Heidelberg
SPIN: 10661523   54/3144-5 4 3 2 1 0 – Printed on acid-free paper

Preface to the Second Edition

Research and development of materials are extraordinarily fast so that any publication introducing advanced materials would not be able to keep up with the progress in the field. Even texts on fundamental principles and basic theories run into the same dilemma. For example, various types of fullerenes and carbon nanotubes were discovered in recent years, but the first edition of this book appeared just too early. Therefore, the newly added Chapter 10 describes recent materials and techniques, which appeared in the past several years. The reader will thus appreciate the overview of developments from the fundamental ideas treated in the virtually unchanged first edition to the latest information in the updating Chapter 10.

Ashiya, Japan   *Francisco Eiichi Fujita*
June, 1998

Preface to the First Edition

In the history of mankind, astronomy and mathematics are supposed to have been developed some thousands years ago from the necessity to support and control agriculture, navigation, architecture, and probably also commerce, in addition to the religious requirements. To produce materials necessary for people's daily life, such as tools, weapons, construction materials, medicines, utensils, and ornaments, the ideas and techniques of chemistry have been widely employed since prehistoric times, too. In ancient times, on the other hand, physics and chemistry were inseparably included in philosophy to understand the material universe, but later, likely along with the development of machines and mechanics, physics became more and more independent of philosophy and also of chemistry, reaching a clear foundation given by Newton's mechanics in 1687. It seems that in the time of the industrial revolution in the eighteenth century, physics was closely connected with mechanical engineering, but less related to the production of materials. For instance, metallurgical and ceramic studies were almost covered by the principles and methods of chemistry. Until recently, physics was supposed to be neither important nor useful in the production of industrial materials but mainly responsible for understanding the nature and behaviour of the substances in the universe.

Today, the role of physics in natural science is essentially unchanged, but the above situation in our modern society is rapidly changing mainly because of the revolutionary introduction of new materials in industrial production.

Hence, physics has become much more involved in the research and development of new materials than before. For instance, when superconductivity was discovered by H. Kamerlingh-Onnes in 1911, nobody, including the discoverer himself, expected the practical use of superconducting materials in our modern life. Today, however, many physicists are engaged in the study and even the production of new superconductors including various kinds of oxides. The invention and improvement of other new materials, such as highly efficient solar batteries, high radiation resistant fusion reactor materials, etc., needs physical ideas and techniques especially based upon solid-state physics.

The increasing necessity of physics can be seen in various aspects of materials science and technology. For instance, in the curriculum of many universities throughout the world, physical metallurgy has been greatly expanded at the expense of courses on manufacturing processes. Regardless of the merits and demerits of such a change, physics is today indispensably connected with materials science and technology.

In June 1989, a small international symposium on "Physics of New Materials" was held in Osaka, Japan, from the above view point of the necessity of basic science, especially solid-state physics, in the research and development of new materials. Encouraged by the success of the symposium, it was planned to publish a new book by a slight expansion of the subjects presented and discussed at the above meeting, which were the electronic structure and properties of transition metal systems, the amorphous solid and liquid structure, the nanophase materials, and the modern microscopic methods in the study of new materials. Four more new subjects were selected; they are the layered compounds and intercalation, the structural phase transformation, the place of atomic order in the physics of materials and metallurgy, and the high-voltage and high-resolution electron microscopy. Under the above eight titles, physical aspects and principles of new materials, such as interesting ferromagnetic alloys, shape memory alloys, atomic-ordered alloys, amorphous alloys, ultrafine particles, layered compounds with and without intercalation, deformable ceramics, and superconducting oxides, will be described and discussed from the view point mentioned above and in terms of thermodynamics, diffraction physics, electron band calculation, theory of phase transitions, and nuclear-physics techniques. In addition to the theoretical treatments, modern experimental techniques useful for the atomic and electronic structure analyses of materials – neutron time of flight, high-voltage and high-resolution electron microscopy, and Mössbauer spectroscopy – will be introduced in connection with the research and development of new materials.

All the contributors are well-known scientists in their respective fields and have participated in the above-mentioned symposium with enough understanding of its purpose and contents. Therefore, they are in full agreement and have readily cooperated with the present editor to publish this volume.

Fukuroi, Japan   *Francisco Eiichi Fujita*
October 1993

Contents

Contributors

Cahn, Robert W., Department of Materials Science and Metallurgy, University of Cambridge, Cambridge, UK

Fujita, Francisco E., Higashiyama 7-12-701, Ashiya 659-0091, Japan

Fujita, Hiroshi, Yamatedai 5-12-22, Ibaragi 567-0009, Japan

Gonser, Ulrich, Fachbereich 12/1 Werkstoffwissenschaften, Universität des Saarlandes, D-66014 Saarbrücken, Germany

Kanamori, Junjiro, Hyakurakuen 3-1-1, Nara 631-0024, Japan

Motizuki, Kazuko, Department of Physics, Okayama University of Science, Ridai-cho 670-0018, Japan

Siegel, Richard W., Materials Science Division, Argonne National Laboratory, Argonne, IL 60439, USA

Sumida, Naoto, Department of Materials Science and Engineering, Faculty of Engineering, Osaka University, Osaka 565, Japan

Suzuki, Kenji, Research and Development Center, Sumitomo Metals Industries, Amagasaki 660-0891, Japan

Suzuki, Naoshi, Department of Material Physics, Faculty of Engineering Science, Osaka University, Toyonaka 560, Japan

1 Introductory Survey

F.E. Fujita

First the historical development of the necessity of physics, especially solid-state physics, in the discovery, analysis, design and production of new materials in modern ages is outlined, together with the definition of the new or advanced materials. The role of physics in the development of new materials is exemplified by the case of shape memory alloys and of strong permanent magnets. A technique to produce nanocrystals (extremely fine crystalline particles) by evaporation, invented by a physicist and their new properties theoretically predicted by other physicists, is also mentioned. In the last part the contents of this volume is introduced with emphasis on the importance of physics.

1.1 New Materials and Necessity of Physics in their Development

Thirty years ago, it was forecasted that our modern society would be supported and operated mainly by three elements of technology which are materials, energy, and information. Today, that prediction seems to be more and more realized. Especially, the rapid rise in the research and development of new materials has not only largely improved our modern life but also controls further expansions of the other two technologies. For instance, to get more solar and nuclear fission energies in the present time and nuclear fusion energy in the near future, the research of materials, such as more efficient solar batteries, light-chemical energy conversion materials, and radiation resistant blanket materials for fusion reactors, is urgently required; and for the exploration of new-type information machines like artificial neurosystems, the invention of new function materials is indispensable. Many other modern technologies also wait for the discovery and invention of new materials.

Needless to say that for the development and application of new materials, the basic research by physical and chemical means is necessary. In the research of solid materials, such as metals, semiconductors, ceramics, and other inorganic chemical compounds, solid-state physics is specially needed as the basis of the study. In the time of the industrial revolution in the eighteenth century, physics was closely related with mechanical engineering, especially in order to understand the mechanics of the motion of matter, the thermodynamics of working gases, and the mechanical behaviour of solid materials[1], while chemistry was connected with the production of steel, ceramics, and other industrial materials by analyzing the structure and chemical reactions of matter. Nevertheless, the bicentenial development of industries and modern science seems to have changed the directions of necessity of physics and chemistry for technologies from the above trend to a new one: Today, physics is deeply concerned with the research and development of new materials. Not only for understanding the mechanism of newly found properties of materials and prediction of new materials but also for their structure analysis and characterization, theories and experimental techniques of physics are necessary. Sometimes, new materials were created from the solid-state physics study as in the case of superconducting materials which will be mentioned below. Also, to find the chemical bonding state or the electronic structure of materials even the techniques of nuclear physics are employed utilizing the hyperfine interactions in the nucleus and electron system, as will be exemplified in the last chapter of this book.

Definition of new materials is not always clear. For instance, the world strongest permanent magnet of Fe–Nd–B alloy is recognized as a new material since its value of the maximum magnetic energy product, $(BH)_{max}$, is almost five times greater than former strong magnets like Alnico 5; but a new stainless steel containing about 30% of Cr, which is almost a thousand times more pit-corrosion resistant than ordinary 13% Cr stainless steel, is not called a new material. Amorphous carbon is certainly not a new material, but when a similar extremely disordered structure was found in silicon sputtered at room temperature and alloys rapidly cooled from the melt, the latter two materials are called new materials. Another example of differentiation between the new and old materials is intercalation, which was already found some forty years ago, for instance, in bromination of graphite, but today other intercalated structures, such as some dichalcogen compounds, are considered as the new materials although their application is scarcely developed yet. The above examples seem to show that the definition of new materials is not always given because of a large progress of a required property, the time of discovery or invention, or the newly found structure but rather due to the exoticism of the discovered property and the expectation of new application.

The new materials introduced in the following sections are all more or less exotic and have the potential for new application, physics of which will be precisely discussed.

1.2 Examples of Physics of New Materials

When the shape memory effect of a Ti–Ni alloy was discovered by *Buhler* et al. [1.1] in 1960, it was an unexpectedly strange phenomenon, the mechanism of

---

[1] The principles of these three subjects were founded by I. Newton's mechanics (1687); R. Boyle (1660) and J. Charles' law (1787); and R. Hooke's (1678), T. Young's (1807), and J. Poisson's (1811) elasticity laws, respectively.

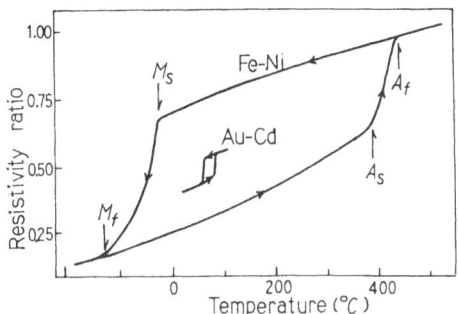

Fig. 1.1. Electrical resistivity changes of Fe–Ni and Au–Cd alloys during martensite transformation and reverse transformation

which was totally unknown. The following metallurgical study revealed that a martensitic transformation took place already before the material was deformed at room temperature and after the deformation, heating to a temperature above the reverse transformation temperature completely erased the history of deformation and recovers the original shape of the material. Hence, it was called the shape memory effect.

Until around 1960, the martensitic transformation was considered to be a typical first order phase transition, in which remarkable discontinuous changes of various parameters associated with the material take place and a large hysteresis of the order of, say, one hundred degrees appears during cyclic transformations, as the electrical resistivity changes of the Fe–Ni system in Fig. 1.1 show. On the other hand, Au–50%Cd alloy was found to exhibit a martensitic transformation with a very small hysteresis of several degrees, as the centre part of the same figure shows [1.2], and the shape memory effect. Many other shape memory alloys were found to be associated with the small hysteresis and the thermoelastic or ferroelastic behaviour. Similar behaviour of structural transition had been known with ferroelectric materials and later with some superconductive intermetallic compounds [1.3]. This structural phase transition of weak first order was precisely studied, especially for the perovskite type ferroelectric materials, in the light of Landau's phenomenological theory of phase transition which will be mentioned in Chap. 6, and this study provided large help in the physical understanding of the mechanism of shape memory effect [1.4]. On the other hand, the metallurgical term, martensitic transformation, has been employed in solid-state physics, opening a new field of the study of phase transition. Such physics underlying the study of the shape memory effect will also be mentioned in Chap. 6.

The research and development of the strong permanent magnet of Fe–Nd–B alloy mentioned before were also closely related with solid-state physics. In the physics of magnetism, it was generally understood that the amount of magnetic moment and/or the exchange interaction of Fe atoms in alloys depends on the Fe-neighbor atomic distance, as Fig. 1.2 shows [1.4]. After the invention of a strong permanent magnet of the Sm–Co system, like $SmCo_5$ type, it was immediately considered by many investigators to replace Sm and Co by another rare earth metal, like Nd, and another magnetic transition metal, like Fe,

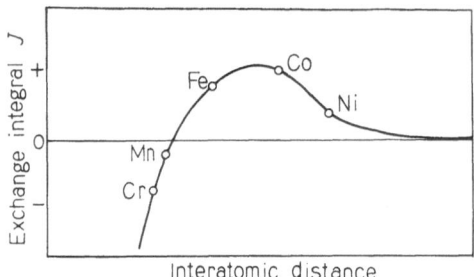

Fig. 1.2. Strength of exchange interaction of Fe atoms in alloys depending on the Fe–neighbour atomic distance

respectively, to obtain the stronger magnet. However, the combination of Nd and Fe was not a successful one, because its Curie temperature was found to be fairly low. *Sagawa* [1.5] tried to increase the Curie temperature of the Nd–Fe alloy by adding an interstitial alloying element, C or B, which was expected to expand the Fe–Fe atomic distance in the structure and, therefore, increase the exchange interaction, as Fig. 1.2 predicted. The Nd–Fe–C alloy was not stable, while the Nd–Fe–B alloy was found to form stable and high Curie temperature compounds, giving rise to the extremely useful magnetic properties. What he finally discovered was that the $Nd_2Fe_{14}B$ compound had a high saturation magnetization I_s, a high Curie temperature T_c and a high anisotropy constant K_1, although the Fe–Fe distances in the structure were not large as compared with those in other rare earth-iron compounds. Nevertheless, it is quite clear that the relation between the Fe–Fe distance and the exchange interaction (and the magnetic moment) mentioned above was the first guiding principle for him to reach the best material to make the strongest permanent magnet. The crystal and electronic structure, especially the spin configuration and interactions associated with the atomic arrangements and neighboring distances in the lattice, were also examined by physical means including neutron diffraction [1.6]. Other important factors to produce a strong permanent magnet with a large coercive force and remanence, such as the domain wall pinning, the uniaxial grain texture, the effective phase mixture, etc., were also thoroughly studied with the aid of theoretical and experimental methods of solid-state physics. As a result, the final product from Nd–Fe–B system exhibited the maximum magnetic energy product, $(BH)_{msax}$, of 405 KJ/m³, which is the world record of today.

There were some other approaches in the research and development of the Nd–Fe–B system alloys. For instance, *Croat* [1.7] obtained a strong permanent magnet of the above system by rapid cooling from the melt, which is a well-known technique to produce amorphous alloys and compounds. The electronic structure of ferromagnetic materials and the structure of amorphous materials will be precisely described in Chaps. 2, 3, respectively.

Another typical example of physics in the research and development of new materials is the "Nanophase materials" mentioned in Chap. 4. Uyeda was the first to produce nanocrystals, i.e., very fine powder crystals of nanometer size, by evaporating a metal in a chamber filled with a low pressure inert gas. The present author remembers that, when Uyeda's first experimental result was presented in a Japanese physics meeting in 1963, it was taken by the audience as just an interesting, exotic display by an experienced physicist. Later, the topological and crystallographical structure of nanocrystals were precisely investigated by Uyeda's group [1.8] using electron microscopy and diffraction. Studies of very fine particles have been carried out by other solid-state physicists. For instance, *Ino* [1.9] clarified by electron microscopy the initial stage of clustering of atoms prior to the formation of a thin film by the vacuum evaporation of a metal onto a substrate. *Kubo* [1.10] studied theoretically the electronic state and spin structure of extremely fine particles, predicting new physical properties of the fine particles. In solid-state physics and chemistry, the problems of isolated clustered atoms, which could be regarded as the very smallest nanocrystals, have opened a new field of study in which the molecular beam and mass separation techniques are employed. However, it is extremely hard to put the particles of such a small size together to make a bulk material for any practical use as a new material. On the other hand, a technique to collect and compact the nanocrystals made by evaporation was developed and gave us a material with new properties unavailable from the ordinary bulk materials. Other possible techniques to obtain the nanocrystal aggregates, i.e., the nanophase materials, involve making an amorphous material first then let it crystallize into a material with an extremely fine grain size of the order of nanometer. An application of it is already found in soft magnetic materials [1.11, 12].

The above mentioned techniques will be mentioned in Chap. 4 together with the new properties and applications of nanophase materials. It is a worth while observation to note that the new properties of nanophase materials mainly arise from the existence of crystal boundaries, which is one of the interesting physics of new materials.

1.3 Brief Introduction of the Contents

Physics of new materials is not a synonym of applied physics of materials. For the research and sometimes the development of new materials, the basic solid-state physics is necessary to find or predict the useful new properties, the mechanism of their functioning, and the structure of new materials, as are exemplified in the last paragraph. In the following chapters in this volume, the reader will find how basic physics, especially solid-state physics, is important and useful for the research of new materials, the development of various kinds of materials, and also to recognize how the different materials are linked by the common physical principles and mechanisms.

Other branches of physics, such as nuclear physics, molecular physics, earth physics, fluid dynamics, low temperature physics, quantum electronics, etc., will not be introduced in this book, although they are to some extent related with the materials. But, application of the Mössbauer effect will be rather precisely mentioned in Chap. 9 as a useful nuclear technique to investigate materials. Another useful technique, electron microscopy, will also be introduced in Chap. 8, where only its modern technical branches, the high voltage and high resolution electron microscopy, will be discussed in relation to the investigation and production of new materials.

Needless to say that there are other wide fields of materials science, in which chemistry plays the more dominant role; they are, for instance, highpolymers, biomaterials, and other soft materials. Chemistry of hard materials, such as inorganic compounds, metals, glasses, etc., is important as well. In this book, however, chemistry of new materials will not be discussed, unless it is substantially needed. Likewise, to mention the exploratory engineering or precise application of new materials is not intended in this book.

As introduced in Chap. 2, the study of electronic structure of metal system based upon the band theory is one of the main elements of solid-state physics which is indispensable for understanding the structure and properties of metals, alloys, semiconductors, and even some insulators. In recent years, the uniform electron band theory has been extensively changed or improved by taking into account the local electronic state associated with each atomic site affected by its environment, from which many plausible explanations of the structure and properties of new materials were obtained. The *ab initio* calculation based on the density functional theory and the coherent potential approximation is first described in Chap. 2. This calculation gives quantitative interpretations and predictions of the physical properties of alloys and compounds including special magnetic behaviour, spontaneous lattice distortion or phase transition, conducting, and insulating properties of new materials. For instance, an exact Slater-Pauling curve, which is one of the bases of the research and development of new magnetic materials, is calculated from the above theory. Other typical application of the *ab initio* calculation, for example, the calculation of electronic structure and hyperfine fields associated with various impurity atoms and lattice defects in metals and oxide superconductors, is also introduced and discussed.

The main subject of Chap. 3 "Structure Characterization of Solid State Amorphized Materials by X-ray and Neutron Diffraction", is on the experimental analysis of the structure of dense random assemblies of atoms, such as amorphous solids, and to relate these structures to various new materials especially in view of the short and medium range order in the extensively disordered structure. First mentioned is the principles of diffraction and scattering techniques using the X-rays from synchrotron radiation and the pulsed neutrons from accelarators to find the interference function $S(Q)$ and the correlation function $g(r)$. It shows that the partial correlation functions, the first diffraction peak structure and epi-peak substructure, are important to analyze the chemical and other short range or more extended order structure. Mechanical alloying, mechanical glinding, melt-quenching, and vapour quenching techniques and their products are compared, and the successive stages working from the perfect crystalline order to the real amorphous disorder is discussed in the framework of diffraction physics. Another new technique of conversion of organic polymer to amorphous ceramics will also be mentioned in a similar way.

The "Nanophase materials" introduced in Chap. 4 are probably one of the newest materials, which have some peculiar and not yet precisely determined characteristics and properties, and they await new applications. The first part of the article describes different ways of preparing the nanometer size particle aggregates. The efficient high speed synthesis and processing to produce the high quality nanophase materials are the key to their wide industrial application. It is pointed out that the structure, stability, and properties of nanophase materials depend largely on how they are made, and the observed essential features and the measured basic data are shown as the fundamentals for application. Finally, the future extension of study and the search for new applications are described.

Another new category of new materials, "Layered Compounds and Intercalation", is theoretically discussed in Chap. 5. Following the introduction of electron theory in Chap. 2, the band structure of a typical intercalated compound of 1T-type TiS_2 is first shown and compared with the experimental results. Comparison between the nonmagnetic and the ferromagnetic states is also described. In the next section, the bonding nature of TiS_2 containing a 3d transition metal is discussed to enable the reader to readily understand the essential nature of intercalation. A different feature of electronic structure of layered compounds is exemplified by $AgTiX_2$, and another interesting layered compound introduced is a hexagonal type TX_2 which can also be intercalated (T: Nb or Ta and X: S or Se). This chapter comprehensively demonstrates the basic relation between the electronic structure and the nature of intercalation.

Structural phase transitions introduced in Chap. 6 are commonly observed among the space substances, planetary substances, minerals, and artificial or industrial materials in our world. In this chapter, first a phenomenological theory of phase transition described by Landau is briefly mentioned, together with the Ehrenfest's well-known definition, for the reader's better understanding of the definition of the order of transition. Then, some examples of phase transitions in new materials will be shown. One is the martensitic transformation of β-phase alloys, which is the basic mechanism of the shape memory effect and the ferroelasticity of alloys. This type of transformation is also closely connected with a new refractory material, i.e., the plastically deformable ceramics such as the partially stabilized zirconia. Another example is the amorphous state produced by rapid quenching of molten metals, which is discussed in relation to Chaps. 3, 4. New phases and new materials produced under high pressure are also an important topic. The phases of diamond and boron nitride are mentioned, and a paradoxical technique for producing diamond thin film without applying high pressure is briefly described.

In Chap. 7, the place of atomic order in the physics of materials and metallurgy is discussed. In the first place, the discovery of long range order (LRO) in the atomic arrangement in a Au–Cu alloy in 1916 by X-ray diffraction and the following studies of LRO are dealt with. In the second, Bragg and Williams' first successful theory (1934) of order–disorder using LRO parameter, other theories using the short range order (SRO) parameter, the theory of critical phenomena, the order of transition, etc., are introduced. Ordering kinetics and related phenomena such as the radiation enhancement of ordering and critical slowing down are discussed, and usefulness of computer simulation to see the kinetics and domain formation is also mentioned. The LRO appears not only in alloys but also in other new materials such as ternary semi-conductors, dielectric crystals, and intermetallic compounds. The objective of this article is to map out the ubiquity of atomic ordering in the physics of solid materials and the extensive potentiality of such ordering for the properties of materials.

To analyze in detail the structure of the materials produced by new ideas and new techniques is always required. Among other available physical means, X-ray diffraction analysis has been most popular and used in almost all fields of science and technology for nearly eighty years. In addition to the diffraction and scattering technique mentioned in Chap. 3, two other tools are introduced and their application to the study of new materials is discussed. One is the modern technique in ordinary transmission electron microscopy mentioned in Chap. 8, and the other is Mössbauer spectroscopy discussed in Chap. 9. In the former, the importance of the high voltage electron microscopy and the high resolution electron microscopy is emphasized in relation to new materials. As the most important physical principle of transmission microscopy, the diffraction con-trast which is extremely sensitive to defect structures in materials, is introduced. On the other hand, by reducing the electron optical aberrations, mainly the spherical aberration, chromatic and stigmatic ones, and selecting the diffracted beams of different orders, the high resolution electron images showing the two dimensional arrangement of atom rows can be composed by the Fourier and its reverse transformation of the incident and diffracted beams automatically done in the electron optical system in the microscope. This is the high resolution electron microscopy to see directly the structure of various materials in atomic scale [1.13]. The microscope with a high voltage of 1 to 2 MV has a large advantage in the observation of thick specimens and in the improvement of resolution due to the short wave length. Another advantage (or sometimes a disadvantage) is the radiation damage of the specimens by the electron bombardment, which could be a model technique to produce new materials, for instance, the amorphization of materials by high energy particle irradiation. Many photographic examples are shown in this article.

The last chapter is on the Mössbauer spectroscopy as an example of modern microscopic methods in the study of materials. Most of the nuclear techniques available today for the analysis of condensed matter came into practical use some thirty years ago. One of the earliest and most useful one is nuclear magnetic resonance, which can reveal the electronic structure of the probe atoms and their environments, especially the near neighbor configuration and magnetic structure, by observing hyperfine interactions. Quite the same or even better information can be obtained by Mössbauer spectroscopy through the hyperfine interaction measurement including the nuclear excited levels. In addition, this method can quantitatively analyze the phases and their crystal or chemical structure. More sophisticated techniques, such as the thermal analysis, the polarization method, etc., and its application to new materials research are also mentioned.

As will be seen in the following chapters, the topics and materials introduced in this volume are not isolated but doubly or multiply connected to each other through different chapters; for instance, the amorphous materials are discussed from the view points of structural analysis by X-ray and neutron diffraction and scattering, thermodynamical theory, hyperfine spectroscopy, and electron microscopy. This means that the present volume has been intended to demonstrate the importance and universality of physics underlying the research and development of new materials.

References

1.1   J.W. Buhler, V.G. Gilfrich, R.C. Wiley: J. Appl. Phys. **34**, 1475 (1963)
1.2   L. Kaufman, M. Cohen: Prog. Met. Phys. **7**, 165 (1958)
1.3   N. Nakanishi: Trans. Jpn. Inst. Metals **6**, 222 (1965)
1.4   K. Shimizu, K. Otsuka: Int'l Metals Rev. **31**, 93 (1986)
1.5   M. Sagawa, S. Fujimura, N. Togawa, H. Yamamoto, Y. Matsuura: J. Appl. Phys. **55**, 2083 (1984)
1.6   J.F. Herbst, J.J. Croat, F.E. Pinkerton, W.B. Yellon: Phys. Rev. B **29**, 4176 (1985)
1.7   J.J. Croat, J.F. Herbst, R.W. Lee, F.E. Pinkerton: J. Appl. Phys. **55**, 2078 (1984)
1.8   K. Kimoto, K. Kamiya, M. Nonoyama, R. Uyeda: Jpn. J. Appl. Phys. **2**, 702 (1963)
1.9   S. Ino: J. Phys. Soc. Jpn. **21**, 346 (1966)
1.10 R. Kubo: J. Phys. Soc. Jpn. **17**, 975 (1962)
1.11 Y. Yoshizawa, S. Oguma, K. Yamauchi: J. Appl. Phys. **64**, 6044 (1988)
1.12 N. Kataoka, A. Inoue, T. Masumoto, Y. Yoshizawa, K. Yamauchi: Jpn. J. Appl. Phys. **28**, L1820 (1989)
1.13 F.E. Fujita, M. Hirabayashi: High-resolution electron microscopy in *Microscopic Methods in Metals*, ed. by U. Gonser, Topics Curr. Phys., Vol. 40 (Springer, Berlin, Heidelberg 1986) pp. 29–71

2 Electronic Structure and Properties of Transition Metal Systems

J. Kanamori

The interrelation between the electronic structure, magnetic, and structural properties of transition metal systems is discussed. Emphasis is placed on the discussion of the quantitative capability of the *ab initio* calculation and the role of electron correlation.

2.1 Background

This chapter is intended to give a perspective of the success and failure of the one electron picture based on the band structure calculation in interpreting and predicting physical properties of metals, alloys, and compounds of transition elements; and the role of electron correlation will be discussed along with it. Discussions which are inevitably confined to selected examples are focused on the future role of the theory in designing 'new materials' and is intended to supplement many excellent references that exist on the subject.

Many properties of metallic systems can be discussed semi-quantitatively by assuming a simple model which characterizes the electronic state in terms of a few parameters. For example, *Friedel* and *Sayer* [2.1] developed a transparent theory of cohesive properties of transition metals which can explain general trends of the crystal structure and the cohesive energy, bulk modulus and other various quantities. *Miedema* [2.2] and *Pettifor* [2.3] developed semi-phenomenological theories which are capable of predicting the cohesive properties of alloys and compounds of wide range. As will be discussed below, most of the alloying effects on the ferromagnetism of iron, cobalt, nickel, and their alloys are also explained by a simple model [2.4]. However, we need a parameter free calculation for quantitative predictions, since *ad hoc* adjustments of parameters to reproduce given experimental data often obscure the real problem behind them. We emphasize also that some conclusions of phenomenological investigations are in contradiction with the results of detailed calculations carried out later. In this chapter, we place emphasis on the *ab initio* calculation based on the density functional theory initially developed by *Hohenberg* and *Kohn* [2.5] and *Kohn* and *Sham* [2.5] which does not contain adjustable parameters.

Systems of transition element atoms are characterized by the atomic d states which make narrow d bands if the interatomic transfer is operative. The one electron approximation or in other words, the band theory, concerns primarily the interatomic transfer and takes into account the effect of the electron–electron coulomb interaction only approximately. As will be mentioned in more detail below, the *ab initio* calculation has yielded satisfactory results for the ground state properties even in those transition metals in which the coulomb interaction gives rise to ferromagnetism. On the other hand, the calculated band width and exchange splitting in a ferromagnetic nickel are larger, respectively, by a factor nearly equal to 2 than the results of photoelectron spectroscopy experiments [2.6]. This discrepancy has been interpreted often as indicating the importance of the electron correlation. We give a discussion below, however, that the inadequacy of the local density approximation used in the calculation in taking into account the electronic structure may be more responsible for it. The competition between the interatomic transfer and the intraatomic electron–electron coulomb interaction is more apparent in the metal–insulator transition found in compounds such as oxides and sulfides of transition elements. In these latter compounds another energy parameter, i.e., the energy cost to transfer an electron from ligand such as oxygen to a transition element atom, comes into the problem. We do not at the present time have a full understanding of the electronic states that a material can realize for various choices of conditions characterized roughly by the band width, the intraatomic coulomb energy, and the charge transfer energy in compounds. The recent discovery of the high T_c superconductivity in oxides containing Cu and other metallic elements has taught us the shallowness of our understanding of the capabilities of materials.

The above discussion indicates that the usefulness of the *ab initio*, one electron calculation is rather limited to metallic systems. Nevertheless, the *ab initio* calculation has established a method of predicting or designing a 'material' nonempirically. The following discussions are intended to give a scope of the approach by indicating its limitation as well as its merits. In Sect. 2.2, we describe briefly the method of calculating the electronic structure and a basic physical picture of the s–d mixing. In Sect. 2.3 we discuss various examples of success and failure of the one electron approximation mostly taken from the cases in ferromagnetic metals and alloys. We present also a discussion of a possible mechanism for enhancing the ferromagnetism of iron, or more precisely iron atom systems as an example of practical applications of the theory. Section 2.4 deals with structural problems of alloys. In Sect. 2.5 we discuss some aspects of electron correlation in high T_c superconducting oxides and related compounds in which the one electron approximation encounter an essential difficulty. Finally, in Sect. 2.6, we discuss future developments of the one electron theory.

2.2 Basic Concepts of Electronic Structure Calculation of Transition Metal Systems

2.2.1 Method of Calculation

The density functional theory developed by *Kohn* et al. [2.5] proves first that the ground state energy corresponds to the minimum of a functional of the density of electrons which is a function of position. Then the electron distribution in the ground state is shown to be determined by solving the Schrödinger equation with an appropriate one electron potential which takes into account the exchange-correlation energy as well as the classical electron–electron and electron–nucleus coulomb interactions. The exchange-correlation potential, which is a functional of the electron density in principle, can be determined only approximately. The approximation which is widely used and concluded to be successful is the local density approximation which expresses the potential at a given location of electron as a function of the electron density at the same site; the function is determined by use of the knowledge of the total energy of the jellium model in which the nuclear charges are smeared out and uniformly distributed in the whole space. For details the reader is referred to reference [2.7]. By calculating the one electron potential self-consistently, one can determine not only the electronic structure for a given lattice configuration of nuclei but also the crystal structure itself and lattice spacings on the basis of the energy minimum principle by comparing the total energy of various configurations. The calculation does not need any input information except for the nuclear charges of constituent atoms. We remark finally that the formalism is extended to the ferromagnetic ground state by introducing the concept of the spin density functional by distinguishing between the densities of electrons with different spin directions.

The *ab initio* calculation was carried out first for pure metals, regular alloys, and stoichiometric compounds. It has been extended to the case of point defects such as impurities and vacancies. In this case the self-consistent determination of the perturbation on the electronic structure and positions of surrounding atoms is one of the central problems. So far the program which extends the self-consistent calculation of the electronic structure up to the fourth neighbors of a point defect at a substitutional site in an otherwise perfect crystal has been developed by *Dederichs* et al. [2.8]. However, to take into account atomic displacements is still a future problem. Also, the electronic structure calculation of interstitial impurities remains at the stage where only the potential at the impurity atom is determined self-consistently, while the potential as well as the atomic positions of surrounding host atoms are assumed to be the same as in a pure crystal. Comparing the results of the calculation within this framework with the hyperfine interaction data of light interstitials such as μ^+, B, C, N in iron, we conclude that the self-consistent determination of lattice relaxation is particularly important for the interstitial case [2.9].

The application to disordered alloys has been made feasible by the coherent potential approximation (CPA) originally proposed by *Soven* [2.10] which introduces the concept of *coherent atom* corresponding to an average of disorderly arranged atoms of various species constituting a given alloy. We solve the impurity problem in which an atom of a given species substitutes a coherent atom in the lattice of the *pure* crystal of *coherent atoms*. The potential of electrons within the impurity atom can be determined self-consistently in the local density approximation mentioned above if the potential of the coherent atom is known; the latter potential which is called *coherent potential* is determined self-consistently by making use of the potentials of component atoms which in turn has been determined in terms of the *coherent potential* by solving the above-mentioned impurity problem. Since the calculation involves the double cycles of achieving the self-consistency first in the local density approximation and secondly in that of the coherent potential, the computational difficulty was solved only by ingenious methods [2.11-13]; *Akai* [2.11, 13] has developed a method which is now recognized as the most efficient one. We shall discuss examples of the calculation later.

There are a variety of methods for the band structure calculation whose detail we shall not discuss in this chapter. The accuracy of the calculation depends on the methods, since they involve more or less simplifying assumptions. In the cases of alloys and impurities in metallic systems the KKR (Korringa-Kohn-Rostoker) method [2.14a] with the muffin-tin potential approximation for the potential is used because it can spare the computational time for other procedures such as CPA. Somewhat more simplified but faster methods such as LMTO (Linearized Muffin-Tin Orbital) [2.14b] have also been developed. In some problems such as those of compounds we may need a more accurate treatment of the potentials than the muffin-tin approximation. The FLAPW (Full Potential Linearized Augmented Plane Wave) method [2.14c] which is capable of taking into account the non-muffin-tin aspect of the potential seems to be the most accurate one at the present time.

2.2.2 s–d Mixing

We discuss here the physical picture of the electronic structure on which the following discussions are based, in particular the s–d mixing which is important in many ways in transition metal systems. The angular symmetries, s, p, d, etc., of wave functions are defined with respect to a given reference point. Suppose we construct a wave function by integrating the Schroedinger equation, assuming a d symmetry with respect to the nuclear position of a given atom. If we continue the integration to the neighboring atom and analyze the angular symmetries of the wave function with respect to its nuclear position, the wave function will turn out to be a mixture of various symmetries. In the case of a transition metal the angular symmetries s, p and d are dominant, though higher harmonics are needed in some cases for detailed quantitative estimation.

Table 2.1 shows that result of a quite instructive calculation based on the KKR–CPA carried out by *Akai* [2.15] for Pd_xNi_{1-x} disordered fcc alloys which are ferromagnetic for $x < 0.95$. In the calculation the observed lattice parameters are used as input information. The table shows the total numbers of states of s, p, f symmetries, and d symmetries, respectively, within the average Wigner-Seitz sphere having an energy below the top of the d band. The numbers are averaged over spin directions, though they depend only slightly on the spin directions. The table shows also the numbers of occupied states below the Fermi

Table 2.1. Examples of calculated numbers of states in $Ni_{1-x}Pd_x$

Concentration	Atom	n_{state} s, p, f	d	$n_{occ\uparrow}$ s, p, f	d	$n_{occ\downarrow}$ s, p, f	d
Ni	Ni	0.74	4.40	0.78	4.52	0.81	3.92
$Ni_{0.5}Pd_{0.5}$	Ni	0.79	4.67	0.80	4.71	0.82	3.82
	Pd	0.66	4.30	0.67	4.36	0.69	4.13
$Ni_{0.1}Pd_{0.9}$	Ni	0.83	4.83	0.82	4.80	0.80	3.73
	Pd	0.70	4.45	0.70	4.37	0.63	4.22

n_{state} denotes the number of states below the top of the d band per spin per atom; $n_{occ\uparrow}$ and $n_{occ\downarrow}$ the number of occupied states

energy. We note first that in pure Ni as well as in Ni–Pd alloys the number of d states per spin per atom below the top of the d band is less than 5. This indicates that we have to be cautious in using the tight binding picture of the d band where the number is equal to 5 if wave functions are naively assumed to be given by linear combinations of atomic d states. The difference is mostly due to the mixing between the s, p, and d states. Thus, the often quoted atomic configuration of a nickel atom in nickel metal, $(3d)^{9.4}$ $(4s)^{0.6}$, merely represents the distribution of electrons in the branches of the band; wave functions of a branch of the d band contain about 10% of s and p symmetry states.

The ferromagnetism of Ni–Pd alloys is characterized by an increase of the magnetic moment of Ni from about 0.6 μ_B to more than 1 μ_B with Pd concentration [2.16]. Since both Pd and Ni have less than one hole in the d band as pure metal, it is difficult with the simple-minded tight binding picture to explain this increase which would require the presence of more than one hole in a Ni atom. The calculation listed in Table 2.1 which agrees with the experimental data shows that the weight of d symmetry states in the d band wave functions increases of Ni atom with Pd concentration. This seems to be primarily due to a decrease of the mixing between Ni d states and Pd s, p states caused by the lattice expansion with Pd concentration. Since the number of occupied s, p, and f states of a Ni atom is almost independent of the Pd concentration, the total number of occupied d symmetry states of a Ni atom is also independent because of the approximate electrical neutrality. An increase of the weight of the d symmetry states in the d band wave functions, therefore, should be compensated by a decrease of the number of occupied d band states. The latter decrease is possible only in the band states of minority spin, since the d band states of majority spin are fully occupied. As a result we can expect an increase of the Ni magnetic moment which can amount to more than 1 μ_B.

The above discussion indicates that the amount of the s–d mixing and consequently the weight of d symmetry states in the d band can be changed by the environment. An increase of the magnetic moment of surface atoms in ferromagnetic transition metals [2.17] and also that of monolayers of transition element atoms deposited on a substratum [2.18] which were predicted by *ab initio* calculations, can be explained by a decrease of the s–d mixing.

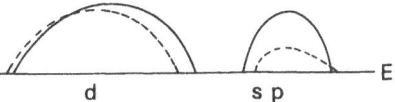

Fig. 2.1. The d states of transition element atoms and s, p states of boron atoms in borides. The full curves represent the states in the absence of the s–d mixing and the dotted ones the d states after the mixing is taken into account

The mixing of atomic d states with s, p states of surrounding atoms can be enhanced if the latter atoms are of nontransition s, p valence elements, since they produce a deeper potential than transition element atoms for s, p symmetry states. A typical example is given by ferromagnetic borides of transition elements M_2B and MB with M = Mn, Fe, Co, etc. [2.19] which can form mixed crystals with different M. The magnetization of the borides follows as a function of the average d electron number, a curve resembling the Slater-Pauling curve of ferromagnetic alloys except for a shift, such that Mn or Fe in the borides corresponds to Co of the latter curve. This can be understood as the result of a large mixing between the d states and the valence states of boron which results in a decrease of the weight of the atomic d states in the d band. Since transition element atoms tend to keep approximately the same number of d electrons in spite of the mixing, an increase of the filling of the d band states are the result of the decrease of the weight which produces a shift of the Slater-Pauling curve. Figure 2.1 illustrates the situation.

2.3 Bulk and Defect Electronic Structure of Ferromagnetic Transition Metal Systems

2.3.1 Calculation for Periodic Systems

Table 2.2 summarizes the results of the *ab initio* calculation for ferromagnetic pure metals carried out by *Moruzzi* et al. [2.20]. We can see that the calculation can satisfactorily reproduce the experimental data on the ferromagnetic state. However, in the calculation the type of the crystal lattice is assumed as input information, though the lattice parameter is determined by the principle of minimum energy. In the case of iron a fcc nonmagnetic state turns out to be lower in energy than the bcc ferromagnetic state as far as the standard method is used for determining the exchange-correlation potential. It is only recently that an extension of the method of determining the one electron potential has been shown to remove the discrepancy [2.21]. The *ab initio* determination of the crystal type is still a subtle problem; since the relevant energy difference is of the order of 10 meV, the difficulty may be understandable.

The Laves phase compounds AB_2 with A = Sc, Y, Zr, Nb and B = Cr, Mn, Fe, Co, Ni crystallize in either or both of the hexagonal C14 and the cubic C15

Table 2.2. The calculated [2.20] and experimental values at 0 K of the lattice constant and spin magnetic moment of ferromagnetic metals

Metal	Lattice constant [a.u.]		Spin magnetic moment (μ_B per atom)	
	Calc.	Exp.	Calc.	Exp.*
bcc iron	5.27	5.40	2.15	2.13
fcc cobalt	6.54	6.68	1.56	1.60
fcc nickel	6.56	6.65	0.59	0.56

*The contribution of the orbital magnetic moment is subtracted

types. Some of them are known to exhibit various types of ordered magnetism; $ScFe_2$ of the C14 structure and both YFe_2 and $ZrFe_2$ of the C15 exhibit ferromagnetism, while $TiFe_2$ of the C14 and YMn_2 of the C15 are antiferromagnetic [2.22]. *Asano* et al. [2.23] carried out a KKR [2.14a] or LMTO [2.14b] *ab initio* calculation to determine the most stable crystal type and the magnetic state for a wide variety of combinations of A and B AB_2. The success of their calculation which concludes correctly most of the stable crystal types as the state of minimum energy with reasonable quantitative agreement with observed lattice parameters is another example of the usefulness of the *ab initio* calculation.

2.3.2 Impurities

Impurities whose atomic number Z is larger than 10 occupy a substitutional site in ferromagnetic transition metals. An *ab initio* calculation in which only the potential within a given impurity atom is determined self-consistently in the local density approximation has been extensively carried out for the cases Z = 1 to 56, i.e., from H to Ba in ferromagnetic iron [2.24]. For some selected cases in ferromagnetic nickel and iron an *ab initio* calculation which determines the electronic structure of impurities and surrounding host atoms of up to fourth neighbors, has been carried out [2.8]. The main purpose of these calculations is to discuss the hyperfine field and nuclear relaxation time of impurity nuclei which reflect the electronic structure of impurities. In the case of impurities of transition elements which can possess a magnetic moment the impurity magnetic moment is also an important subject of the discussion. Furthermore, the result is utilized to construct a theory of the alloying effect on the ferromagnetism.

Figure 2.2 shows an example of the calculated hyperfine field of impurities in ferromagnetic iron and corresponding experimental values for a s, p valence period. The hyperfine field of nontransition element nuclei in ferromagnetic metals arises mainly from the Fermi contact interaction [2.25] whose magnitude is proportional to the difference of the local densities of electrons at the nucleus

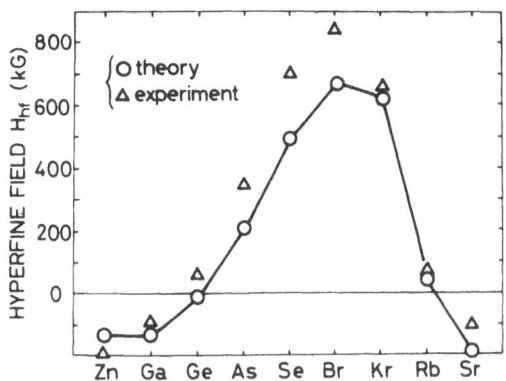

Fig. 2.2. An example of the systematic variation of H_{hf} with impurity valence in ferromagnetic iron [2.9]

between the spin directions. The spin polarization is produced by the mixing between the s states of the impurity and the spin polarized 3d states of the surrounding host atoms. States of other symmetries such as p impurity can also be spin polarized. Their contribution to the hyperfine field which is made either directly through the dipole interaction with nuclear moment or indirectly through the polarization effect on core s states [2.25] is negligibly small in the case of nontransition element impurities which remain approximately non-magnetic. In the case of transition element impurities the contribution is fully taken into account. We can see from Fig. 2.2 that the hyperfine field starts with a negative value at the beginning of a s, p period corresponding to alkali elements and increases with Z to become positive in the middle of the period. Near the end of the period it passes a maximum and drops into a negative value either at the end of the period or at the beginning of the next period.

The systematic variation reflects the evolution of the electronic structure with impurity valence where the s–d mixing discussed in Sect. 2.2.2 plays an important role. Figure 2.3 shows an example of the calculated local density of states at the impurity nuclear position. The local density of states is characterized by the presence of a dip at an energy E_{dip} whose position does not change with the impurity valence. The position of E_{dip} for majority spin states lies lower than that for minority spin states; the difference corresponds roughly to the exchange splitting. *Terakura* [2.26] has shown that E_{dip} corresponds to the boundary between the bonding states and the antibonding states between the impurity s and host d states. Suppose we integrate the Schroedinger equation starting near the nuclear position of a host transition element atom, assuming a d symmetry appropriate for the bonding with the impurity s state. The tail of the d orbital which is dependent on the energy parameter will change its sign somewhere; if the energy corresponds to the top of the d band, the sign change takes place near the surface of the atomic sphere of the transition element atom. With an energy somewhat below the top of the d band, it will take place at the nuclear position of the impurity. Then the mixing with the impurity s state will vanish approximately because it is orthogonal to it. Thus the energy where this

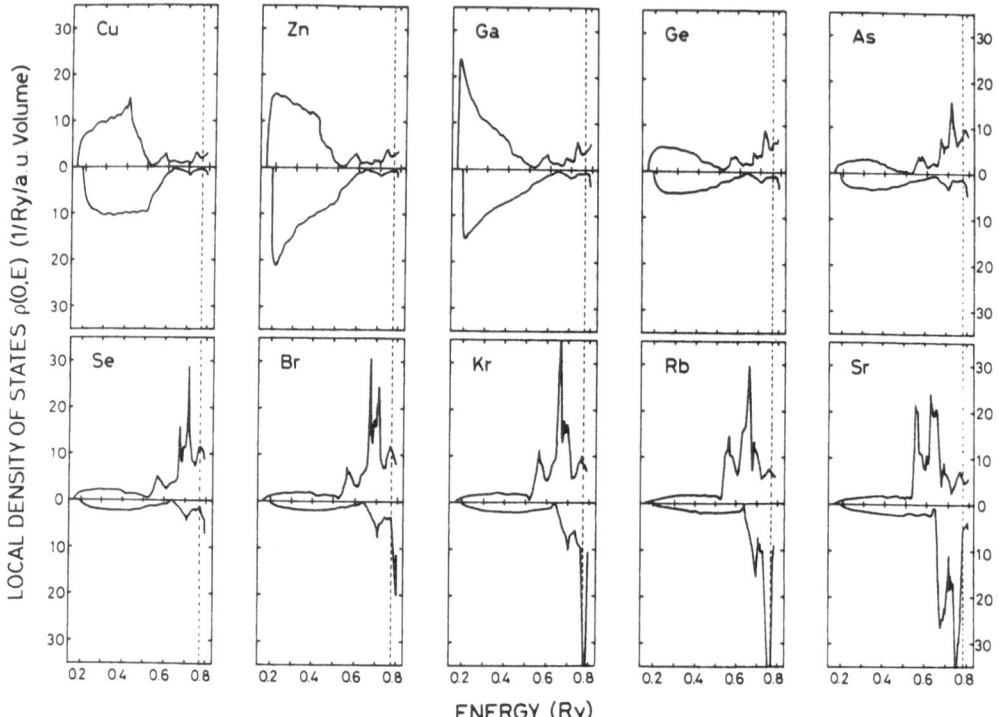

Fig. 2.3. The calculated local density of states at the nuclear position for the impurities in iron whose H_{hf}'s are shown in Fig. 2.2. The upper and lower parts represent the majority and minority spin states, respectively

situation is realized corresponds to E_{dip}. We can see also that the wave function is of the bonding nature for $E < E_{dip}$, since the host d and the impurity s are mixed with the same sign, while it is of the antibonding type for $E > E_{dip}$.

The systematic variation of the hyperfine field can be understood on the basis of the above mentioned picture of the s–d mixing [2.27]. We can suppose that the effective energy levels of the impurity valence states of s, p symmetries lie high above the top of the d band at the beginning of a given s, p period. With increasing nuclear charge through the period, the levels come down and pass finally the d band at the end of the period to become part of the core states in the next period. The bonding states below E_{dip} make a negative contribution to the hyperfine field in total, while the occupied antibonding states between $E_{dip} < E < E_F$ make a positive contribution. The bonding states of minority spin have larger amplitudes at the impurity site than those of majority spin, the former d orbitals having a larger tail extending towards the impurity because of a relatively shallower potential raised by the exchange potential inside the transition element atom. The resulting larger mixing with the impurity s orbital explains the negative sign of the contribution to the hyperfine field. The positive sign of the contribution of the antibonding states is due to the fact that E_{dip} of majority spin states lies lower in energy than that of minority spin; the number of the occupied antibonding states, therefore, is larger for majority spin than for minority spin. The systematic variation of the hyperfine field in a s, p period can be understood as the result of an increase of the positive contribution of the antibonding states whose amplitude at the impurity nucleus increases with the increase of the nuclear charge.

We need to take into account the change of the electronic structure of surrounding host atoms in some problems; the change of the isomer shift of ^{57}Fe neighboring to impurities in iron is a typical example, since it reflects the change of the electronic structure caused by the presence of an impurity. *Akai* et al. [2.28] carried out an *ab initio* calculation of the electronic structure of surrounding host Fe atoms as well as that of a given impurity which can reproduce the systematic variation of the isomer shift with impurity atomic number quite satisfactorily.

In relation to the hyperfine interaction in metals, we may add that the *ab initio* calculation is extended by *Dederichs* et al. [2.29] to the electric field gradient at the nucleus which gives rise to the electric quadrupole interaction. Since the calculation applies mostly to the systems of nontransition elements of the hcp crystals, we omit further discussions.

2.3.3 Disordered Alloys

It is well known that the average magnetization per atom of ferromagnetic disordered Fe–Co, Fe–Ni, and Ni–Co alloys falls as a function of average d electron number approximately on a common curve which is called the Slater-Pauling curve. Other combinations such as Fe–Mn, Fe–Cr, Ni–Mn, Ni–Cr, etc., give rise to various branches of the curve. *Friedel* [2.30] gave a general and quite insightful argument on the mechanism of the alloying effects on Ni for small concentration of other transition elements, introducing for the first time the concept of the virtual bound state for the d states of impurities. His discussion was extended to concentrated alloys including Fe base alloys by *Hasegawa* and *Kanamori* [2.4] through the use of a simple model Hamiltonian and the coherent potential approximation. The Hamiltonian is an extension of the Hubbard model of pure metal which will be given explicitly in the next Sect. 2.3.4. The KKR–CPA *ab initio* calculation [2.31] which can reproduce the experimental data quite well without any adjustable parameters has been carried out recently.

The CPA calculation [2.4] elucidated the mechanism underlying the various features of the alloying effect on the ferromagnetism. We discuss here the case of bcc Fe–Co alloys [2.14c] as a typical example. The average magnetization of Fe–Co bcc alloys increases with Co concentration in spite of the fact that Fe atoms are replaced by Co atoms having smaller magnetic moment. In fact the neutron diffuse scattering measurement shows that the average atomic magnetic moment of Co remains to be around 1.7 μ_B, while the average Fe moment increases. Furthermore, the Mössbauer and NMR experiments indicate that it is the magnetic moment of Fe atoms neighboring Co which is increased. The increase overcompensates the decrease due to the substitution of some Fe atoms by Co.

The underlying mechanism can be understood by recalling the following simple quantum mechanical principle. Suppose two atoms having different energy levels: If we allow an electron transfer between them, two energy levels repel each other such that the level of higher energy moves towards higher energy and the lower one towards lower energy. If these atoms are embedded in a metal, each energy level will broaden into an energy band due to the electron migration. However, the fact that the band of higher energy is shifted towards higher energy and vice versa remains true. If the Fermi level lies midway in the band as is illustrated in Fig. 2.4, the atom whose level is of higher energy will lose its electron, because the portion of unoccupied states increases.

In bcc Fe–Co alloys, the energy levels of minority spin states of Co atoms lie lower in energy as a whole than those of Fe atoms; incidentally, those of the majority spin states of Fe and Co atoms are approximately of the same energy. The difference between the two spin directions arises from the exchange potential whose spin dependent part may be expressed roughly as

$$E_{ex} = \pm U(n\downarrow - n\uparrow) \tag{2.1}$$

with $U > 0$, where the upper and lower signs apply to the majority and minority spins, respectively; $n\uparrow$ ($n\downarrow$) stands for the average number of electrons of majority (minority) spin at a given atom. A little consideration leads us to the understanding that the deeper potential due to a larger nuclear charge in Co is compensated for majority spin by the exchange potential which is smaller in absolute magnitude than that in Fe because of a smaller magnetic moment, while the difference is enhanced for minority spin. The above mentioned level repulsion applies, therefore, to the minority spin states in Fe–Co alloys.

As a result Fe atoms tend to lose a certain amount of minority spin electrons due to the mixing between Fe and Co atoms; this loss is compensated by an increase of the majority spin electrons which is brought in by the decrease of $n\downarrow$. The increase of the majority spin electrons is possible because Fe atoms have some vacant majority spin states above the Fermi level. Consequently, the magnetic moments of Fe atoms surrounding Co atoms increase. *Hamada* [2.32]

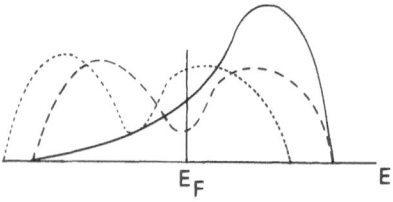

Fig. 2.4. Alloying effect on the density of states. The dotted and broken curves represent the minority spin d bands of cobalt and iron, respectively. In Fe–Co alloys the local density of states of minority spin of Fe atoms is changed into the full curve

carried out a more detailed calculation on the basis of the tight binding model which concluded an approximate proportionality of the increase of the Fe magnetic moment to the number of neighboring Co atoms.

The coherent potential approximation has been successfully applied to various problems of disordered systems other than those pertaining to ferromagnetism. We mention here the vacancy generation in TiO and other compounds which was discussed by *Huisman* et al. [2.33]. Vacancies in solids are usually generated by the mechanism of lowering the free energy at finite temperature by increasing the entropy. In such a case the vacancy concentration will decrease with decreasing temperature. In certain cases such as TiO, however, the vacancy concentration is quite high even at low temperatures. It was discussed by *Goodenough* [2.34] and others [2.33] that the generation of vacancies in such systems would be energetically favorable. *Huisman* et al. [2.33] have studied the situation by carrying out a CPA calculation on the tight binding model and also another more detailed calculation on the electronic structure to conclude that the generation of vacancies in the oxygen sublattice costs indeed a very small positive energy or even a negative one in the case of TiO, while in the isomorphous compound TiC it cost definitely a positive energy in agreement with experiment. The lowering of the generation energy of vacancies in TiO is mainly caused by the appearance of new electronic states associated with the defects below the Fermi energy.

2.3.4 Failure of the *ab initio* Calculation

The success of the *ab initio* calculations some of which have been discussed so far leads us to the expectation that the band structure obtained by it would reproduce the excitation spectra obtained by photoemission or by inverse photoemission; the former introduces a hole and the latter an electron to the ground state if temperature is low enough [2.6]. As was mentioned in Sect. 2.1, however, the measured band width and exchange splitting in ferromagnetic nickel are smaller by 40–50% than the calculated values, while reasonable agreements are obtained for both the quantities in the case of iron and for the d band width in copper. The case of cobalt seems to be intermediate, the experimental data being smaller than the calculation by about 20%.

Since the validity of the density functional theory is limited in principle to the ground state properties, the discrepancies may not hurt its usefulness, though certainly it casts some shadow on it. The discrepancy in the case of nickel has been ascribed to the many-body effect which the *ab initio* calculation might fail to take into account [2.35, 36]. The theories are based on the so-called Hubbard model defined by the tight binding Hamiltonian given by

$$H = \Sigma_{i,j,\sigma} t_{ij} a_{i\sigma}^{+} a_{j\sigma} + U\Sigma_{i} a_{i\uparrow}^{+} a_{i\uparrow} a_{i\downarrow}^{+} a_{i\downarrow}, \tag{2.2}$$

where i and j specify the lattice sites; the first term represents the band energy; the second the intraatomic coulomb energy; $\sigma = \uparrow$ or $\downarrow$ denotes the spin direction. The model can be extended to the case of multi-orbitals per atom if necessary.

The discussion of the many body effects starts with the assumption that the *ab initio* band structure calculation corresponds to the Hartree-Fock approximation on the model Hamiltonian (2.2) which replaces the interaction term by

$$U\Sigma_i \langle a_{i-\sigma}^+ a_{i-\sigma} \rangle a_{i\sigma}^+ a_{i\sigma} \tag{2.3}$$

with $\langle n_{i-\sigma} \rangle = \langle a_{i-\sigma}^+ a_{i-\sigma} \rangle$ representing the self-consistently calculated average value. In the case of nickel, we take usually the hole picture, assuming 0.6 holes per atom on the average. The many-body theories [2.35, 36] incorporate two and three particle correlations by going beyond the Hartree-Fock approximation. Since holes of parallel spin will not occupy the same atom in the model (2.2) by the exclusion principle, the exchange splitting or in other words the energy difference of the energy between two holes of opposite spins in the same band state is determined by $U(\langle n\uparrow \rangle - \langle n\downarrow \rangle)$ in the Hartree-Fock approximation with $\uparrow$ and $\downarrow$ denoting majority and minority spin, respectively.

When the coulomb interaction U is large compared with the band energy measured by t, the motion of two holes in energetically low lying states will occur in the following way. Suppose two holes of opposite spins are traveling in a crystal: As far as they are in different atoms, they do not interact with each other in the model given by (2.2). If they are going to enter in an atom simultaneously, they will change their motion to avoid each other to keep the lower energy. Thus the effective number of holes of opposite spin which is seen by a given hole at an atom and consequently, the exchange splitting itself will be reduced from the value in the Hartree-Fock approximation. In the case of high electron density we should take into account more than two particle correlations. It is shown by a sophisticated argument [2.35, 36], however, that the exchange splitting in the ferromagnetic state of nickel is reduced mostly due to the two particle correlation mentioned above. The band width problem is more subtle, since the two and three body correlations do not give rise to a reduction of the band width in the single orbital model, while one can expect a reduction in the multi-orbital model [2.36].

We point out here that no discussion has been given on the fact that the photoelectron spectroscopy measurements are consistent with the *ab initio* calculation in the case of iron and copper. There is a possibility that the many-body effects discussed so far for the Hubbard model are taken into account to a large extent in the *ab initio* calculation based on the local density approximation. The present author [2.37] has recently pointed out that there is a possible mechanism for the reduction of both the band width and the exchange splitting which is operative only in the case of nickel. When a hole is created in a d state of a given atom in the photoemission experiment, it can be shielded by increasing d electrons in other d states in the case of iron, while it is shielded by electrons in s and p states in the case of nickel. It is argued that the shielding charge in the latter case inevitably extends to neighboring atoms, which can not be taken into account in the local density approximation. It is argued further that the shielding may cause the reduction of both the band width and the exchange splitting. In the case of iron, on the other hand, the local density approximation may be assumed to be capable of taking into account the essential part of the electron correlation. We refer the details of the argument to [2.37].

Certainly, there should be a limit to the capability of the *ab initio* calculation based on the density functional theory. However, the many-body effect is not the only answer for the discrepancy with experiment. We need to examine the interplay between the electronic structure and electron correlation carefully. We may quote the case of the energy gap in Si as another example showing the importance of the interplay, though it is not a transition metal system. The energy gap obtained by a straightforward application of the local density approximation is about a half of the observed energy gap of about 1 eV. It is argued that an improvement on the local density approximation based on a careful treatment of the electronic structure can remove the discrepancy to a large extent [2.7, 38].

2.3.5 Enhancement of Ferromagnetism in Iron by Nonmagnetic Atoms

We discuss the possibility of enhancing the ferromagnetism of iron by making either alloys or compounds with nontransition elements. One of the motivations is given by the celebrated new permanent magnet material $Nd_2Fe_{14}B$ recently found by *Sagawa* et al. [2.39] in which some of Fe atoms possess the magnetic moment nearly equal to 3 μ_B. Another example is $Fe_{16}N_2$ found by *Kim* and *Takahashi* [2.40] where the average Fe moment may amount to 3 μ_B. In the case of $Nd_2Fe_{14}B$ it is often argued that boron is an ingredient stabilizing the crystal structure and does not contribute directly to the ferromagnetism. The present author has presented a discussion that boron plays an essential role in enhancing the ferromagnetism [2.41]. The discussion which can be extended to other cases may serve in finding new ferromagnets containing nontransition elements such as boron, carbon, nitrogen, etc., as will be discussed below.

The local electronic structure of Fe atoms surrounding a boron resembles that of Co atoms, because the mixing with s and p states of boron shifts the d bands of both spin directions towards lower energy due to the level repulsion mentioned above and the decrease of the weight of d symmetry states. As was discussed in Sects. 2.2.2 and 2.3.2, 2s and 2p states of boron are situated above the d band. These pseudo-cobalt Fe atoms then enhance the ferromagnetism of more distant Fe atoms. This picture is consistent with the magnetic moment distribution found by neutron and Mössbauer experiments on $Nd_2Fe_{14}B$ [2.42].

There is experimental evidence for the importance of boron. Related compounds R_2Fe_{17} which have a similar atomic arrangement of Fe atoms in spite of the absence of boron, have smaller magnetizations and lower T_c. Moreover, an unpublished *ab initio* calculation made by Asano shows that the electronic structure of the compounds is quite similar to that of $R_2Fe_{14}B$; the only conspicuous difference is the fact that the d band states of majority of the spins of Fe atoms are almost entirely below the Fermi level in the latter compound, while some portion of them remains above the Fermi level in the former. The difference may be ascribed to the downward shift of the d band due to the presence of boron in the latter.

2.4 Structural Problems

Structural problems of metallic systems constitute a field in which the calculation of the electronic structure is becoming predominantly important. As was mentioned in Sect. 2.3.1, the determination of the crystal structure which a given material should take as the state of minimum energy can be sometimes beyond the accuracy of the calculation. On the other hand, we now have many examples in which the calculation has proved that the electronic structure is a key to the understanding of an observed structural phase transition. In some cases, it can reproduce the phase diagram of a given alloy system quite satisfactorily. In the following, we discuss some examples which may reflect the present status of the field.

2.4.1 Methods of Calculating Phase Diagram of Alloy Systems

The phase diagram of an alloy system is usually characterized by phase transitions which may be divided into two categories, those accompanied by a change of the basic lattice and those of the order–disorder type on the same basic lattice. In the Cu–Au system, for example, we have only the phase transitions of the latter type on a fcc lattice. We may also include some cases of the segregation in the latter where component metals have the lattice of the same type; Cu–Ag is a typical example. We briefly summarize the methods of calculation in the following, referring details to literature [2.43].

In order to construct a phase diagram, we need to compute the entropy of atomic disorder on a given lattice at finite temperatures besides that associated with lattice vibrations which can be taken into account by standard methods if necessary. For simple models, we may apply various sophisticated methods to the calculation of the total free energy. In the practical problems of alloys where effective interatomic interactions are of extended range, the cluster variation method proposed by *Kikuchi* [2.44] as early as 1951 and the Monte-Carlo simulation calculation [2.45] are the practicable methods at the present time. The cluster variation method starts with selected clusters of lattice sites, for example, in the case of the fcc lattice the tetrahedron consisting of mutually nearest neighboring sites, the octahedron whose three axes connect the second neighboring sites and subclusters of them. In a general atomic arrangement of a given alloy we will find clusters of various atomic configurations and we can classify each cluster by its atomic configuration. An atomic arrangement can be characterized by the number of the clusters of each class. The entropy is expressed as a function of these numbers of the clusters. With the energy computed from an assumed model for interatomic interactions we obtain the free energy which is to be minimized by varying the numbers of the clusters.

There are two types of the approaches to obtain the interatomic interactions on a given lattice from the electronic structure calculation; the interactions between nuclear charges are included in any case. In the first approach which was initiated by *Gautier* et al. [2.46], we start with the disordered state of alloys whose electronic state is calculated by the coherent potential approximation. The interatomic interactions are obtained as a perturbation energy produced by a concentration variation in the disordered state. The method was successfully applied to transition metal alloys in the framework of the tight binding approximation.

In the second approach we deduce the interatomic interaction from the energies of various regular arrangements of atoms computed by the *ab initio* calculation. The method was first proposed by *Connolly* and *Williams* [2.47] and applied extensively by *Terakura* et al [2.48] and *Takizawa* et al. [2.49] to Cu–Au, Au–Ag, Ag–Cu, and other alloys. A similar method is applied by *Ferreira* et al. [2.50] to semiconducting compounds.

2.4.2 Ordering on fcc Lattice

Before going into the discussion of the electronic structure calculation, we briefly mention the analysis of ordered structures by use of the phenomenological energy expression for alloys $A_{1-x}B_x$ on a given lattice

$$E = \Sigma_k V_k p_k + \dots , \tag{2.4}$$

where p_k is the total number of kth neighboring pairs of B (or A) atoms and V_k is the corresponding interaction constant defined by $V_k = V_k^{AA} + V_k^{BB} - 2V_k^{AB}$ in terms of the interaction constants of AA, BB, and AB pairs. The energy expression can comprise three and four body interactions if necessary. With fixed interaction constants a set of ordered structures appear as the ground state at special concentrations x_i ($i = 1, 2, \dots$) and the ground state for a general concentration x falling on an intermediate range $x_i < x < x_{i+1}$ is given by a two-phase mixture of the ordered states appearing at x_i and x_{i+1}. The special concentrations are determined by the interaction constants and lattice geometry.

A method for the analysis of the ordered states with a given set of interaction constants was developed by the present author and *Kaburagi* [2.51–53]. Figure 2.5 shows the result for fcc lattice in the case where only V_1 and V_2 are taken into account. Figure 2.6 shows examples of relevant ordered structures. Table 2.3 lists typical ordered alloys on fcc which are classified on the basis of the analysis [2.51].

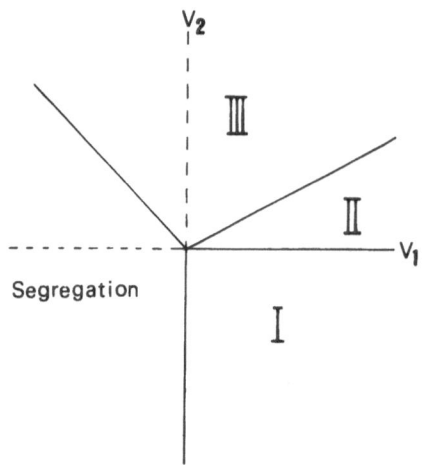

Fig. 2.5. The regions in V_1–V_2 plane where the ordered alloys listed in Table 2.3 and Fig. 2.6. appear

Table 2.3. Examples of ordered fcc alloys [2.51]

Type I min. p_1, max. p_2

Cu_3Au, $CuAu_3$, Cu_3Pd, Ag_3Pt, Au_3Pd, Ni_3Pt, Ni_3Fe, $NiFe_3$, Ni_3Mn, Fe_3Pt, $FePt_3$, Fe_3Ni, Mn_3Pt, $MnPt_3$, Pt_3Co, Pd_3Fe, Ir_3Cr, Pd_3MnI, $CuAuI$, $NiPt$, $NiMn$, $FePt$, $PdFe$

Type II min. p_1, min. p_2 under the cond. of min. p_1

Ni_4Mo, Ni_4W, Au_4V, Au_4Cr, Au_4Mn, Ni_3V, Pd_3V, Pd_3Nb, Au_5Mn_2, Ni_2V, Ni_2Cr, Pd_2V, Pt_2V,

Type III min. p_2, min. p_1 under the cond. of min. p_2

$CuPt$

There is a systematic trend for the occurrence of characteristic ordered alloys. Alloys of the CuAuI and Cu_3Au types (type I) characterized by the condition of minimum p_1 and maximum p_2 are found in the cases where the average number of d electrons is relatively high. On the other hand those of type II satisfying the condition of minimum p_1 and minimum p_2 under the condition of minimum p_1 are found in the cases of somewhat lower d electron number. We can also note that intermediate structures which do not satisfy the minimum or maximum condition for p_2 are found in Mn–Pd and other systems near the boundary values of the average d electron number. This phenomenological conclusion [2.52] was later supported by the electronic structure calculation made by *Bieber* et al. [2.54]; they concluded for example, that the energy difference between the $L1_2$ (Cu_3Au) type (type 1: min. p_1 and max. p_2) and the DO_{22} type (type 2: min. p_1 and min. p_2) changes its sign when the average d electron number per atom is near 7. In fact, Ni_3Mn is of the type I while Ni_3V is of the type II, for example.

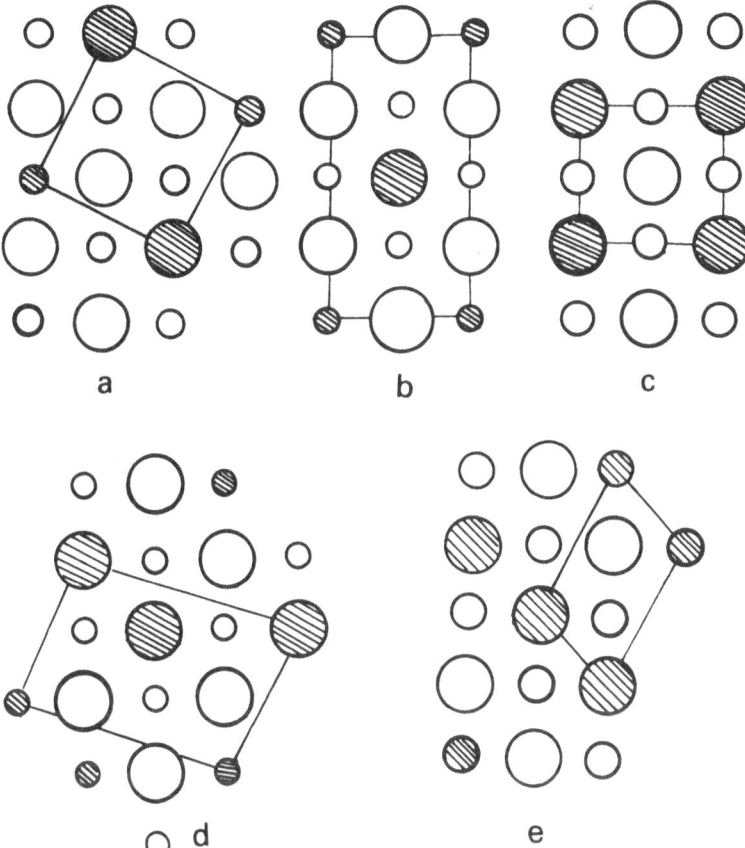

Fig. 2.6. Structures of some ordered alloys listed in Table 2.3. Each figure represents the projection on the (0 0 1) plane; large and small circles correspond to those at (x, y, n) and $(x, y, n + 1/2)$ sites with integer n, respectively. Hatched circles indicate the sites occupied by minority component atoms; all the sites in the direction [0 0 1] are occupied by atoms of the same species; lines indicate a unit cell. (a) A_4B of type II, (b) A_3B of type II, (c) A_3B of type I, (d) A_5B_2 of type II, (e) A_2B of type II

Vacancies on the anion lattice of compounds such as $TiO_{0.5}$ [2.53, 55, 56] is another example of orderings. Part of their ordering patterns can be understood with models of extended range pair interactions such as that given by (2.4).

2.4.3 Examples of *ab initio* Calculations

As was mentioned at the end of Sect. 2.4.1 *Terakura* et al. [2.48] and *Takizawa* et al. [2.49] have carried out extensive investigations on fcc alloys such as Au–Ag, Ag–Cu, Au–Cu, and also A–B with A = Ni or Pd or Pt and B = Cu or Ag or Au which are based on the *ab initio* density functional calculation of the electronic structure of regular alloys and the cluster variation calculation of the phase diagram through the *Connolly* and *Williams* [2.47] approach. They investigate not only fcc base ordered structures but also bcc base structures in some cases, though we shall not discuss the latter possibility here. The Cu–Ag system is a well-known example of segregation in which no ordered structure is found while the Cu–Au system is characterized by the appearance of Cu_3Au, CuAuI, and $CuAu_3$ ordered structures which satisfy the condition of minimum p_1 and maximum p_2. Alloys of compositions near 50–50 have a modulated structure called CuAuII in a narrow temperature range above CuAuI. The Au–Ag system makes a solid solution in the whole concentration range; the transition to ordered structures seem to occur at a temperature too low to be observed.

The calculation made by *Terakura* et al. [2.48] can reproduce quantitatively well the phase diagram and heat of solution. It is an interesting question why the Cu–Ag system cannot make a solid solution. The calculation of *Terakura* et al. [2.48] shows that the underlying mechanism cannot be ascribed to the electron affinity in contradiction to *Miedema*'s phenomenological discussion [2.2]. A clear-cut understanding of the mechanism is not available at the present time. Since the calculation can reproduce the observed results well, the underlying mechanism should be found within the calculation.

The Cu–Pt system is another interesting example. The structure of ordered alloys near $Cu_{0.5}Pt_{0.5}$ is the only example which satisfies the condition of minimum p_2 and minimum p_1 under the condition of minimum p_2 (type III). *Takizawa* et al. [2.49] have shown that indeed the structure is of minimum energy for this system.

The calculations carried out by *Terakura* et al. [2.48] include not only the pair interactions but also several more-than-two particle interactions when they deduce the interaction constants for the phase diagram calculation based on the cluster variation method. However, they conclude that the pair interactions are predominantly large in favor of the phenomenological analysis mentioned in Sect. 2.4.2.

2.4.4 Lattice Distortion

Many lattice distortions which are not associated with a rearrangement of atoms originate in the electronic structure. The cubic to tetragonal transition of ferromagnetic Fe–Pd alloys found by *Oshima* et al. [2.57] is a typical example. The system is fcc or bcc disordered alloys which make a transition from cubic to tetragonal with $c/a > 1$ with decreasing temperature at the composition near $Fe_{0.7}Pd_{0.3}$. The transition conserves approximately the volume and is almost of the second kind. A sketch of the phase diagram is given in Fig. 2.7.

Akai [2.58] has carried out a KKR–CPA calculation which shows that the transition is of electronic origin. Moreover, he has concluded that the tendency towards the tetragonality is a general one to be expected in transition metal

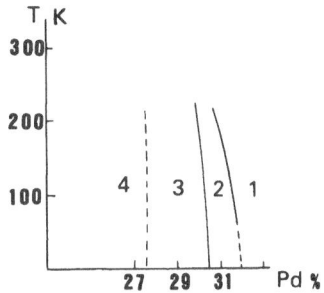

Fig. 2.7. A sketch of the phase diagram $Fe_{1-x}Pd_x$. Phase 1. is *fcc*, 2. *fct* (*tetragonal*), 3. *bct*, and 4. *bcc* [2.57]

alloys of similar compositions, Fe–Ni and Fe–Pt. This discussion is in agreement with the observation that the elastic constant $c_{11}-c_{12}$ decreases near the martensitic transformation even in the cases where the transition to a tetragonal symmetry is not observed.

Akai's calculation concerns the stability of the cubic phase at absolute zero with varying alloy concentration. The axial pressure which drives away from or back to the cubic symmetry is calculated for the system assuming an axial ratio slightly off the value 1 (0.98 is the assumed value). The contribution of Fe atoms is of the sign causing the instability, while that of Pd atoms is of the sign favoring the cubic symmetry. It seems that the alloying effect distorts the local density of states of minority spin at Fe atoms to cause the band Jahn–Teller effect. In order to produce a sufficient distortion of the local density states we need a certain amount of Pd concentration, while the larger the Pd concentration is, the larger their stabilizing contribution. Thus the transition to the tetragonality will occur in a narrow range of the Pd concentration if any. Otherwise, we shall observe a lowering of $C_{11}-C_{12}$ there. Akai assumes that the sign of $c/a - 1$ is determined by the higher order terms of the lattice strain which is not effective at the assumed lattice distortion in the calculation of the axial pressure.

2.5 Limitation of the One Electron Theory

Blinded or encouraged by the success of the one electron theory in the field of metals and alloys, we tend to extend the discussion to compounds presumptuously. As far as so-called intermetallic compounds such as those of the Laves phase are concerned, the calculation seems to yield a meaningful picture of the basic electronic structure. However, oxides which are in the limelight at present time because of high T_c superconductivity have confronted the theory in such a way that it is difficult to credit the *ab initio* calculation with a merit in predicting or interpreting the electronic properties.

"Why is NiO insulating?" is a classical question raised by *Mott* [2.59]. The *ab initio* calculation seems to fail to yield the insulating phase [2.60]. The problem was discussed with important progress, albeit intermittently, before the advent of the high T_c superconductivity. First the concept of the Mott insulator was established. Suppose two Ni^{2+} ions are sufficiently distant from each other in a crystal of NiO. If the energy increase in the process described by

$$Ni^{2+}(d^8) + Ni^{2+}(d^8) \Rightarrow Ni^+(d^9) + Ni^{3+}(d^7)$$

is larger than the energy gain due to the migration of d^7 and d^9 states, i.e., the band energies, one can conclude that the insulating state is stable. We can show easily that the energy associated with the above mentioned process is roughly given by the intraatomic coulomb energy between two d electrons by simply comparing the number of intraatomic electron pairs before and after the process.

If this picture were justified for the actual nickel oxides, we could confine ourselves to the d states. However, *Fujimori* et al. [2.61] has pointed out that the process

$$Ni^{2+}(d^8) + Ni^{2+}(d^8) \Rightarrow Ni^+(d^9) + Ni^{2+}(d^8)L(-1)$$

can cost a smaller energy than the process discussed above, where $L(-1)$ means a hole in the ligand, i.e., surrounding oxygen ions. The energy associated with this process is the energy cost, Δ, to transfer an electron from oxygen to Ni^{2+}. The insulating phase is stabilized if Δ is larger than the energy gain due to the migration of the states d^9 and $L(-1)$.

Sawatzky and *Allen* [2.62] generalized Fujimori's argument to derive the phase diagram sketched in Fig. 2.8 in terms of the intraatomic coulomb interaction parameter U and the charge transfer energy Δ. There is a hyperbola-like boundary for the insulating antiferromagnetic state of Ni compounds, where NiO seems to be well inside it. However, NiS seems to be near the boundary, since under pressure or by raising temperature it undergoes an insulator to metal transition [2.63]. The photoemission experiment carried out by *Fujimori* et al. [2.64] indicates that the spectra change only slightly through the transition in contradiction with the result of the electronic structure calculation based on the one electron approximation. We can suspect that the metallic phase near the boundary is a highly correlated electronic state which is remote from that obtained by the one electron approximation.

The high T_c Cu compounds seem to be close to the boundary between the insulating antiferromagnetic phase and metallic one in the stoichiometric composition [2.65]. It may not be very surprising that extra holes and electrons brought in the system by nonstoichiometry will be in a highly correlated state

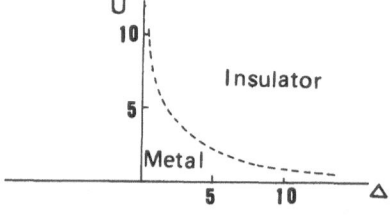

Fig. 2.8. A sketch of the Sawatzky-Allen diagram. The boundary curve is the result of an unpublished calculation made by M. Takahashi and J. Kanamori. U and Δ are measured in the unit of the matrix element t of the transfer between Ni and its ligand

for which the *ab initio* calculation of one electron states is not suitable. We do not know yet the extent of diversification of the manifestation of the electron correlation near the boundary between the insulating and metallic phases. The high T_c superconductivity has been found so far in Cu compounds but not in Ni and other transition element compounds. This may be related to the fact that more-than-one holes (electrons) occupy the same atom simultaneously in the latter cases. In relation to it we may remark that the heavy fermion behavior of conduction electrons found in actinide and rare-earth compounds is another manifestation of highly correlated metallic states. In transition metal systems a Laves compound $Y_{0.97}Sc_{0.3}Mn_2$ [2.66] is found to exhibit a similar behavior. The interplay between the electronic structure and electron correlation seems to be the field which awaits further exploration.

2.6 Future Development

We have discussed the significance of the electronic structure calculation, particularly that of the *ab initio* calculation for transition metal systems. The success of the calculation is quite useful particularly for the cases where the underlying mechanism of the experimental findings is obscure. If the *ab initio* calculation is successful, we may hope that we can find the mechanism by an analysis of the calculation. In some cases, however, we encounter a difficulty to draw an unambiguous physical picture from it. We may say in such cases that there remains a gap between the calculation and theoretical understanding. Thus, approaches based on simplified models will not lose their significance in the future provided that they can give a transparent physical picture.

Nevertheless, the *ab initio* calculation will be ever important for metallic systems. It is reaching the stage of applying it to the material design. Also, it can replace experiment in the problems where extreme conditions not realizable in laboratory should be simulated as in planetary science. However, we need to extend its capability to aperiodic systems. *Car* and *Parrinello* [2.67] have developed a method to combine the electronic structure calculation with the molecular dynamics which can treat even a system *in motion*. Being based on the pseudopotential approach which is not appropriate for transition elements, the method is not applicable to our problems at the present time. *Akai* et al. [2.68] have recently proposed an alternative approach applicable to transition metal systems based on the full potential KKR method [2.69]. We can hope that the method can take into account, for example, the lattice relaxation around an impurity in the near future.

It is a future problem to develop a tool for exploring the electronic states of insulating and nearly insulating compounds; we possess at the present time no general method as powerful as the *ab initio* calculation of the metallic state. The problems mentioned in Sect. 2.5 produce at the present time an insurmountable barrier for the one electron approximation.

References

2.1 J. Friedel: Ann. Phys. (France) **1**, 257 (1976)
     J. Friedel, C.M. Sayers: J. Physique **38**, 697 & L263 (1977)
2.2 A.R. Miedema: Philips Techn. Rev. **36**, 217 (1976)
2.3 D.G. Pettifor: Solid State Commun. **28**, 621 (1978); Solid State Physics **40**, 43 (1987)
2.4 H. Hasegawa, J. Kanamori: J. Phys. Soc. Jpn. **31**, 382 (1971); J. Phys. Soc. Jpn. **33**, 1599, 1607 (1972)
2.5 P. Hohenberg, W. Kohn: Phys. Rev. **136**, B864 (1964)
     W. Kohn, L.J. Sham: ibid. **140**, A1133 (1965)
2.6 D.E. Eastman, F.J. Himpsel, J.A. Knapp: Phys. Rev. Letters **44**, 95 (1980)
     D.E. Eastman, J.F. Janak, A.R. Williams: J. Appl. Phys. **50**, 7423 (1979)
2.7 R.O. Jones, O. Gunnarsson: Rev. Mod. Phys. **61**, 689 (1989)
2.8 N. Stefanou, A. Oswald, R. Zeller, P.H. Dederichs: Phys. Rev. **B35**, 6911 (1987)
     B. Drittler, N. Stefanou, S. Blugel, R. Zeller, P.H. Dederichs: Phys. Rev. **B40**, 8203 (1989)
2.9 M. Akai, H. Akai, J. Kanamori: J. Phys. Soc. Jpn. **56**, 1064 (1987)
2.10 P. Soven: Phys. Rev. **156**, 809 (1967)
2.11 H. Akai: Physica B **86–88**, 539 (1977)
     H. Akai, J. Kanamori: J. Phys. Soc. Jpn. **51**, 1176 (1982)
2.12 G.M. Stokes, W.M. Temmerman, B.L. Gyorffy: Phys. Rev. Lett. **41**, 339 (1978)
2.13 H. Akai: *Electronic Structure and Lattice Defects in Alloys* ed. by R.W. Siegel, F.E. Fujita (Trans. Tech. Pub., Switzerland), Materials Science Forum **37**, 211 (1989)
2.14 W. Kohn, N. Rostoker: Phys. Rev. **84**, 1111 (1954)
     O.K. Andersen: Phys. Rev. **B12**, 3060 (1975)
     H.J.F. Jansen, A.J. Freeman: Phys. Rev. **B30**, 561 (1984)
2.15 H. Akai: J. Phys. Soc. Jpn. **51**, 468 (1951)
2.16 J.W. Cable, H.R. Child: Phys. Rev. **B1**, 3809 (1970)
2.17 A.J. Freeman: J. Mag. Mag. Mat. **35**, 31 (1983)
2.18 S. Blügel, M. Weinert, P.H. Dederichs: Phys. Rev. Lett. **60**, 1077 (1988)
2.19 M.C. Cadeville, A.J.P. Meyer: Compt. Rend. **255**, 3391 (1962)
2.20 V.L. Moruzzi, J.F. Janak, A.R. Williams: *Calculated Electronic Properties of Metals* (Pergamon, New York 1978)
2.21 P. Bagno, O. Jepson, O. Gunnarsson: Phys. Rev. **B40**, 1997 (1989)
2.22 Y. Nakamura, M. Shiga, S. Kawano: Physica B **120**, 212 (1983)
     R. Ballou, J. Deportes, R. Lemaire, Y. Nakamura, B. Ouladiaf: J. Mag. Mag. Mat. **70**, 129 (1987)
2.23 S. Asano, S. Ishida: J. Mag. Mag. Mat. **70**, 39 (1987)
2.24 H. Akai, M. Akai, J. Kanamori: J. Phys. Soc. Jpn. **54**, 4246, 4257 (1985)
2.25 A.J. Freeman, R.E. Watson: *Magnetism IIA* ed. by G.T. Rado, H. Suhl (Academic, New York 1965) 168
2.26 K. Terakura: J. Phys. F **7**, 1773 (1977)
2.27 H. Katayama-Yoshida, K. Terakura, J. Kanamori: J. Phys. Soc. Jpn. **49**, 972 (1980)
2.28 H. Akai, S. Blügel, R. Zeller, P.H. Dederichs: Phys. Rev. Lett. **56**, 2407 (1986)
2.29 P. Blaha, K. Schwarz, P.H. Dederichs: Phys. Rev. **B37**, 2792 (1988)
2.30 J. Friedel: Nuovo Cimento Suppl. **2**, 287 (1958)
2.31 H. Akai, P.H. Dederichs, J. Kanamori: J. Physique **49**, C8-23 (1988)
2.32 N. Hamada: J. Phys. Soc. Jpn. **46**, 1759 (1979)
2.33 L.M. Huisman, A.E. Caelsson, C.D. Gelatt Jr., H.Ehrenreich: Phys. Rev. **B22**, 991 (1980)
2.34 J.B. Goodenough: Phys. Rev. **B5**, 2764 (1972)
2.35 A. Liebsch: Phys. Rev. Lett, **43**, 1431 (1979); Phys. Rev. **B23**, 5203 (1981)
2.36 J. Igarashi: In *Electron Correlation and Magnetism in Narrow-Band Systems* ed. by T. Moriya, Springer Ser. Solid-State Sci. Vol. 29 (Springer, Berlin, Hedelberg 1981) p. 109
     J. Kanamari, A Kotani (eds.): *Care-level Spectroscopy in Condensed Systems*, Springer Ser. Solid-State Sci., Vol. 81 (Springer, Berlin, Heidelberg 1988) p. 168

2.37 J. Kanamori: Prog. Theoret. Phys. **84**, Supplement 101, to be published
2.38 O. Gunnarsson K. Schonhammer: Phys. Rev. Lett. **56**, 1968 (1986)
2.39 M. Sagawa, S. Fujimura, N. Togawa, H. Yamamoto, Y. Matsuura: J Appl. Phys. Pt, II **55**, 2083 (1984)
2.40 T.K. Kim, M. Takahashi: Appl. Phys. Lett. **20**, 492 (1972)
2.41 J. Kanamori: Proc. 10th Intl. Workshop on *Rare-Earth Magnets and Their Applications* (Soc. Nontraditional Technology, Tokyo 1989) 1
2.42 D. Givord, H.S. Li, F. Tasset: J. Appl. Phys. **57**, 4100 (1985)
2.43 D. de Fontaine: *Electronic Structure and Lattice Defects in Alloys*, ed. by R.W. Siegel, F.E. Fujita (Trans. Tech. Pub., Switzerland 1989) Materials Science Forum **37**, 25; Solid State Physics **34**, 74 (1979)
2.44 R. Kikuchi: Phys. Rev. **81**, 988 (1991)
2.45 K.K. Mon, K. Binder: Phys. Rev. **B42**, 675 (1990)
       K. Binder, D.W. Heermann: *Monte Carlo Simulation in Statistical Physics*, 2nd edn. Springr Ser., Solid-State Sci., Vol. 80 (Springer, Berlin, Heidelberg 1992)
2.46 F. Gautier, F. Ducastelle, J. Giner: Phil. Mag. **31**, 1373 (1975)
2.47 J.W. Connolly, A.R. Williams: Phys. Rev. **B27**, 5169 (1983)
2.48 K. Terakura, T. Oguchi, T. Mohri, K. Watanabe: Phys. Rev. **B35**, 2169 (1987)
2.49 S. Takizawa, K. Terakura, T. Mohri: Phys. Rev. **B39**, 5792 (1989)
2.50 L.G. Ferreira, A.A. Mbaye, A. Zunger: Phys. Rev. **B37**, 10547 (1988)
2.51 J. Kanamori: Prog. Theoret. Phys. **35**, 16 (1966)
       M. Kaburagi, J. Kanamori: Prog. Theoret. Phys. **54**, 20 (1975)
       J. Kanamori: J. Phys. Soc. Jpn. **53**, 250 (1984)
2.52 J. Kanamori, Y. Kakehashi: J. Physique **38**, C7-274 (1977)
2.53 J. Kanamori: Proc. Int'l. Symp. *Modulated Structures* (American Inst. Physics, New York 1979) p. 117
2.54 A. Bieber, F. Ducastelle, F. Gautier, G. Treglia, P. Turchi: Solid State Commun. **45**, 585 (1983)
2.55 C.H. de Novion, V. Moisy-Maurice: J. Physique **38**, C7-211 (1977)
2.56 J.P. Landesman, P. Turchi, F. Ducastelle, G. Treglia: Mat. Res. Soc. Symp. Proc., Vol. 21 (Elsevier, Amsterdam 1984) p. 363
2.57 R. Oshima, K. Kosuga, M. Sugiyama, F.E. Fujita: Sci. Rep. RITU A29 Suppl. 1, 67 (1981)
2.58 H. Akai: *Electronic Structure and Lattice Defects in Alloys*, ed. by R.W. Siegel, F.E. Fujita (Trans. Tech. Pub., Switzerland); Materials Sci. Forum **37**, 211 (1989)
2.59 N.F. Mott: *Metal-Insulator Transitions* (Taylor and Francis, London 1974)
2.60 K. Terakura, T. Oguchi, A.R. Williams, J. Kubler: Phys. Rev. **B30**, 4734 (1984)
2.61 A. Fujimori, F. Minami, S. Sugano: Phys. Rev. **B29**, 5225 (1984)
2.62 A. Sawatzky, J.W. Allen: Phys. Rev. Lett. **53**, 2339 (1984)
2.63 J.T. Sparks, T. Komoto: Phys. Lett. **25**, 398 (1967)
       S. Anzai, K. Ozawa: J. Phys. Soc. Jpn. **24**, 271 (1968)
2.64 A. Fujimori, K. Terakura, M. Taniguchi, S. Ogawa, S. Suga, M. Matoba, S. Anzai: Phys. Rev. **B37**, 3109 (1988)
2.65 A. Fujimori, E. Takayama-Muromachi, Y. Uchida, B. Okai: Phys. Rev. **B35**, 8814 (1987)
       A. Fujimori, E. Takayama-Muromachi, Y. Uchida: Solid State Commun. **63**, 857 (1987)
       A. Congiu Castellano, M. De Santis, P. Delogu, A. Gargano, R. Giorgi: Solid State Commun. **63**, 1135 (1987)
2.66 M. Shiga, H. Wada, Y. Nakamura, J. Deportes, K.R.A. Ziebeck: J. Physique **49**, C8-241 (1988)
2.67 R. Car, M. Parrinello: Phys. Rev. Lett. **55**, 2471 (1985)
2.68 H. Akai, B. Drittler, P.H. Dederichs: *Molecular Dynamics Simulations*, ed. by F. Yonezawa, Springer Ser. Solid-State Sci., Vol. 103 (Springer, Berlin, Heidelberg 1992)
2.69 B.H. Drittler: KKR-Greensche Funktionmethode für das volle Zellpotential, Dissertation Tech. Univ. Aachen (1990)
2.70 A. Svane, O. Gunnarsson: Phys. Rev. Lett. **65**, 1148 (1990)
       Z. Szotek, W.M. Temmerman, H. Winter: Phys. Rev. **B 47**, 4029 (1993)
2.71 T. Kotani: Phys. Rev. **B 50**, 14816 (1994)
2.72 M. Takahashi, J. Kanamori: J. Phys. Soc. Jpn. **60**, 3154 (1991)

2.73 H. Akai: In *Interatomic Potential and Structural Stability*, ed. by K. Terakura, H. Akai, Springer, Ser. Solid-State Sci., Vol. 114 (Springer, Berlin Heidelberg 1992) pp. 42–53

T. Otsubo, Y. Nakayama, I. Minami, M. Tanigaki, S. Fukuda, A. Kitagawa, M. Fukuda, K. Matsuta, Y. Nojiri, H. Akai, T. Minamisono: Hyperfine Interact. **80**, 1051 (1993)

2.74 P. H. Dederichs, R. Zeller, H. Akai, H. Ebert: J. Magn. Magn. Mater. **100**, 241 (1991)

2.75 H. Akai, P. H. Dederichs: Phys. Rev. **B 47**, 8739 (1993)

M. Schroter, H. Ebert, H. Akai, P. Entel, E. Hoffmann, G. G. Reddy: Phys. Rev. **B 52**, 188 (1995)

H. Akai: In *Computational Physics as a New Frontier in Condensed Matter Research*, ed. by H. Takayama (Jpn. Phys. Soc., Tokyo 1995) p. 96

2.76 J. M. D. Coey, Hong Sun: J. Magn. Magn. Mater. **87**, L251 (1990)

2.77 M. Komuro, Y. Kozono, M. Hanazono, Y. Sugita: J. Appl. Phys. **67**, 5126 (1990)

2.78 H. Akai, M. Takeda, M. Takahashi, J. Kanamori: Solid State Commun. **94**, 509 (1995)

2.79 M. N. Baibich, J. M. Broto, A. Fert, F. Nguyen Van Dau, F. Petroff, P. Etienne, C. Creuzet, A. Friederich, J. Chazelas: Phys. Rev. Lett. **61**, 2472 (1988)

Proc. 1st Int'l,Symp. on Metallic Multilayers, Kyoto, March (1993), ed. by S. Maekawa, H. Fujimori, T. Shinjo, R. Yamamoto (North Holland, Amsterdam 1993)

2.80 S. Blugel: Ground state properties of ultrathin magnetic films, Habilitation, Thesis, RWTH Aachen (1995)

3 Structure Characterization of Solid-State Amorphized Materials by X-Ray and Neutron Diffraction

K. Suzuki

Amorphous solids have attracted strong attention in modern materials science for the following two reasons: The first is due to a wide range of utilization of amorphous solids as engineering materials of metals, ceramics, and organic polymers in high-tech fields. The other is a fundamental point of view approach to materials physics and chemistry as opposed to that of the perfect crystal. The properties of amorphous solids are essentially structure sensitive, since the amorphous solid remains in the non-equilibrium state and has a great degree of freedom in the atomic motion and configuration.

The traditional procedures [3.1] for preparing amorphous solids such as melt-quenching, vapor-deposition, and precipitation from solutions are based on the strategy that a metastable solid-state with disordered atomic arrangement is frozen by the rapid removal of the kinetic energy of atoms moving violently in the liquid and gas phases.

Currently, great attentions have been paid to mechanical alloying, firing of metal-organic polymers, and hydrogen induced amorphization as a solid-state process for the preparation of amorphous solids [3.2–4], where atoms in a stable equilibrium crystal are excited and frozen into an energized metastable solid state with disordered atomic structures. It is noteworthy that the solid-state amorphization process is an approach from the opposite direction against the conventional processes.

This chapter describes new aspects of amorphous solid physics focused on the atomic-scale structural evolution during the mechanical solid-state amorphization of Ni–V (miscible system) [3.5] and Cu–Ta (immiscible system) [3.6] metallic alloys, the medium-range structure of Ni-based binary [3.7] and ternary [3.8] amorphous alloys induced by the frustration of chemical bonds among constituent atoms, the nanometer-scale hybrid structure of amorphous Si–C–O–Ti fibers prepared by firing polytitanocarbosilane as an organic precursor [3.9, 10] and the structural contrast for hydrogen induced amorphization between YFe_2 and YNi_2 C15-type Laves phase intermetallic compound [3.11].

3.1 New Generation Scattering Experiments

Neutron and X-ray scattering are still one of the most powerful techniques for characterizing the atomic-scale structure of amorphous solids, although they supply only the one-dimensional information on the radial average of pair correlations. In the last decade, the new generation of radiation sources for pulsed neutron [3.12, 13] and synchrotron radiation [3.14] based on accelerators have reached the stage of routine use in condensed materials research.

In particular, the utilization of pulsed neutron source generated by nuclear spallation reaction has brought the remarkable progress of experimental performances as follows:

(a) enormous extension of the dynamic range for momentum and energy transfer during scattering event,
(b) great improvement in signal-to-noise ratio because of the off-source during measuring time,
(c) attainment of high resolution in both the space and time correlation,
(d) versatile use of time-of-flight spectroscopic technique and so on.

Accelerator-based pulsed sources have a relatively high flux for epithermal neutrons in the range of wavelength less than 0.5 Å. This makes the total structure factor $S(Q)$ measured up to a very high scattering vector value Q beyond 50 Å^{-1}. Furthermore, the use of short wavelength neutrons in epithermal energy region remarkably reduces a doubt about the validity of the static approximation, but makes it more difficult to correct the inelastic scattering effect such as the Placzek correction.

By extending the $S(Q)$ measurement until a very high scattering vector value of $Q > 30$ Å^{-1}, the local coordinations in various different kinds of amorphous solids have been determined to be rather close to those found in crystalline counterparts [3.8, 13, 15]. The structure characterizations described in this chapter were mainly carried out by using the spallation pulsed neutron source KENS installed in the National Laboratory for High Energy Physics, Tsukuba.

3.2 Mechanical Alloying and Mechanical Disordering

3.2.1 Mechanical Amorphization of Ni–V Miscible System

So far ball/rod [3.16] milling techniques have been commonly used for processing amorphous metallic alloys through the solid-state amorphization, in which there are two different kinds of thermodynamic process, that is to say, mechanical alloying (MA) and mechanical disordering (MD), depending on the adopted starting materials. The MA process synthesizes amorphous alloy powders by reacting elemental crystalline powders through long distance solid state chemical diffusion, being often accompanied with the negative heat of formation. In the MD process, however, crystalline alloy or compound powders are transformed into amorphous solid powders by destroying the periodical long-range order of atomic arrangement without any changes of chemical composition due to long distance solid-state diffusion, being usually accompanied with positive heat of formation.

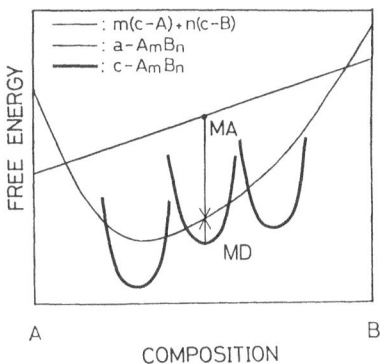

Fig. 3.1. Free energy changes for mechanical alloying (MA) and mechanical disordering (MD)

The MA and MD processes are reactions going to thermodynamically opposite directions of each other, as illustrated in Fig. 3.1. Interesting points are whether the atomic scale structures of MA and MD amorphous alloys still have significant differences between them, and how they are compared with those of melt-quenched (MQ) and vapor-quenched (VQ) amorphous alloys. Since the MA and MD process are slowly carried out as a function of milling time, we can directly follow the topological process of crystal-to-amorphous solid structure transition by neutron diffraction [3.17, 18]. This is another interesting point in this kind of study.

The topological structure of amorphous metals is approximately described in terms of the dense random packing of spheres (DRP), which is exclusively constructed from tetrahedral structure units [3.19]. In the fcc or bcc crystalline lattice, however, both the tetrahedral and octahedral structure units are included with a ratio of 2-to-1. This implies that octahedral structure units must be preferentially destroyed and modified into tetrahedral structure units during the solid-state mechanical amorphization process.

Fukunaga et al. [3.5] have directly observed the topological change from fcc crystal to amorphous solid in the MA process of 4Ni (fcc) + 6V(bcc) → Ni_4V_6 (amorphous) by total neutron scattering. The total neutron structure factor $S(Q)$ of Ni–V binary alloys is described as a weighted sum of three partial structure factors $S_{ij}(Q)$ (i, j = Ni, V) as follows:

$$S(Q) \frac{1}{\langle b \rangle^2} [c_{Ni}^2 b_{Ni}^2 S_{NiNi}(Q) + 2c_{Ni}c_V b_{Ni}b_V S_{NiV}(Q) + c_V^2 b_V^2 S_{VV}(Q)],$$

where c_i and b_i are chemical composition and the coherent scattering length, respectively, for neutron of ith atom, and $\langle b \rangle = c_{Ni}b_{Ni} + c_V b_V$.

As the value of b_V (= 0.0382×10^{-12} cm) is negligibly small compared with that of b_{Ni} (= -1.03×10^{-12} cm), the observed contribution to $S(Q)$ is almost entirely due to $S_{NiNi}(Q)$. Therefore, we can selectively find how the Ni fcc lattice is modified into an amorphous structure during the MA process. Figure 3.2 shows the variation in $S(Q)$ for the MA process of $4Ni + 6V \rightarrow Ni_4V_6$ as a function of milling time. Before starting the MA process, Bragg peaks in $S(Q)$

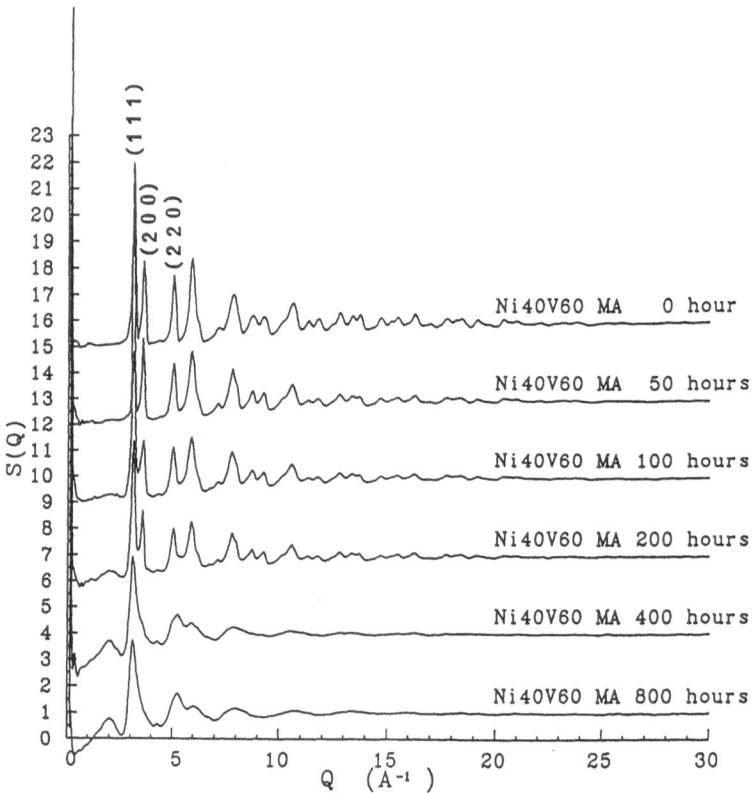

Fig. 3.2. Neutron total structure factors $S(Q)$ for MA process of 4Ni (fcc) + 6V (bcc) → Ni_4V_6 (amorphous) as a function of milling time

appear only from the Ni fcc crystal. These Bragg peaks are quickly broadened and modified with increasing milling time.

The pre-peak located at $Q = 1.9$ Å^{-1} appears slightly just after 100 h of milling time and becomes very obvious with further increase in milling time. This kind of pre-peak is commonly found in $S_{NiNi}(Q)$ of sputter-deposited Ni-42 at % V amorphous alloy and melt quenched Ni-60 at % Ti and Ni-50 at % Zr amorphous alloys [3.7]. *Steeb* and *Lamparter* [3.20] have noticed that the pre-peak is definitely associated with the medium-range ordering of minority atoms with small diameters.

When milling time is beyond 400 h, the (2 0 0) reflection peak is particularly lost and vibration in the high Q region disappears in $S(Q)$. As shown in Fig. 3.3, reduction in the (2 0 0) peak height suggests that the octahedral structure unit in fcc lattice is preferentially destroyed during MA process [3.18]. This situation is clearly demonstrated in the neutron total radial distribution function $4\pi r^2 \rho g(r)$ defined as the Fourier transform of $S(Q)$ truncated at $Q_{max} = 30$ Å^{-1}.

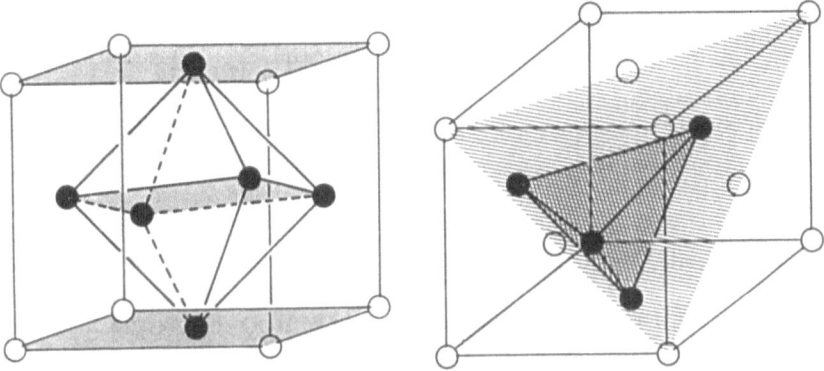

Fig. 3.3. Tetrahedral and octahedral structure units in fcc lattice

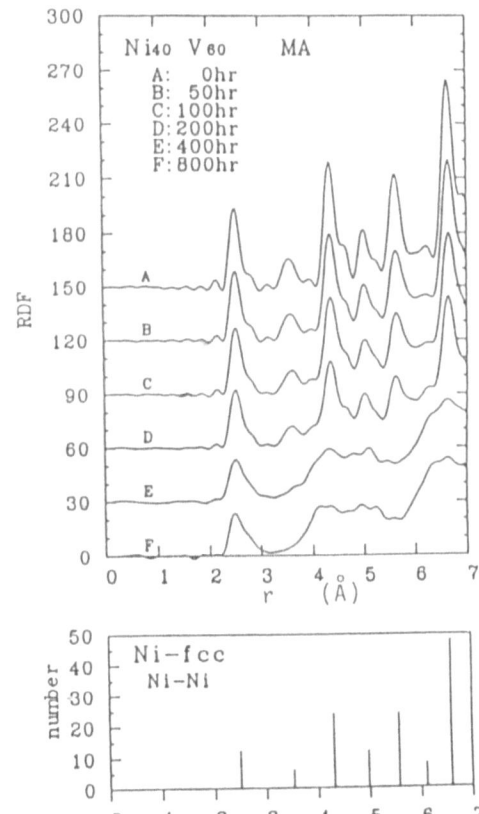

Fig. 3.4. Neutron total radial distribution functions $4\pi r^2 \rho g(r)$ for MA process of 4Ni (fcc) + 6V (bcc) → Ni_4V_6 (amorphous) as a function of milling time

Figure 3.4 shows a characteristic variation in $4pr^2\rho g(r)$ as a function of milling time. The second peak located around $r = 3.5$ Å and the fifth peak around $r = 5.6$ Å are drastically lost with increasing milling time. This behaviour directly corresponds to the preferential disappearance of octahedral structure units in Ni fcc lattice, because the second and fifth neighbor atoms occupy the key sites constructing an octahedral structure unit as shown in a model of Fig. 3.5.

The neutron total structure factors $S(Q)$ for MA Ni–V amorphous alloys obtained after 400 h of MA time are compared as a function of V content in Fig. 3.6 [3.21, 22]. Because of a negligibly small value of the neutron scattering length for a V atom, the $S(Q)$ in Fig. 3.6 is nearly equal to the partial structure factor $S_{NiNi}(Q)$ for Ni–Ni pair in MA Ni–V amorphous alloys. The overall behaviors of $S(Q)$ are almost similar over the wide range of alloy composition from 45 to 75 at % V. Figure 3.7 shows the neutron total radial distribution functions $4pr^2\rho g(r)$ for MA Ni–V amorphous alloys, which are derived as the Fourier transform of $S(Q)$ truncated at $Q_{max} = 23$ Å^{-1}. The $4\pi r^2\rho g(r)$ in Fig. 3.7 are also very similar to each other. This behaviour suggests that the Ni–Ni correlation is rigidly preserved in a whole range of the chemical composition of MA Ni–V amorphous alloys.

In order to evaluate the chemical short-range order persisting in MA Ni–V amorphous alloys, we obtain the coordination numbers of Ni atoms around a Ni atom by calculating the area under the first peak of $4pr^2\rho g(r)$ as shown in Fig. 3.7. The results are shown in Fig. 3.8, in which shaded area corresponds to

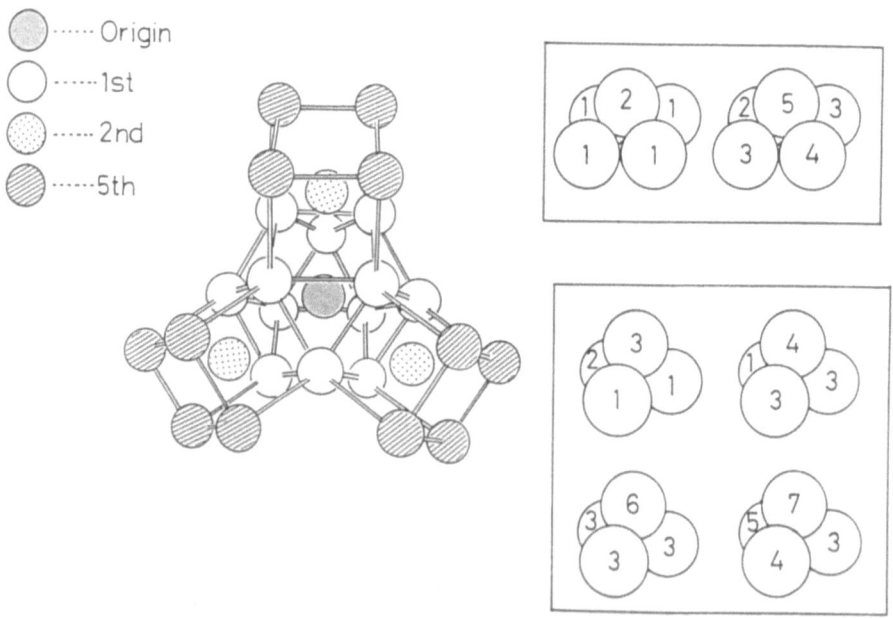

Fig. 3.5. Atomic arrangement around an atom in fcc lattice

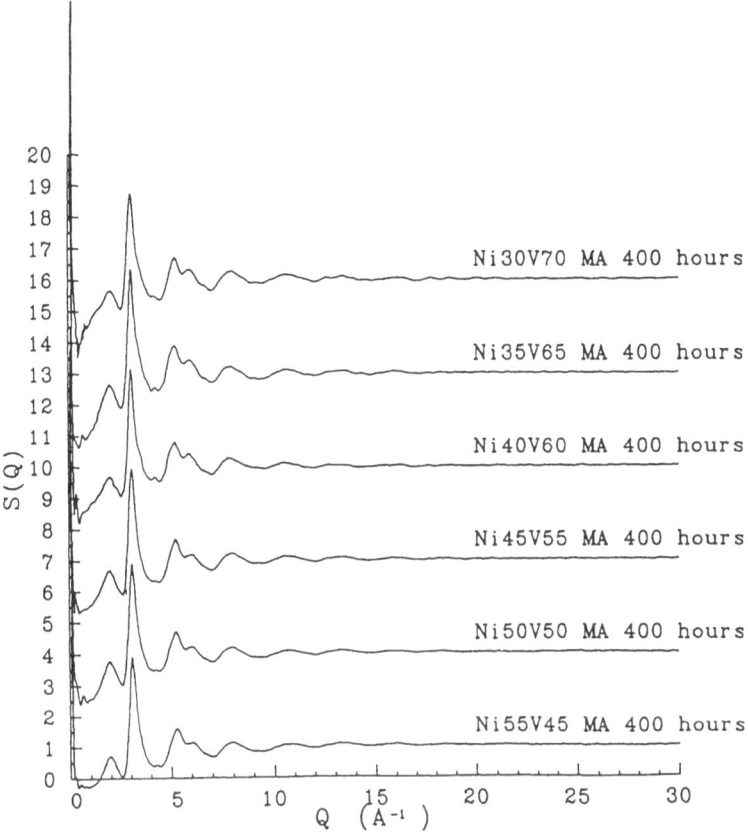

Fig. 3.6. Neutron total structure factors $S(Q)$ of Ni–V amorphous alloys prepared by MA process for 400 h as a function of V content

the possible maximum and minimum contributions of Ni–V pair correlation to the first peak area in the radial distribution function.

It is noteworthy that the range of chemical composition where Ni–V amorphous alloys can be formed by MA process exactly coincides with that of crystalline σ-phase in Ni–V binary system. The crystalline σ-phase (space group; $P4_2/mnm$) has a definite degree of the chemical order among five different atom sites in its unit cell. We suppose a chemically disordered σ-phase in which Ni and V atoms are randomly distributed in all sites of the unit cell. The coordination numbers of Ni–Ni pair correlation in both the chemically ordered and disordered σ-phases are also plotted as a function of V content in Fig. 3.8.

A comparison in Fig. 3.8 shows that the coordination numbers of Ni–Ni pair correlation in MA Ni–V amorphous alloy satisfactorily follow those in chemically ordered σ-phase. This means that chemical short-range structure of Ni–V

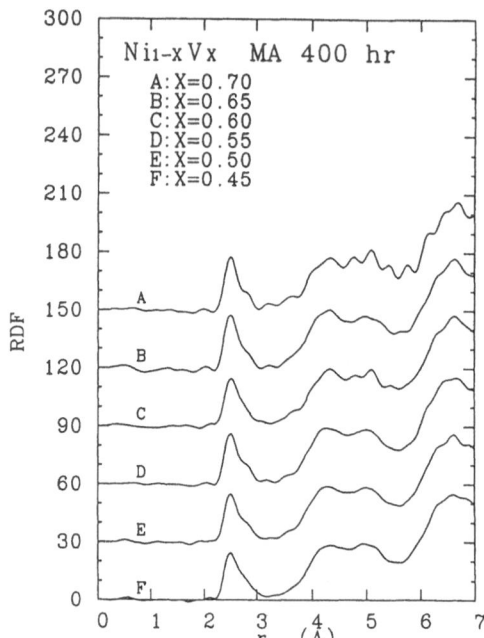

Fig. 3.7. Neutron total radial distribution functions $4\pi r^2 \rho g(r)$ of Ni-V amorphous alloys prepared by MA process for 400 h as a function of V content

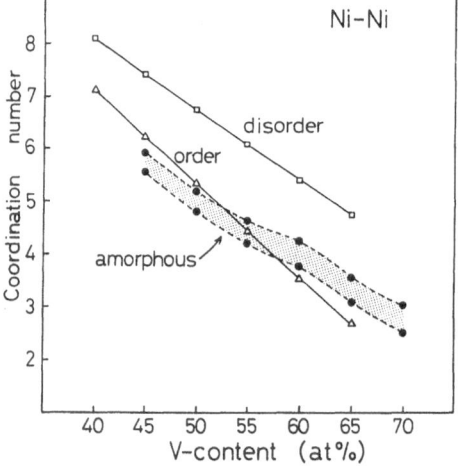

Fig. 3.8. Coordination numbers of Ni atoms around a Ni atom in MA amorphous alloys; ordered σ phase and disordered σ phase of Ni-V binary system as a function of V content

amorphous alloys prepared by MA process is rather similar to that of crystalline σ-phase in the thermodynamic equilibrium state.

During the MD process of Ni_4V_6 crystalline σ-phase compound, the observation of neutron $S(Q)$ in Fig. 3.9 shows that drastic decrease happens in the height of the first sharp diffraction peak (FSDP) corresponding to the (1 0 1)

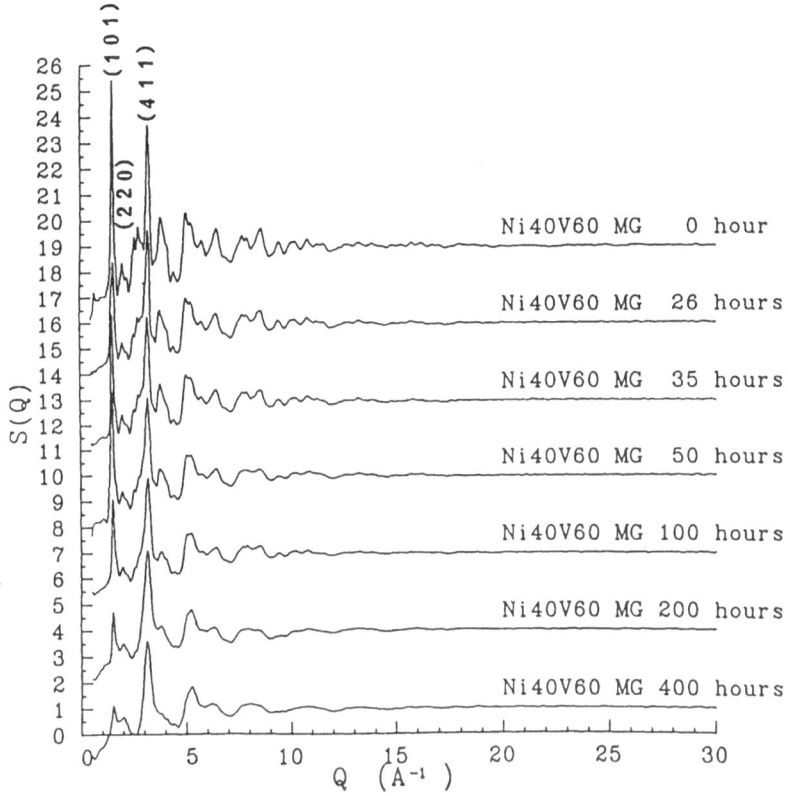

Fig. 3.9. Neutron total structure factors $S(Q)$ for MD process of Ni_4V_6 (crystalline σ phase) $\rightarrow Ni_4V_6$ (amorphous) as a function of milling time

Bragg peak located around $Q = 1.52 \text{ Å}^{-1}$ and there is essentially no change in the shape of pre-peak corresponding to the (2 2 0) peak located around $Q = 2.00 \text{ Å}^{-1}$ [3.22]. Furthermore, we can find that higher index peaks appearing in the range of $Q > 5 \text{ Å}^{-1}$ do not follow significant modifications during MD process. As shown in Fig. 3.10 it is difficult to find dramatic variations in the short-range structure corresponding to the first peak and the medium-range structure around 4 Å in the second peak in the neutron total radial distribution functions. However, we can find significant evolutions in the third peak profile around 5 Å.

The above behavior can be understood in terms of the characteristic variation in the medium-range structure of mutual connection between tetrahedral structure units, because crystalline σ-phase structure is constructed only from tetrahedral structure units. The corner and edge sharing between tetrahedral structure units in the ctystalline σ-phase structure are predominantly destroyed and modified into the face sharing in the amorphous solid structure [3.18, 22]. This speculation is based on the fact that the (1 0 1) Bragg peak corresponds

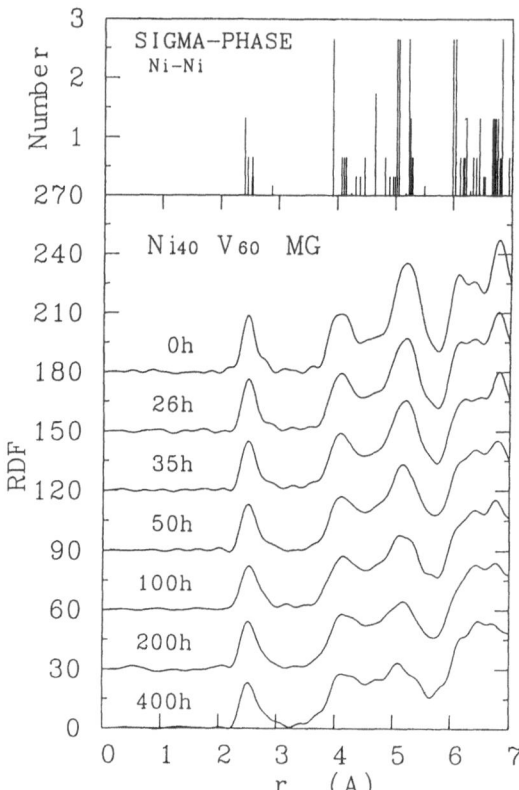

Fig. 3.10. Neutron total radial distribution functions $4\pi r^2 \rho g(r)$ for MD process of Ni_4V_6 (crystalline σ phase) $\rightarrow Ni_4V_6$ (amorphous) as a function of milling time

well to the interatomic distance formed by the corner and edge sharing between tetrahedral structure units, while the (220) Bragg peak is originated from short atomic distances in the close packing formed by face sharing of tetrahedral structure units.

3.2.2 Mechanical Amorphization of Cu–Ta and Cu–V Immiscible Systems

It is well known that the driving force for solid-state amorphizing reaction by chemical interdiffusion is a negative heat of formation between constituent elements [3.23]. However, *Fukunaga* and others have found the amorphous alloy formation by MA process in Cu–Ta [3.6] and Cu–V [3.24] binarty systems, which are immiscible even in the liquid state and characterized by the positive heat of mixing. This means that the MA process accompanied with energizing excitation of constituent atoms does not always need the negative heat of formation.

The atomic-scale structure of Cu–Ta and Cu–V amorphous alloys prepared by MA process was carefully examined by *Fukunaga* et al. [3.6, 24] using

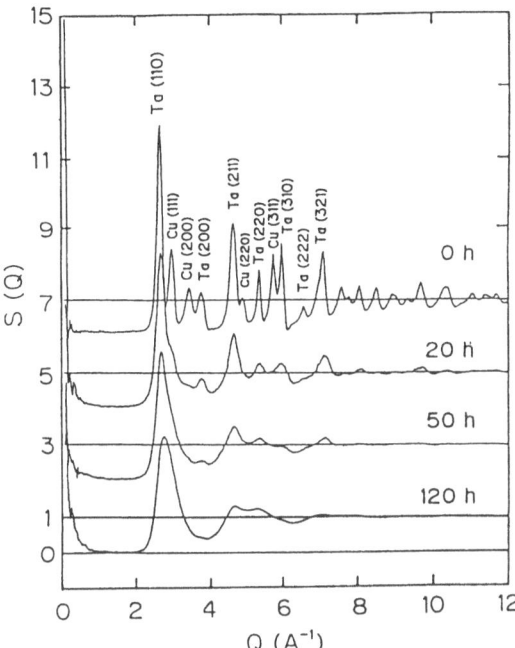

Fig. 3.11. Neutron total structure factors $S(Q)$ for MA process of 3Cu (fcc) + 7Ta (bcc) → Cu_3Ta_7 (amorphous) as a function milling time [3.5]

neutron total scattering. Figure 3.11 shows the neutron total structure factors $S(Q)$ for the MA process of 3Cu (fcc) + 7Ta (bcc) → Cu_3Ta_7 (amorphous) as a function of milling time. The Ta (1 1 0) and Cu (1 1 1) reflection peak become broad with increase of milling time, while (2 0 0) peaks for both Cu fcc and Ta bcc crystal are disappearing and the contribution of small angle scattering is very obvious in the range of $Q < 1$ Å^{-1} with proceeding of amorphization. The neutron total radial distribution functions $4\pi r^2 \rho g(r)$, defined as the Fourier transform of the $S(Q)$ truncated at $Q_{max} = 12$ Å^{-1}, are shown in Fig. 3.12. We can find that the second neighbor atoms in Cu fcc and Ta bcc crystal are selectively shifted with increase of milling time.

The structural evolutions in the MA process of Cu (fcc) + V (bcc) → CuV (amorphous) were also observed as a function of milling time using neutron total scattering by *Fukunaga* et al. [3.24]. As shown in Figs. 3.13, 14, the behavior of $S(Q)$ and $4\pi r^2 \rho g(r)$ is entirely similar to that observed for the MA process of 4Ni (fcc) + 6V (bcc) → Ni_4V_6 (amorphous). Even in immiscible systems, amorphous alloys are formed with atomic-scale mixing in an excited state by the MA process in which octahedral structure units are predominantly modified into tetrahedral structure units.

However, the atomic mixing in both the Cu_3Ta_7 and CuV amorphous alloys are not uniform in the short- and medium range length, because the small angle scattering intensity appears and rather increases in the range $Q < 1$ Å^{-1} in neutron $S(Q)$ with proceeding of MA process. In case for the MA process of

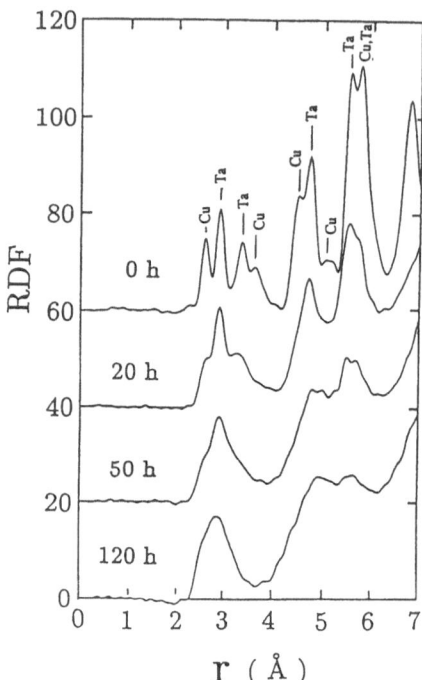

Fig. 3.12. Neutron total radial distribution functions $4\pi r^2 \rho g(r)$ for MA process of 3Cu (fcc) + 7Ta (bcc) → Cu_3Ta_7 (amorphous) as a function of milling time [3.6]

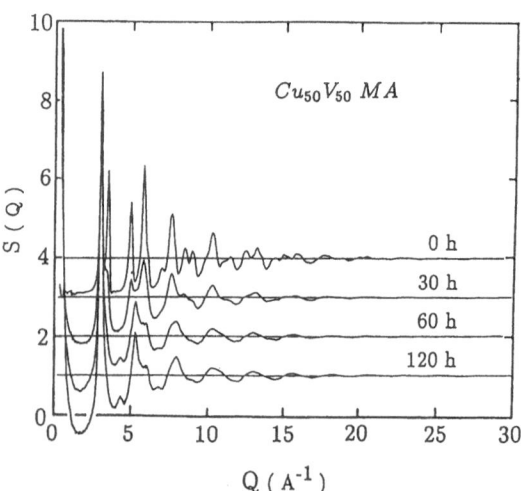

Fig. 3.13. Neutron total structure factors $S(Q)$ for MA process of Cu (fcc) + V (bcc) → CuV (amorphous) as a function of milling time [32.4]

Cu + V → CuV, new peaks are growing around $Q = 4.3$ and 6.7 Å^{-1} in the $S(Q)$, as shown in Fig. 3.13.

These peaks are contributions from Bragg peaks of (200) and (310) planes in a metastable Cu–V bcc solid solution. The MA Cu–V alloy is expected to include both the amorphous phase and bcc crystal. Such coexistence of two

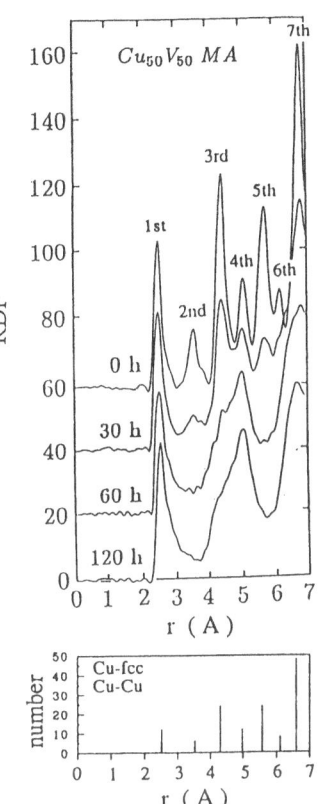

Fig. 3.14. Neutron total radial distribution functions $4\pi r^2 \rho g(r)$ for MA process of Cu (fcc) + V (bcc) → CuV (amorphous) as a function of milling time [3.24]

metastable phases provides a continuous crystallization over a broad temperature range, as confirmed by a DSC measurement by *Fukunaga* et al. [3.24].

3.3 Medium-Range Structure of Metallic Amorphous Alloys

3.3.1 Pre-peak in the Structure Factor of Binary Amorphous Alloys

Small peaks well defined are added around the Q range of $1-2$ Å^{-1} lower than the main peak position in $S(Q)$ of net-work glasses and compound forming liquid alloys. These peaks are often called the first sharp diffraction peak (FSDP) [3.25] and so far is known to originate from the packing of structure units in medium-range length scale.

In case of amorphous metallic alloys, where the dense random packing of spheres is a first step approach to the atomic scale structure, we often observe the pre-peaks in $S(Q)$ [3.8]. The category of the pre-peak [3.26] should be different from that of the FSDP. *Fukunaga* et al. [3.7] have observed that Ni–Ni correlations give rise to the pre-peak around $Q = 1.9 \, \text{Å}^{-1}$ in $S_{NiNi}(Q)$ of Ni–V, Ni–Ti, and Ni–Zr amorphous alloys.

As shown in Fig. 3.15, the pre-peak increases with decreasing Ni content. The Fourier transform of the pre-peak in $S_{NiNi}(Q)$ into r-space shows that the pre-peak in $S_{NiNi}(Q)$ contributes predominantly to the second peak located around $r = 4.1 \, \text{Å}$ in $G_{NiNi}(r)$. In fact, Fig. 3.16 shows that the relative height of the second peak to the first peak in $G_{NiNi}(r)$ increases with decreasing Ni content,

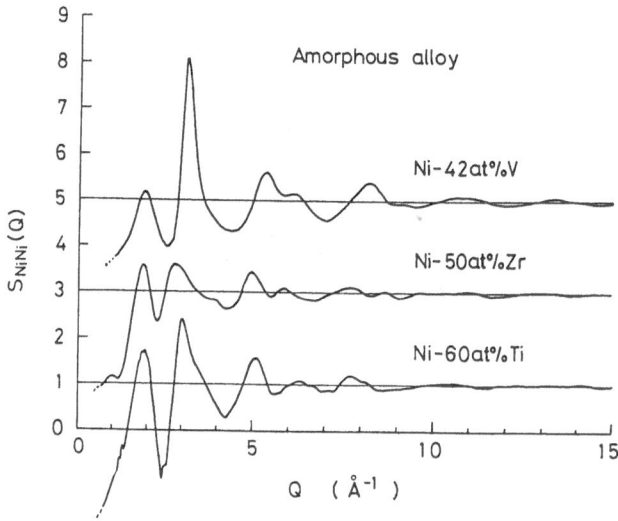

Fig. 3.15. Ni–Ni partial structure factors $S_{NiNi}(Q)$ of Ni_6V_4, Ni_5Zr_5, and Ni_4Ti_6 amorphous alloys

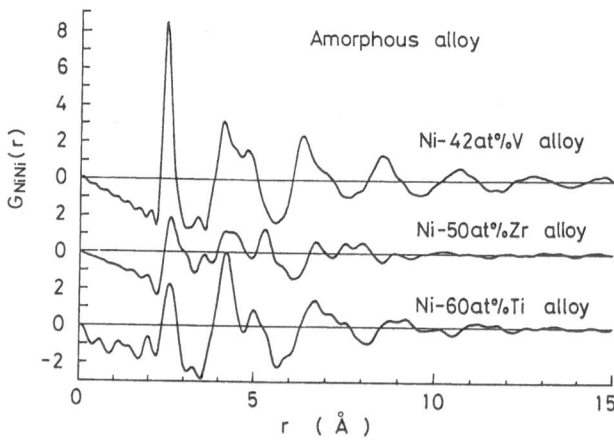

Fig. 3.16. Ni–Ni partial reduced radial distribution functions $G_{NiNi}(r)$ of Ni_6V_4, Ni_5Zr_5, and Ni_4Ti_6 amorphous alloys

corresponding to the decrease of the pre-peak in $S_{NiNi}(Q)$ as shown in Fig. 3.15. As pointed by *Steeb* and *Lamparter* [3.20], the pre-peak is definitely associated with the medium-range ordering of Ni atoms, which is expected to be closely related to the mutual connection among tetrahedral structure units in the amorphous alloy. It is interesting to note that the Ni–Ni pre-peak appears even in a MA and MD NiZr and NiV amorphous alloys, irrespective of preparation methods.

3.3.2 Chemical Frustration in Ternary Amorphous Alloys

Current progress in small angle scattering experiments has revealed that amorphous alloys have a uniform density distribution but contain inevitably the concentration fluctuation over a broad range from several to thousand Ångstroms. The thermodynamical calculation made by *Fujita* [3.27] clearly shows the formation of stable clustering of atoms even in a liquid alloy. Such a concentration fluctuation is a key factor for controlling the properties of multi-component amorphous alloys. Even in a binary alloy, the short-range and medium-range fluctuations of the chemical concentration are deeply correlated to its crystallization process and thermal stability.

Particularly, the local concentration fluctuation induces a drastic modification in the properties of ternary amorphous alloys. For example, Pd–Cu–Si alloys can be vitrified in a bulk state by cooling the melt in water, if several atomic percents of Cu are added into Pd–Si amorphous alloys [3.24]. We know empirically that magnetic, mechanical, and chemical properties are greatly improved in ternary amorphous alloys, compared with binary amorphous alloys, by adding small amounts of a third element.

The chemical bond between A–B unlike atoms is uniquely formed in an A–B binary alloy by mixing A and B metals. Therefore, the concentration fluctuation in the A–B binary alloy is only the variation in the chemical composition A-to-B from site to site in alloy. In an A–B–C ternary alloy, however, there is a freedom for an A atom to choose A–B bond or A–C bond. This is an essentially different behavior in ternary alloys, which can not be found in binary alloys. The essentially same behaviors are expected in quaternary and higher multi-component alloys, which have greater choice for selecting the chemical fluctuation to relax some frustration in the bonding formation among constituent elements.

Evidences for the chemical frustration in ternary amorphous alloys mentioned above have been experimentally observed in Ni–Zr–V [3.25], Ti–Zr–Cu [3.26], and Ti–Zr–Si [3.27] alloy systems by neutron total scattering using the spallation pulsed neutron source KENS.

Figures 3.17, 18 show the neutron $S(Q)$ and $g(r)$ observed for $(Ni_{0.5}Zr_{0.5})_{1-x}$ V_x amorphous alloys prepared by melt-quenching. We find no significant changes in the overall behaviors for both the $S(Q)$ and $g(r)$ with adding V atoms until 25 at %. With increasing V content, however, the pre-peak located around $Q = 1.9$ Å^{-1} decreases and small angle scattering takes place in a low scattering

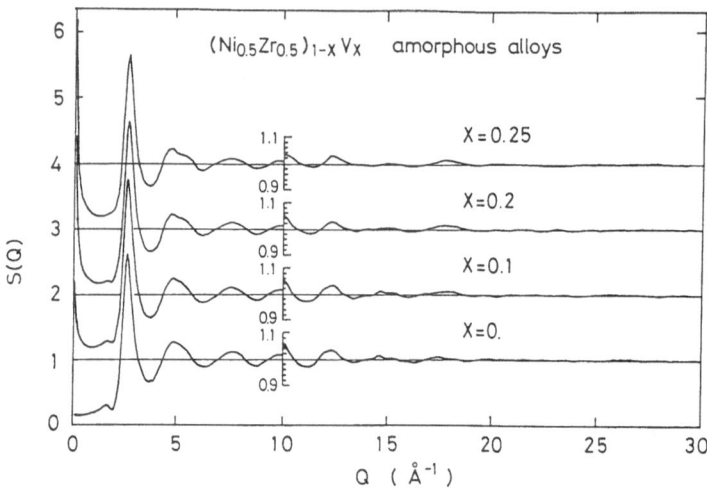

Fig. 3.17. Neutron total structure factors $S(Q)$ of $(Ni_{0.5}Zr_{0.5})_{1-x}V_x$ amorphous alloys as a function of V content

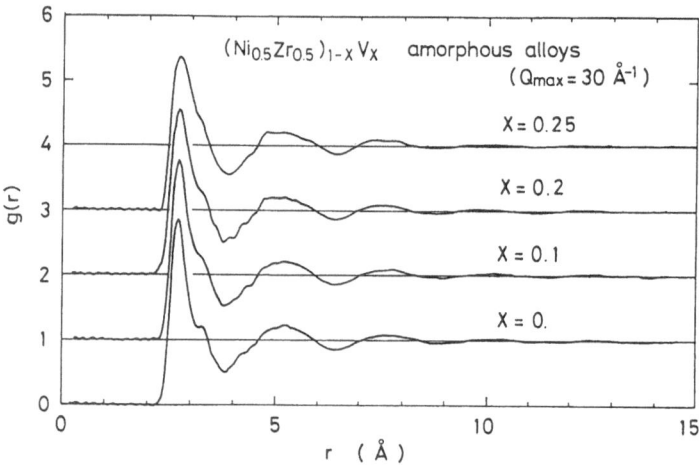

Fig. 3.18. Neutron total pair distribution functions $g(r)$ of $(Ni_{0.5}Zr_{0.5})_{1-x}V_x$ amorphous alloys

vector region of $Q < 1 \text{ Å}^{-1}$ in $S(Q)$. The $g(r)$ shows that the first peak located around $r = 2.70 \text{ Å}$ lowers the height and broadens the width, and the shoulder appearing around $r = 3.20 \text{ Å}$ is more smoothly joined to the first peak, with increasing V content.

As a V atom has a negligibly small cross section for coherent neutron scattering, compared with those of Ni and Zr atoms, the first peak in $g(r)$ is not apparently contributed from V atoms, but constructed only from Ni–Ni, Ni–Zr, and Zr–Zr pair distribution functions. *Fukunaga* and colleagues have separated

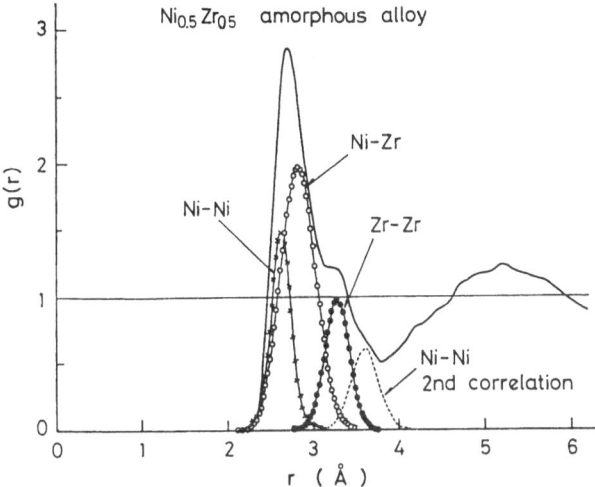

Fig. 3.19. Ni–Ni, Ni–Zr, and Zr–Zr partial pair distribution functions of NiZr amorphous alloy

the three partial pair distribution functions of MQ (melt-quenched) NiZr amorphous alloy as shown in Fig. 3.19. By comparing Fig. 3.18 with Fig. 3.19, the behavior of the $g(r)$ shown in Fig. 3.18 can be interpreted by considering that part of Zr atom sites is occupied with the V atoms added.

The *Cargill* and *Spaepen* [3.32] chemical short-range order parameters η_{NiZr} for Ni–Zr–V amorphous alloys were calculated as a function of V content by using the three partial pair distribution functions separated for each alloy. As shown in Fig. 3.20, the value of η_{NiZr} increases with increasing V content in all the alloys. This implies that the addition of V atoms enhances the chemical

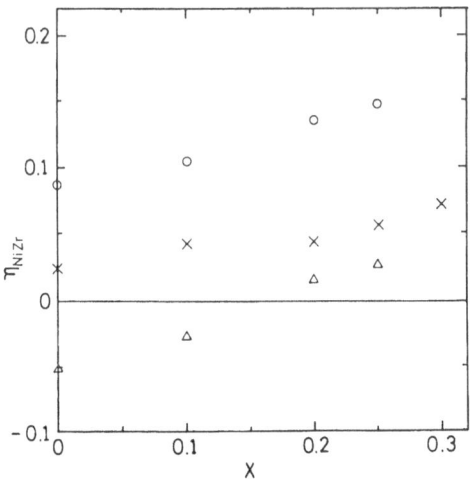

Fig. 3.20. Cargill-Spaepen chemical short-range order parameter η_{NiZr} for Ni–Zr unlike atom pair in $(Ni_{0.33}Zr_{0.67})_{1-x}V_x$ ($\triangle$), $(Ni_{0.5}Zr_{0.5})_{1-x}V_x$ ($\bigcirc$), and $(Ni_{0.67}Zr_{0.33})_{1-x}V_x$ ($\times$) ternary amorphous alloys

short-range order between Ni and Zr atoms in Ni–Zr–V ternary amorphous alloys. However, the small angle scattering observed in the $S(Q)$ suggests that the distribution of V atoms is not uniform but locally fluctuated in Ni–Zr–V ternary amorphous alloys.

If the fluctuated domain is assumed to be spherical then the radius of the fluctuated domain is expected to be about 4.5 Å. Figure 3.21 is a schematic model of the microscopic phase structure for Ni–Zr–V ternary amorphous alloys in which granules of Ni–Zr amorphous alloys are formed with the radius of several Ångstroms and V atoms precipitate predominantly in the boundary among the granules.

In Ni–Zr binary amorphous alloys, as shown by no observation of small angle scattering, Ni and Zr atoms are distributed uniformly but form the chemical short-range structure having a well-defined local coordination. When V atoms are added into Ni–Zr binary alloys, some chemical fluctuation takes place because of the formation of the hierarchy in the bonding force among Ni, Zr, and V atoms. This is an origin for creating a nanometer composite structure in ternary amorphous metallic alloys.

Such a fluctuation in the chemical short-range structure also takes place in crystalline alloys. However, the drastic appearance of the chemical fluctuation such as observed in amorphous alloys is difficult in crystalline alloys, because atomic sites are regularly restricted at lattice sites in crystalline alloys. Since the topological limit of atomic sites is relaxed in amorphous alloys, the nature of

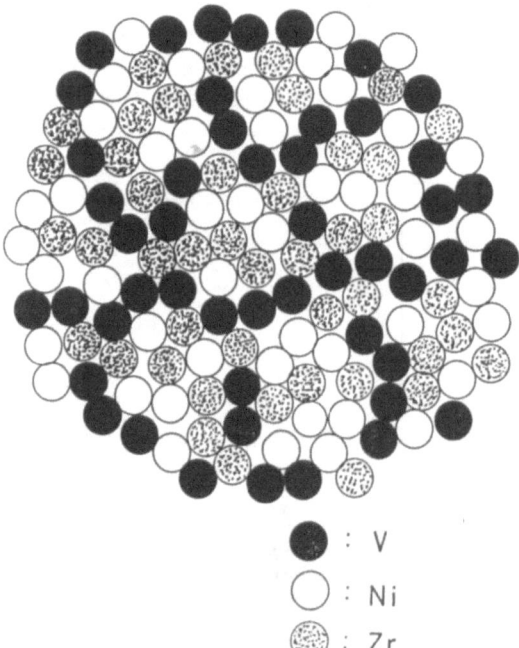

● : V

○ : Ni

◎ : Zr

Fig. 3.21. A schematic model for nanometer-scaled composite of "poly-amorphous" solid structure

chemical bonding of atoms is straightly emphasized in amorphous alloys. The microscopic composite structure inherently existing in multicomponent amorphous alloys may be assigned as "polyamorphous" [3.8], corresponding to "polycrystalline".

3.4 Conversion of Organic Polymers to Amorphous Ceramics

Recently the conversion of metal–organic polymers to amorphous covalent ceramics by firing the precursors in an inert atmosphere has been considered as a fifth route [3.29] for preparing amorphous materials following the conventional methods of melt-quenching, vapor-deposition, precipitation from solutions, and solid-state mechanical amorphization. This method has been often called "Yajima process", because the late Prof. S. Yajima at Tohoku University succeeded first in the preparation of SiC fibers by firing polycarbosilane as an organic precursor [3.34]. At the present time "Yajima process" has been extensively applied not only to the SiC system but also to SiC–Ti [3.35], Si_3N_4 [3.36], and BN [3.37] systems.

In the first stage of the "Yajima process", the fibers of metal–organic polymers such as polycarbosilane are formed by melt-spinning and then stabilized by oxidizing in air at the temperature of about 200°C. The second stage is the firing of the oxidized polycarbosilane fiber in an inert atmosphere around the temperature of 1200°C.

Suzuya et al. [3.9] have measured the small angle X-ray scattering (SAXS) profiles of Si–C–O–Ti amorphous fibers as prepared by firing polytitanocarbosilane near 1200°C. Figure 3.22 is the log–log plot of SAXS intensities as a function of the scattering vector Q. In the low scattering vector region of $Q < 0.1$ Å^{-1}, we can obviously find the anisotropic SAXS intensities between the scattering vectors parallel (Q_1) and perpendicular (Q_2) toward the long fiber axis, respectively. The SAXS intensity for Q_2 is quite large compared with that for Q_1. This implies, as illustrated schematically in Fig. 3.23 [3.9], that the spatial fluctuation extends to the order of 100 Å in the cross-section of a single fiber. Such a distinct anisotropic structure of the fibers is caused by the formation of oxygen-rich interfaces between filaments constructing a single fiber when polytitanocarbosilane fibers are partially oxidized during curing process.

The anisotropic SAXS intensities are drastically reduced after annealing the fibers above 1300°C. This fact suggests that SiO_2 fine particles precipitate from the oxygen-rich interface boundaries surrounding each filament in the fiber and then the anisotropic structure extending perpendicular to the fiber axis is destroyed to result in the isotropic structure.

As found in Fig. 3.22, the SAXS intensities for both the Q_1 and Q_2 directions completely coincide to be isotropic in the high scattering vector region of $Q > 0.2$ Å^{-1} and have the peak maxima around $Q = 0.2$ Å^{-1} originating from interparticle correlations. The particles are uniformly distributed with the

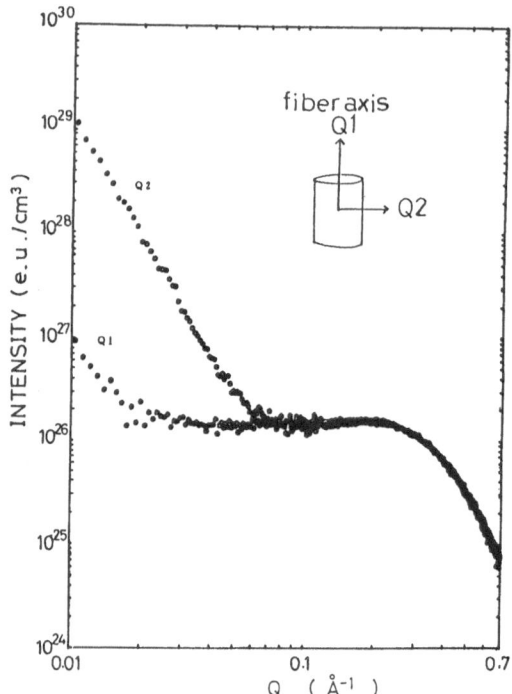

Fig. 3.22. Anisotropic intensities of small angle X-ray scattering (SAXS) for Si–C–O–Ti amorphous fibers prepared from polytitano-carbonsilane

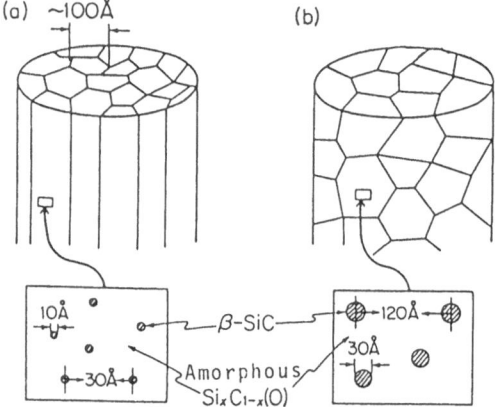

Fig. 3.23. Structure models of Si–C–O–Ti amorphous fibers; (a) firing polytitanocarbosilane fibers at 1200°C in argon gas and (b) annealing Si–C–O–Ti amorphous fibers at 1300°C for 1 h in argon gas

average interparticle distance of about 30 Å. The Guinier plot provides that the radius of particles is about 5 Å^{-1}. As the Porod plot is satisfied in the high scattering vector region of $Q > 0.5$ Å^{-1}, the particles are precipitated with the boundaries sharply discontinuous against the matrix.

Based on the absolute value of the SAXS intensities, we conclude that the precipitated particles are mainly composed of SiC. The SiC particles drastically

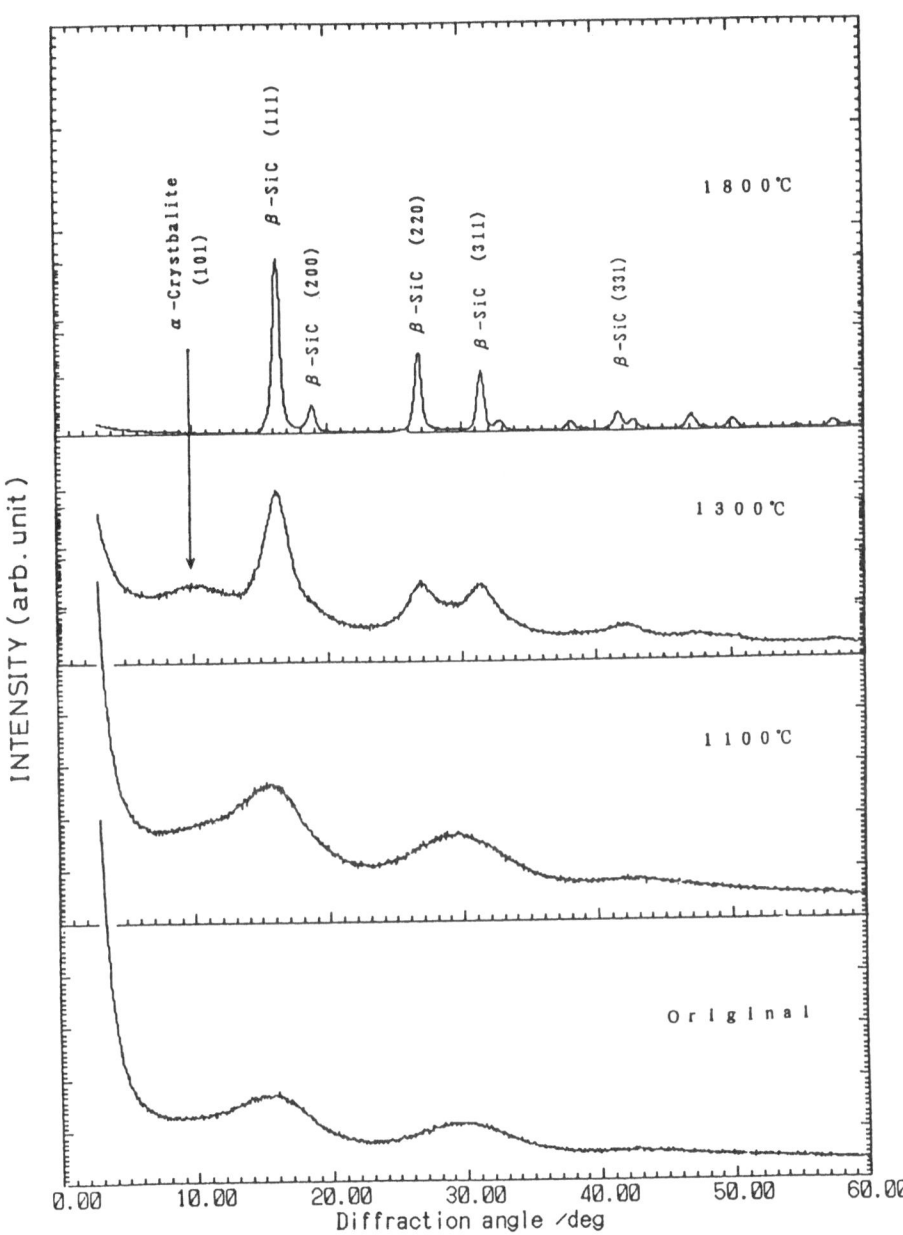

Fig. 3.24. X-ray diffraction patterns of Si–C–O–Ti fibers annealed in argon gas for 1 h at various temperatures

grow up to the radius of about 15 Å and are distributed with the large interparticle distance of about 120 Å after annealing the fiber above 1300°C.

Figure 3.24 shows the wide angle X-ray diffraction patterns of Si–C–O–Ti fibers as a function of annealing temperature. When the annealing temperature is beyond 1300°C, a broad Bragg peak corresponding to the (1 0 1) reflection of α-crystoballite SiO_2 appears around $2\theta = 10°$ (incident X-ray: MoKα) and the second peak located around $2\theta = 29°$ are separated into two peaks to indicate clearly the precipitation of microcrystalline β-SiC. The annealing at 1800°C provides the emission of CO gas and accelerates exclusively the growth of β-SiC crystals. These observations confirm that the isotropic SAXS intensities are contributed from β-SiC particles.

The matrix of Si–C–O–Ti amorphous fibers consist predominantly of amorphous Si_xC_{1-x} (O) including excess carbon content. It is quite difficult to neglect the possibility that a part of oxygen is dissolved into the matrix. However, large majority of oxygen participates in Si–O bonds at the boundaries between filaments and forms α-crystoballite SiO_2 particles by annealing. Recent studies on Raman scattering [3.35], EXAFS [3.39], and XPS [3.40] of Si–C fibers supports the structure model, as illustrated in Fig. 3.23, of Si–C–O–Ti amorphous fibers obtained by the SAXS measurement.

3.5 Hydrogen-Induced Amorphization

Hydrogen-induced amorphization is currently known as a solid-state method for synthesizing amorphous alloys [3.41]. In this method, amorphous metallic hydrides are formed by charging hydrogen atoms into crystalline intermetallic compounds. This phenomenon was first observed for the hydride reaction of Zr_3Rh (crystal) → $Zr_3RhH_{5.5}$ (amorphous) by Yeh et al. [3.42].

Although both of the YFe_2 and YNi_2 crystalline intermetallic compounds have the C15-type Laves phase structure, YFe_2 forms crystalline YFe_2D_x ($x = 1.7 \sim 4.3$) and YNi_2 is transformed into amorphous YNi_2D_x ($x = 1.7 \sim 3.8$) by the reaction with D_2 gas under pressure of 5.0 MPa at room temperature. However, crystalline YFe_2D_x is transformed into amorphous $YFeD_x$ by heating it up to about 200°C in a D_2 gas atmosphere. It is interesting to examine such a contrasting behavior between YFe_2 and YNi_2 crystalline intermetallic compounds from the structural point of view.

X-ray total structure factors $S_x(Q)$ for C15-YFe_2, c (crystalline)-$YFe_2D_{4.3}$, and a (amorphous)-$YFe_2D_{3.4}$ are compared in Fig. 3.25 [3.11]. We can find that the C15-type Laves phase structure certainly remains in $S_x(Q)$ of c-$YFe_2D_{4.3}$ in which Bragg peaks shift toward the low Q side corresponding to the lattice expansion due to interstitial occupation of D atoms. Heating c-$YFe_2D_{4.3}$ at 190°C in a D_2 gas atmosphere for 30 min results in the amorphization to a-$YFe_2D_{3.4}$. On the other hand, as shown in Fig. 3.26, C15-YNi_2 is simultaneously amorphized into a-$YNi_2D_{3.8}$ together with a hydride reaction with D_2

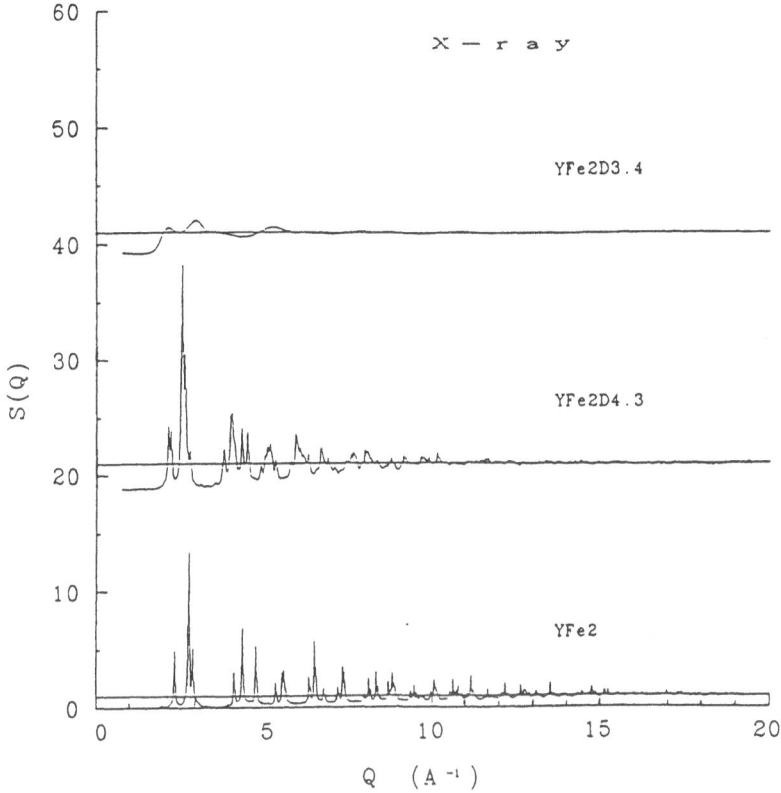

Fig. 3.25. X-ray total structure factors $S_x(Q)$ of C15-YFe$_2$, crystalline YFe$_2$D$_{4.3}$, and amorphous YFe$_2$D$_{3.4}$

gas at room temperature. After heating at 190°C in a D$_2$ gas atmosphere, the overall profile of $S_x(Q)$ does not show significant changes but the small ripples on the broad peaks in $S_x(Q)$ disappear [3.11].

Figures 3.27, 28 are the X-ray total radial distribution functions RDF$_x(r)$ obtained by Fourier transforming the $S_x(Q)$ shown in Figs. 3.25, 26. The first peaks in RDF$_x(r)$ for c-YFe$_2$D$_{4.3}$ and a-YNi$_2$D$_{3.6}$ have nearly the same profile which are broad with a shoulder on the low r side but are located at different positions. The first peak position in RDF$_x(r)$ of c-YFe$_2$D$_{4.3}$ is higher than that of a-YNi$_2$D$_{3.6}$. The peak profiles in RDF$_x(r)$ are quite different between a-YFe$_2$D$_{3.4}$ and a-YNi$_2$D$_{3.6}$ but the peak positions are located at nearly same distances. The peak located around $r = 2.5$ Å corresponds to Fe–Fe or Ni–Ni correlation, and the peak appearing around $r = 3.0$ Å describes Fe–Y or Ni–Y correlation.

Atomic distances and coordination numbers for each atom–atom correlation are summarized in Tables 3.1, 2. Based on these tables, we can conclude that the topological short-range structure of metallic atoms in a-YFe$_2$D$_x$ is con-

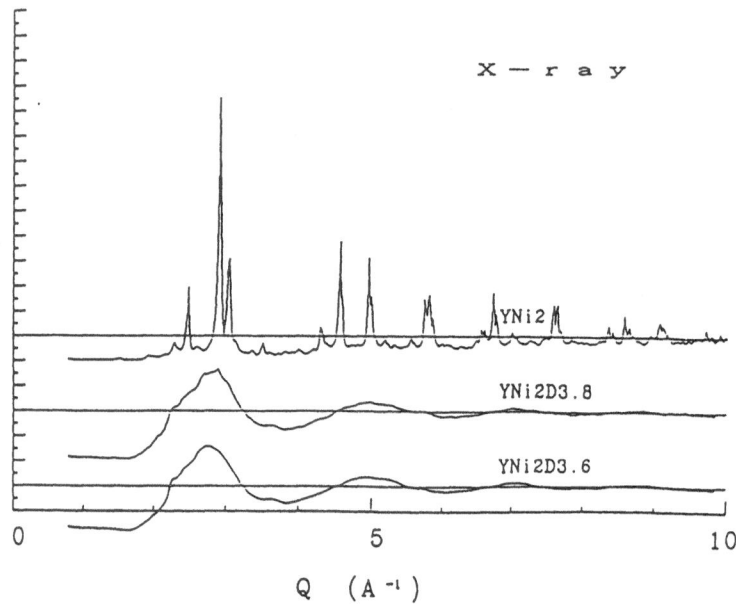

Fig. 3.26. X-ray total structure factors $S_x(Q)$ of C15-YNi$_2$, amorphous YNi$_2$D$_{3.8}$, and amorphous YNi$_2$D$_{3.6}$

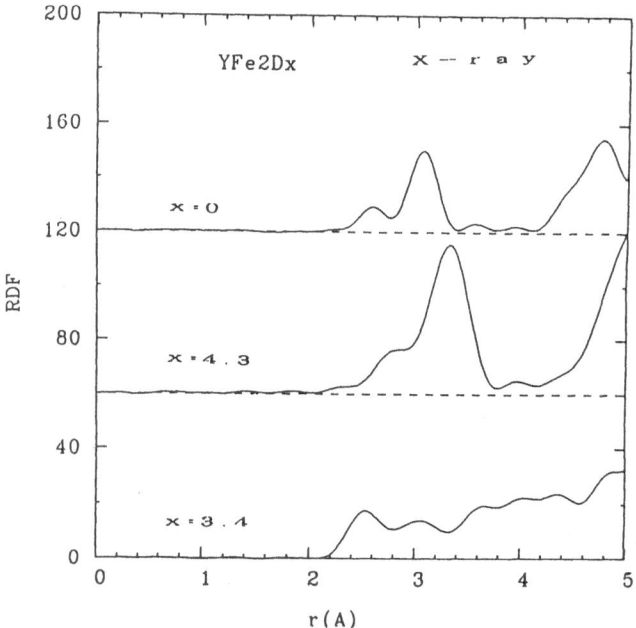

Fig. 3.27. X-ray total radial distribution functions RDF$_x(r)$ of C15-YFe$_2$, crystalline YFe$_2$D$_{4.3}$, and amorphous YFe$_2$D$_{3.4}$

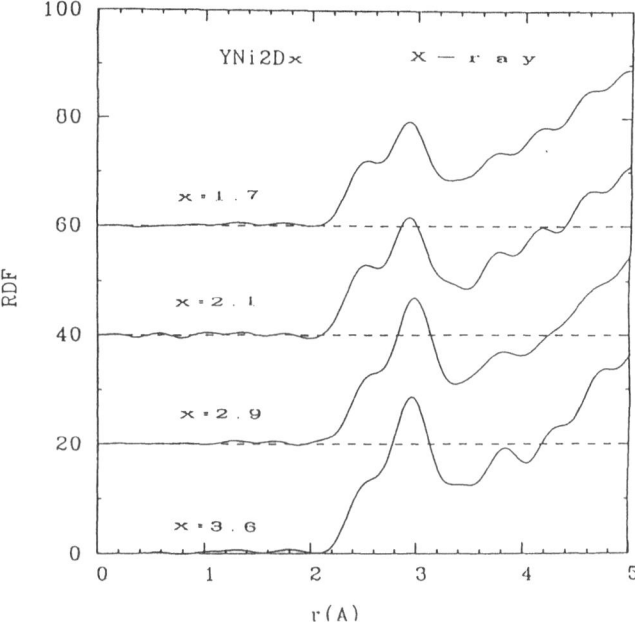

Fig. 3.28. X-ray total radial distribution functions $RDF_x(r)$ of C15-YNi$_2$ and amorphous YNi$_2$D$_{3.6}$

Table 3.1. Atomic distances [d (Å)] and coordination numbers [n] for metal atom–metal atom correlations in crystalline and amorphous YFe$_2$D$_x$. L.P. means C15-type crystalline Laves phase and G.R. is the Goldschmidt radius

		Fe–Fe		Fe–Y		Y–Y	
	x	d	n	d	n	d	n
Cryst.	4.3	2.80	6.5 ± 0.6	3.3	5.7 ± 0.3	3.45	3.5 ± 0.5
Amor.	3.4	2.54	7.7 ± 0.7	3.02	1.7 ± 0.2	3.69	10.5 ± 0.5
Amor.	2.5	2.53	7.7 ± 0.7	3.00	2.0 ± 0.2	3.69	11.0 ± 0.5
Amor.	2.1	2.53	7.8 ± 0.7	2.97	1.9 ± 0.2	3.65	10.9 ± 0.5
Amor.	1.7	2.50	8.4 ± 0.7	2.95	2.0 ± 0.2	3.60	10.6 ± 0.5
Cryst.	L.P.	2.60	6.0	3.04	6.0	3.18	4.0
	G.R.	2.52		3.06		3.60	

$d \pm 0.03$ Å

siderably modified from that in C15-YFe$_2$ by hydride reaction. In particular, Fe–Fe and Fe–Y distances are reduced while Y–Y distance expands in a-YFe$_2$D$_x$ compared with those in C15-YFe$_2$. The coordination number of Fe–Fe correlation slightly increases while that of Fe–Y correlation drastically decreases to one-half to one-third and that of Y–Y correlation increases to two to three times.

Table 3.2. Atomic distances [d (Å)] and coordination numbers [n] for metal atom–metal atom correlations in amorphous Y_2NiD_x crystalline and alloys. L.P. means C15-type crystalline Laves phase and G.R. is the Goldschmidt radius

		Ni–Ni		Ni–Y		Y–Y	
	x	d	n	d	n	d	n
Amor.	3.6	2.57	6.0 ± 0.6	2.96	5.0 ± 0.5	3.82	7.9 ± 0.8
Amor.	2.9	2.56	6.0 ± 0.6	2.97	5.0 ± 0.5	3.78	8.0 ± 0.8
Amor.	2.1	2.52	5.9 ± 0.6	2.92	5.0 ± 0.5	3.78	8.1 ± 0.8
Amor.	1.7	2.51	6.2 ± 0.6	2.91	4.4 ± 0.5	3.73	7.9 ± 0.8
	L.P.	2.54	6.0	2.98	6.0	3.11	4.0
	G.R.	2.48		3.04		3.60	

$d \pm 0.03$ Å

In contrast to it, Ni–Ni and Ni–Y distances in a-YNi_2D_x show little change from those in C15-YNi_2 but Y–Y distance is much stretched. The coordination number of Ni–Ni correlation remains almost the same and that of Ni–Y correlation slightly decreases while that of Y–Y correlation increases. The structural changes during the reaction of C15-$YNi_2 \to$ a-YNi_2D_x are considerably small compared with those of C15-$YFe_2 \to$ a-YFe_2D_x.

We can expect that c-YFe_2D_x preserves the metallic atom structure which is topologically similar with those in C15-YFe_2 because D atoms occupy interstitial sites in C15-YFe_2. However, Fe–Y bond is broken, and Fe–Fe and Y–Y bonds are further formed during hydride reaction. On the other hand, the Ni–Ni bond has essentially no effect and Ni–Y bond is partly broken in the hydride reaction of C15-$YNi_2 \to$ a-YNi_2D_x. Further information about the Y–Y bond in a-YNi_2D_x is also observed, but its degree is less than in a YFe_2D_x.

The neutron total $RDF_n(r)$ for C15-YFe_2, c-$YFe_2D_{4.3}$, and a-$YFe_2D_{3.4}$ are shown in Fig. 3.29 [3.11]. The peak appearing around $r = 1.73$ Å in $RDF_n(r)$ shows the formation of the D–Fe correlation in c-$YFe_2D_{4.3}$. The peak around $r = 2.28$ Å corresponds to the D–Y correlation in the hydrides. The peak height for D–Fe correlation is drastically reduced in a-$YFe_2D_{3.4}$ compared with c-$YFe_2D_{4.3}$.

The D–Y correlation is buried in a broad peak appearing around $r = 2$–3 Å because of overlapping with the Fe–Fe correlation. Therefore, the separation of the Fe–Fe correlation from the D–Y correlation is necessary to obtain the coordination number of the D–Y correlation. By the normalization of X-ray $RDF_x(r)$ to neutron $RDF_n(r)$ using the weighting factor ratio of X-ray scattering length to neutron scattering length, the contribution of the Fe–Fe correlation was subtracted from the neutron $RDF_n(r)$ to isolate the D–Y correlation [3.43]. The atomic distances and coordination numbers for the D–Y correlation obtained are summarized as a function of D content together with those for the D–Fe correlation in Fig. 3.30.

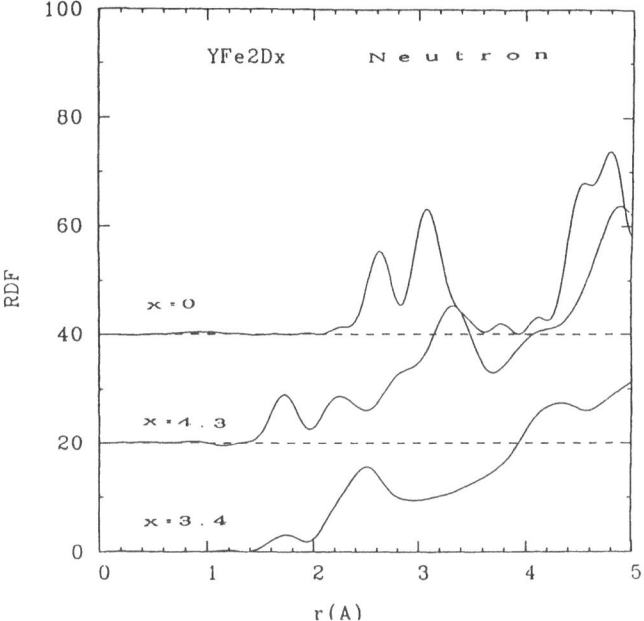

Fig. 3.29. Neutron total radial distribution functions $RDF_n(r)$ of C15-YFe_2, crystalline $YFe_2D_{4.3}$, and amorphous $YFe_2D_{3.4}$

Almost all D atoms in c-$YFe_2D_{4.3}$ occupy the 4-fold tetrahedral site, of which the average chemical structure is described by $[Fe_2Y_2]$. In fact, the C15-type Laves phase structure contains two different kinds of tetrahedral sites, $[Fe_2Y_2]$ and $[Fe_3Y]$. D atoms prefer to occupy the interstice in $[Fe_2Y_2]$, but a part of D atoms also sit in the interstice in $[Fe_3Y]$. However, D atoms in a-YFe_2D_x certainly occupy the interstices in $[Y_4]$ and $[FeY_3]$ consisting of higher content of Y atoms. This means that the hydride reaction causes drastic modification of the short-range structure of metallic atoms, which corresponds well to the result obtained by X-ray diffraction.

The neutron $RDF_n(r)$ of C15-YNi_2 and a-YNi_2D_x are shown in Fig. 3.31 [3.43]. The D–Ni correlation in a-YNi_2D_x appears as a fairly large peak located around $r = 1.68$ Å in $RDF_n(r)$ compared with that in a-YFe_2D_x. The peak for the D–Y correlation appearing at $r = 2.3$–2.4 Å is overlapped with that for the Ni–Ni correlation located around $r = 2.5$–2.6 Å in $RDF_n(r)$. Therefore, the atomic distance and coordination number of the D–Y correlation were estimated by subtracting the contribution of the Ni–Ni correlation, which was obtained by normalizing X-ray $RDF_x(r)$ to neutron $RDF_n(r)$ from the second peak in neutron $RDF_n(r)$.

Results are shown in Fig. 3.32. The total coordination numbers of metallic atoms surrounding a D atom are nearly 4.5 and do not depend on the D content. This means that great parts of D atoms occupy the 4-fold tetrahedral site. It is

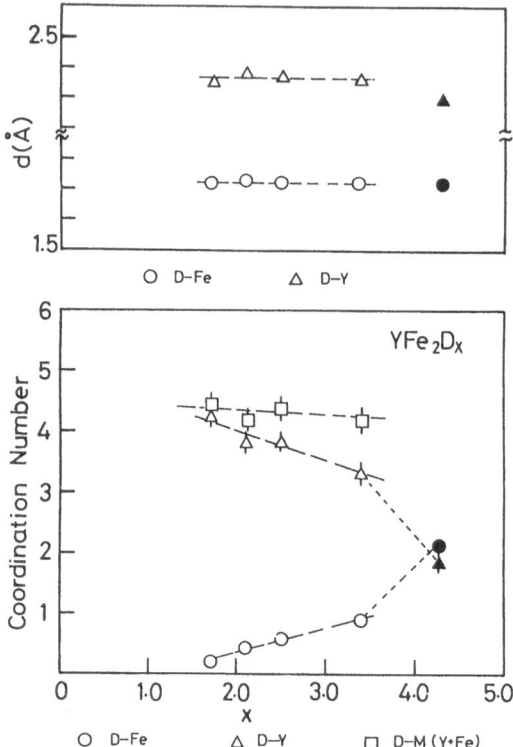

Fig. 3.30. Atomic distances and coordination numbers of D-Y correlation in crystalline $YFe_2D_{4.3}$ and amorphous YFe_2 as a function of D content

noteworthy that the chemical structure of the tetrahedral sites in a-YNi_2D_x has a fairly high Ni content compared with that of a-YFe_2D_x. Even in amorphous state, D atoms prefer to occupy the interstices in [YNi_2] and/or [YNi].

Why such a difference takes place between YFe_2 and YNi_2 intermetallic compounds? This mainly originates from the fact that the Ni-Y bond ($H = -1.17$ kJ/mol YNi_2) is more stable than the Fe-Y bond ($H = -4.8$ kJ/mol YFe_2) [3.44] and the atomic size of a Fe atom is larger than that of a Ni atom. D atoms are well known to form a very stable chemical bond with Y atoms ($H = -80$ kJ/mol D). Preferential occupation of D atoms in the tetrahedral sites consisting of higher content of Y atoms can be understood, since the atomic size of a Y atom is large and the strain due to interstitial D atoms is still small. With increasing D content, D atoms gradually have to occupy the tetrahedral site containing Ni or Fe atoms. The average size of tetrahedral sites available for D atom occupation is large enough in C15-YFe_2 to lead to a little strain in c-$YFe_2D_{4.3}$. However, c-$YFe_2D_{4.3}$ is further stabilized by transforming into a-$YFe_2D_{3.6}$ with the formation of a stable D-Y bond at the sacrifice of breaking a-Y-Fe bond. On the other hand, D atoms can not sit stably in the tetrahedral sites in C15-YNi_2 because the average size of tetrahedral sites in C15-YNi_2 is

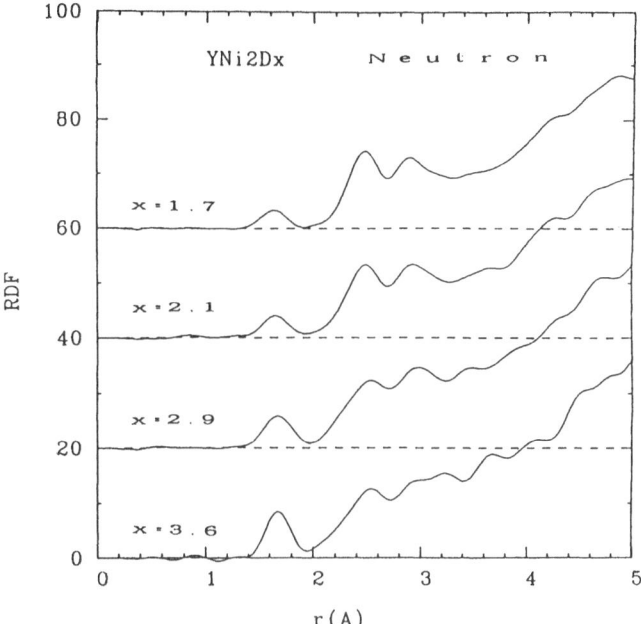

Fig. 3.31. Neutron total radial distribution functions $RDF_n(r)$ of C15-YNi_2 and amorphous YNi_2D_x

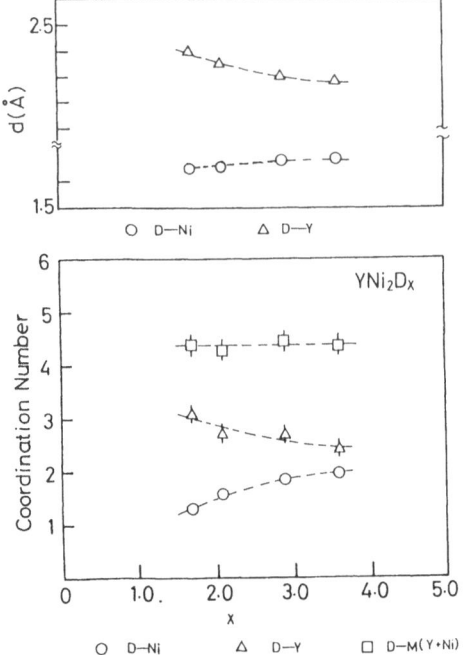

Fig. 3.32. Atomic distances and coordination numbers for D–Y correlation in amorphous YNi_2D_x as a function of D content

too small for D atoms to occupy without great accumulation of strain. Therefore, the crystal-to-amorphous solid structure transition is simultaneously induced with hydride reaction by forming a stable D–Y bond. Because the Ni–Y bond is fairly stable, the Ni–Y correlation is well preserved in a-YNi_2D_x.

References

3.1   Proc. 4th Int'l Conf. on Rapidly Quenched Metals, ed. by T. Masumoto, K. Suzuki (The Japan Institute for Metals, Sendai, Japan 1982)
3.2   Proc. Conf. on Solid State Amorphizing Transformations, ed. by R.B. Schwarz, W.L. Johnson (North-Holland, Amsterdam 1988); in J. Less-Common Metals **140** (1988)
3.3   P.H. Shingu (ed.): Mechanical Alloying (Trans. Tech. Pub., Zurich 1992); in Materials Science Forum **88/90** (1992)
3.4   Proc. 7th Int'l Conf. on Liquid and Amorphous Metals, ed. by H. Endo (North-Holland, Amsterdam 1990); in J. Non-Crystalline Solids **117/118** (1990)
3.5   · T. Fukunaga, Y. Homma, M. Misawa, K. Suzuki: J. Non-Crystalline Solids **117/118**, 721 (1990)
3.6   T. Fukunaga, K. Nakamura, K. Suzuki, U. Mizutani: J. Non-Crystalline Solids **117/118**, 700 (1990)
3.7   T. Fukunaga, S. Urai, N. Watanabe, K. Suzuki: J. Phys. F **16**, 99 (1988)
3.8   K. Suzuki: J. Non-Crystalline Solids **117/118**, 1 (1990)
3.9   K. Suzuya, T. Kamiyama, T. Yamamura, K. Okamura, K. Suzuki: J. Non-Crystalline Solids **150**, 167 (1992)
3.10 K. Suzuya, K. Shibata, K. Okamura, K. Suzuki: J. Non-Crystalline Solids **150**, 255 (1992)
3.11 K. Suzuki, X. Lin: J. Alloys and Compounds **193**, 7 (1993)
3.12 Proc. Workshop on Research Opportunities in Amorphous Solids with Pulsed Neutron Sources, ed. by D.L. Price (North-Holland, Amsterdam 1985); in J. Non-Crystalline Solids **76** (1985)
3.13 K. Suzuki: Glasses in *Method of Experimental Physics – Neutron Scattering*, Vol. 23 B, ed. by D.L. Price, K. Sköld (Academic, New York 1987) Chap. 12, p. 243
3.14 A. Bienenstock: J. Non-Crystalline Solids **106**, 17 (1988)
3.15 K. Suzuki: J. Non-Crystalline Solids **95/96**, 15 (1987)
3.16 M.S. El-Eskandarany, K. Aoki, K. Suzuki: J. Less-Common Metals **167**, 113 (1990)
3.17 K. Suzuki: J. Non-Crystalline Solids **112**, 23 (1989)
3.18 K. Suzuki: J. Phys. Condens. Matter **3**, F39 (1991)
3.19 J.L. Finney, J. Wallace: J. Non-Crystalline Solids **43**, 165 (1981)
3.20 S. Steeb, P. Lamparter: J. Non-Crystalline Solids **61/62**, 237 (1984)
3.21 T. Fukunaga, Y. Homma, M. Misawa, K. Suzuki: Mat. Sci. Eng. A **134**, 987 (1991)
3.22 Y. Homma, T. Fukunaga, M. Misawa, K. Suzuki: Materials Science Forum **88/90**, 339 (1992)
3.23 R.B. Schwarz, W.L. Johnson: Phys. Rev. Lett. **51**, 415 (1983)
3.24 T. Fukunaga, M. Mori, K. Inou, U. Mizutani: Mat. Sci Eng. A **134**, 863 (1991)
3.25 S.C. Moss, D.L. Price: Random packing of structure units and the first sharp diffraction peak in glasses, in *Physics of Disordered Materials*, ed. by D. Adler, H. Fritzsche, S.R. Ovshinsky (Plenum, New York 1985) p. 77
3.26 D.L. Price, S.C. Moss, R. Reijers, M.-L. Saboungi, S. Susman: J. Phys. Condens. Matter **1**, 1005 (1989)
3.27 F.E. Fujita: J. Non-Crystalline Solids **106**, 286 (1988)
3.28 Y. Nishi, K. Suzuki, T. Masumoto: J. Jpn. Inst. Metals **44**, 1336 (1980)
3.29 T. Fukunaga, M. Ishii, M. Misawa, K. Suzuki: Chemical short-range order of Ni–Zr–V ternary alloys, in *KENS REPORT VII*, ed. by N. Watanabe, M. Arai, H. Asano, Y. Endoh (National Laboratory for High Energy Physics, KEK, Tsukuba 1987/88) p. 59

3.30 T. Fukunaga, M. Misawa, K. Suzuki: Atomic rearrangement of Ti–Zr–X (X = Cu or Ni) ternary alloys, in *KENS REPORT VII*, ed. by N. Watanabe, M. Arai, H. Asano, Y. Endoh (National Laboratory for High Energy Physics, KEK, Tsukuba 1987/88) p. 62

3.31 T. Fukunaga, S. Shibuya, M. Misawa, K. Suzuki: J. Non-Crystalline Solids **95/96**, 263 (1987)

3.32 G.S. Cargill III, F. Spaepen: J. Non-Crystalline Solids **43**, 91 (1981)

3.33 G.D. Soraru, F. Babonneau, J.D. Mackenzie: J. Non-Crystalline Solids **106**, 256 (1988)

3.34 S. Yajima, J. Hayashi, M. Omori: Chem. Lett. **931** (1975)

3.35 T. Yamamura, T. Ishikawa, M. Shibuya, T. Hisayuki, K. Okamura: J. Mater. Sci. **23**, 2589 (1988)

3.36 D. Seyferth, G.H. Wiesemann, C. Prud'homme: J. Am. Chem. Soc. **66**, C13 (1983)

3.37 B.A. Bender, R.W. Rice, J.R. Spann: Ceram. Eng. Sci. Proc. **6**, 1171 (1985)

3.38 Y. Sasaki, Y. Nishina, M. Sato, K. Okamura: J. Mater. Sci. **22**, 443 (1987)

3.39 C. Laffon, A.M. Flank, P. Lagarde, M. Laridjani, R. Hagege, P. Olry, J. Cotteret, J. Dixmier, J.L. Miquel, H. Hommel, A.P. Legrand:: J. Mater. Sci. **24**, 1503 (1989)

3.40 L. Porte, S. Sartre: J. Mater. Sci. **24**, 271 (1989)

3.41 K. Aoki, X,-G. Li, T. Aihara, T. Masumoto: Mater. Sci. Eng. A **133**, 316 (1991)

3.42 X.L. Yeh, K. Samwer, W.L. Johnson: Appl. Phys. Lett. **42**, 242 (1983)

3.43 X Lin: Dissertation, Tohoku University (1990)

3.44 C. Colinet, A. Pasturel, K.H. Bushow: J. Appl. Phys. **62**, 3712 (1987)

3.45 T. Fukunaga: Physica B 213/214, 518–522 (1995)

3.46 T. Fukunaga: J. Jpn. Soc. Powder & Powder Metallurgy **43**, 738–741 (1996)

3.47 M.S. El-Eskandarany, K. Aoki, K. Suzuki: J. Appl. Phys. **71**, 2924 (1992)

3.48 K. Suzuki, K. Shibata, H. Mizuseki: J. Non-Crystalline Solids **156/158**, 58 (1993)

3.49 H. Mizuseki, K. Shibata, K. Suzuki: Physica B **213/214**, 538–540 (1995)

3.50 T. Otomo, M. Arai, K. Shibata, H. Mizuseki, K. Suzuki: Physica B **213/214**, 544–546 (1995)

3.51 K. Suzuki: J. Non-Crystalline Solids **192/193**, 1–8 (1995)

3.52 T. Kamiyama, Y. Wang, K. Suzuki, M. Shibuya, T. Yamamura: J. Jpn. Soc. Powder & Powder Metallurgy **41**, 795 (1994)

4 Nanophase Materials: Synthesis, Structure, and Properties

R.W. Siegel

In the past few years, atom clusters with average diameters in the range of 5–50 nm of a variety of materials, including metals and ceramics, have been synthesized by evaporation and condensation in high-purity gases followed by consolidation in situ under ultrahigh vacuum conditions to create nanophase materials. These new ultrafine-grained materials have properties that are often significantly different and considerably improved relative to those of their coarser-grained counterparts. The property changes result from their small grain sizes, the large percentage of their atoms in grain boundary environments, and the interactions between grains. Since their properties can be engineered during the synthesis and processing steps, cluster-assembled nanophase materials appear to have great technological potential beyond the current scientific interest in their grain-size dependent properties. Recent research on nanophase materials is reviewed and their future is considered.

4.1 Background

Nanophase materials are one of the broad class of nanostructured materials artificially synthesized with microstructures modulated in zero to three dimensions on length scales less than 100 nm that it has become possible to create over the past few years. The various types of nanostructured materials share three features: atomic domains spatially confined to less than 100 nm, significant atom fractions associated with interfacial environments, and interactions between their constituent domains. Nanostructured materials thus include zero-dimensionality atom clusters and cluster assemblies, one-dimensionally modulated multilayers, and their three-dimensional analogues, nanophase materials. While the man-made synthesis of these materials is rather recent, it appears that some nanostructured materials have been with us since before 'us' existed.

Evidence from the earliest meteorites that have been found and studied suggests that such materials have been part of our universe from its beginnings; it has been suggested that primordial materials with nanometer-scale phase structures condensed from our solar nebula [4.1–3]. It is only now, billions of years later, that the planned man-made synthesis of nanostructured materials has begun. Among these efforts, the synthesis of nanometer size atom clusters of metals and ceramics by means of the gas-condensation method, followed by their in situ consolidation under high-vacuum conditions, has resulted in a new methodology for synthesizing ultrafine-grained materials by means of which the creation of new levels of property engineering may become possible through the sophisticated control of scale, morphology, interaction, and architecture.

Atom clusters in the nanometer size regime, containing hundreds to tens of thousands of atoms, can be produced in sufficient numbers, by means of either physical or chemical processes, that they can be assembled into materials, which can be studied by a variety of conventional experimental methods. These materials have the potential for incorporating and taking advantage of a number of size-related effects in condensed matter ranging from electronic effects (so-called "quantum size effects") caused by spatial confinement of delocalized valence electrons and altered cooperative ("many-body") atom phenomena, such as lattice vibrations or melting, to the suppression of such lattice-defect mechanisms as dislocation generation and migration in confined grain sizes. The possibilities to assemble size-selected atom clusters into new materials with unique or improved properties will likely create a revolution in our ability to engineer materials with controlled optical, electronic, magnetic, mechanical, and chemical properties for many future technological applications.

This chapter reviews some of the research that has been accomplished in the area of cluster-assembled nanophase materials over the past few years. During this period, increasing interest has focused on synthetic nanostructured materials in anticipation that their properties will be different from, and often superior to, those of conventional materials that have phase or grain structures on a coarser size scale [4.4]. This interest has been stimulated not only by the considerable recent effort and success in synthesizing a variety of zero-dimensionality quantum well structures and one-dimensionally modulated, multi-layered materials with nanometer scale modulations, but also by the possibilities for synthesizing three-dimensionally analogous, bulk nanophase materials via the assembly of clusters of atoms [4.5].

Control of the size or sizes and morphologies of the phase domains or granular entities (e.g., clusters) being assembled is of primary importance in any of the methods for the synthesis of nanostructured materials. Beyond this, chemical control of the phases and cleanliness of the interfaces between phases or grains must be controlled as well. In bulk cluster-assembled nanophase materials, such control appears to be readily available. Before proceeding with a description of their synthesis and processing, it is useful to list some of the unique advantages of the assembly of nanophase materials under controlled atmospheres from gas-condensed clusters; they are as follows [4.6]:

(1) The < 100 nm sizes of the atom clusters and their surface cleanliness allow conventional restrictions of phase equilibria and kinetics to be overcome during material synthesis and processing; this results from the combination of short diffusion distances, high driving forces, and uncontaminated surfaces and interfaces available with cluster 'building blocks'.

(2) The large fraction of atoms residing in the grain boundaries and interfaces of these materials allow for interface atomic arrangements to constitute significant volume fractions of material, and thus novel materials properties may result from such 'defect' atomic environments.

(3) The reduced size scale and large surface-to-volume ratios of the individual nanophase grains can be predetermined and can alter and enhance a variety of physical and chemical properties.

(4) A wide range of materials can be produced in this manner, including metals and alloys, intermetallic compounds, ceramics, and semiconductors; and such materials can be synthesized to contain crystalline, quasicrystalline, or amorphous structures.

(5) The extensive possibilities for reacting, coating, and mixing in situ various types, sizes, and morphologies of clusters create significant future potential for the synthesis of a variety of new multicomponent composites with nanometer-sized microstructures and engineered properties that can be both multifunctional and hierarchical.

4.2 Synthesis and Processing

The synthesis of ultrafine-grained materials by the in situ consolidation in vacuum of nanometer size gas-condensed ultrafine particles or atom clusters [4.7] was first applied to metals [4.8, 9]. By consolidating clusters in this manner, materials with a large fraction of their atoms in grain boundaries could be formed, as shown schematically in Fig. 4.1. This method was subsequently applied to the synthesis of nanophase ceramics [4.10, 11]. The study of the gas condensation of ultrafine particles or atom clusters has, however, had a rather longer history stretching back to the formation and use of 'smokes', such as

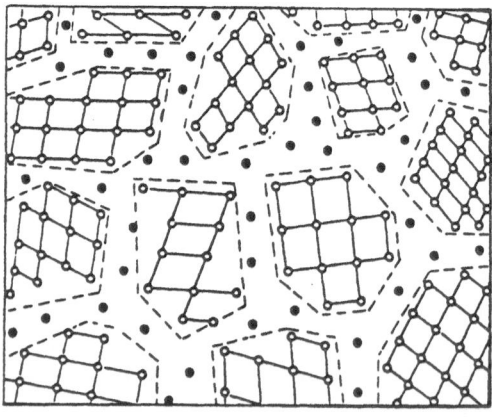

Fig. 4.1. Schematic representation of a cluster-assembled nanophase material. The black circles represent relaxed atoms in the grain boundaries between the grains formed by the consolidated clusters in which the white circles represent atoms in regular lattice positions. The dashed lines are meant only to guide the eye; no chemical differences between the atoms are implied. After [4.9]

carbon or bismuth 'blacks', for a variety of applications [4.12]. Scientific research into the controlled production of ultrafine particles by means of the gas-condensation method has been more recent [4.13–16], but still preceeded cluster assembly and therefore provided an important basis for this work. In addition, of course, the previous knowledge of powder metallurgy and ceramics generated over an even longer period, provided much needed background information for the work on nanophase materials to progress. The application of these ideas in recent years [4.17–21] to the synthesis of a variety of nanophase metals and ceramics has built upon this broad scientific and technological base.

Early research on the atom clusters formed via the gas-condensation method [4.13–15] defined most of the important parameters that control the sizes of the clusters formed in the conventional gas-condensation method that has been used to synthesize nanophase materials. The pioneering work of *Uyeda* and coworkers [4.13] demonstrated that a wide range of metallic ultrafine particles (between 10 and 100 nm diameter) could be condensed in a low pressure Ar atmosphere and that their sizes could be controlled by varying the gas pressure in the range of about 1–30 torr (0.13–4 kPa). This work and many of the later developments in this area of research have been recently reviewed [4.16]. Subsequent work by *Thölen* [4.15] extended this method to a number of additional metals and to a study of the nucleation, growth, and coalescence of the clusters. However, the most detailed study to date of the conventional gas-condensation process for forming ultrafine metal particles or clusters via condensation in naturally convecting inert gases (He, Ar, or Xe) was carried out by *Granqvist* and *Buhrman* [4.14]. Some of their results will be discussed further below, but it was these early studies that elucidated the essential parameters (primarily type of gas, gas pressure, and evaporation rate) that control the formation of gas-condensed atom clusters which have enabled the synthesis of cluster-assembled nanophase materials in recent years.

The basic aspects of the generation of atom clusters via gas condensation can be described using the conceptual model shown in Fig. 4.2. A precursor material, either an element or compound, is evaporated in a gas maintained at a low pressure, usually well below one atmosphere. The evaporated atoms or molecules lose energy via collisions with the gas atoms or molecules and undergo a homogeneous condensation to form atom clusters in the highly supersaturated

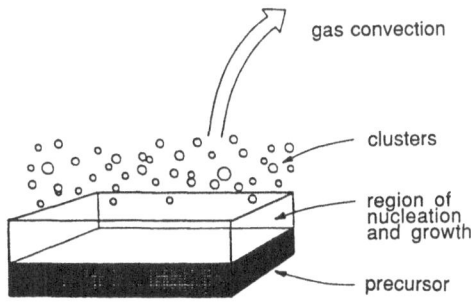

Fig. 4.2. Conceptual model for cluster formation via gas condensation. After [4.14]

vicinity of the precursor source. In order to maintain small cluster sizes, by minimizing further atom or molecule accretion and cluster–cluster coalescence, the clusters once nucleated must be removed rapidly from the region of high supersaturation. Since the clusters are already entrained in the condensing gas, this is readily accomplished by setting up conditions for moving this gas. Such gas motion has generally been driven by natural convection under the combined action of gravity and the temperature difference between the precursor source and a cooled thermophoretic cluster collection surface. However, a forced gas flow can also be used, with significant advantages in terms of both cluster size control and process efficiency.

It should be emphasized that there are thus only three fundamental rates, that function relative to one another, which essentially control the formation of the atom clusters in the gas-condensation process [4.6]. They are (1) the rate of supply of atoms to the region of supersaturation where condensation occurs, (2) the rate of energy removal from the hot atoms via the condensing medium, the gas, and (3) the rate of removal of the clusters once nucleated from the supersaturation region. Other factors can also affect the clusters finally collected, particularly those that result in significant cluster–cluster coalescence, but these three rates represent the core of the process. Accordingly, the smallest cluster sizes for a given precursor are obtained for a low evaporation rate and condensation in a low pressure of a light inert gas, such as He. These conditions lead to a lower supersaturation of precursor atoms in the gas, slower removal of energy from the evaporated atoms (via the lighter gas atoms at lower pressure), and more rapid convective gas flow owing also to the lower gas pressure. The rapid gas flow is significant, since it guarantees a shorter dwell time of the gas-entrained condensed clusters in the supersaturated region in which, if they remained, they would grow further. The effects of these controllable experimental parameters (gas type and pressure and evaporation rate) on the average cluster sizes produced under conditions of natural convective gas flow are shown in Figs. 4.3(a), (b).

A typical apparatus [4.22] which uses such a process for the synthesis of nanophase materials via the in situ consolidation of gas-condensed clusters is shown schematically in Fig. 4.4. It consists of an ultrahigh-vacuum (UHV) system fitted with two resistively-heated evaporation sources (which operate in the manner of Fig. 4.2), a cluster collection device (liquid-nitrogen filled cold finger) and scraper assembly, and in situ compaction devices for consolidating the powders produced and collected in the chamber. Before making the powders, the UHV system is first evacuated by means of a turbomolecular pump to below 10^{-5} Pa and then back-filled with a controlled high-purity gas atmosphere at pressures of about a few hundred Pa. For producing metal powders this is usually an inert gas, such as He, but it can alternatively be a reactive gas or gas mixture if, for example, clusters of a ceramic compound are desired.

The clusters that are collected via thermophoresis on the surface of the cold finger form very open, fractal structures. A view of such a fractal collection of

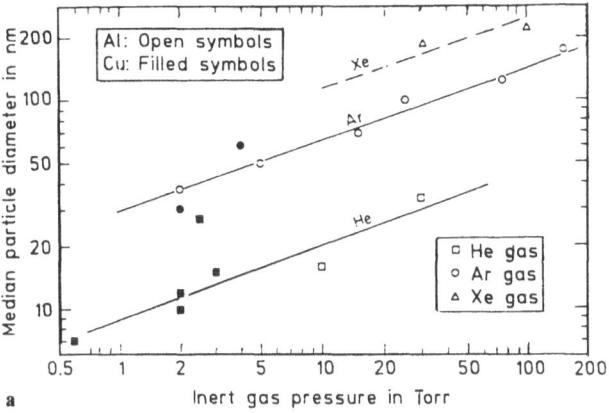

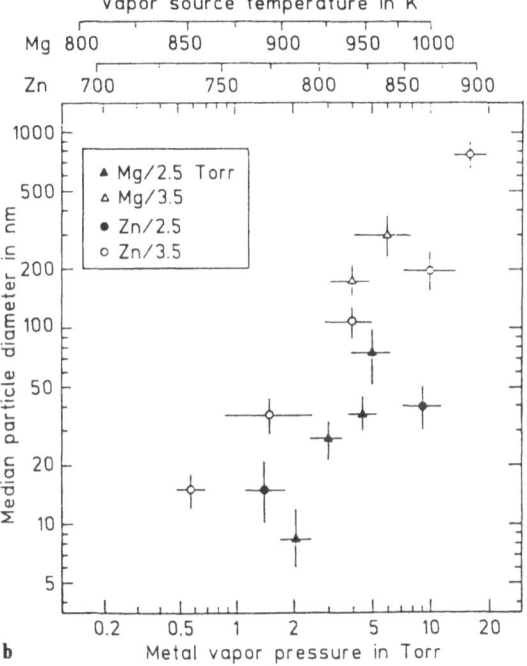

Fig. 4.3. (a) Median particle diameter versus pressure of He, Ar, or Xe gas for clusters of Al and Cu formed via gas-condensation; the straight lines only serve to guide the reader. (b) Median particle diameter versus vapor pressure at the metal surface (or source temperature) for Mg and Zn evaporated and gas-condensed in two different Ar pressures (2.5 and 3.5 torr). From [4.14]

nanophase TiO_2 taken from the cold finger, as seen by transmission electron microscopy, is shown in Fig. 4.5. The clusters are held on the collector surface rather weakly, via Van der Waals type forces, and are easily removed from this collection surface by means of a Teflon scraper. Upon removal, the clusters fall like "snow" from the surface and are funneled into a set of compaction devices,

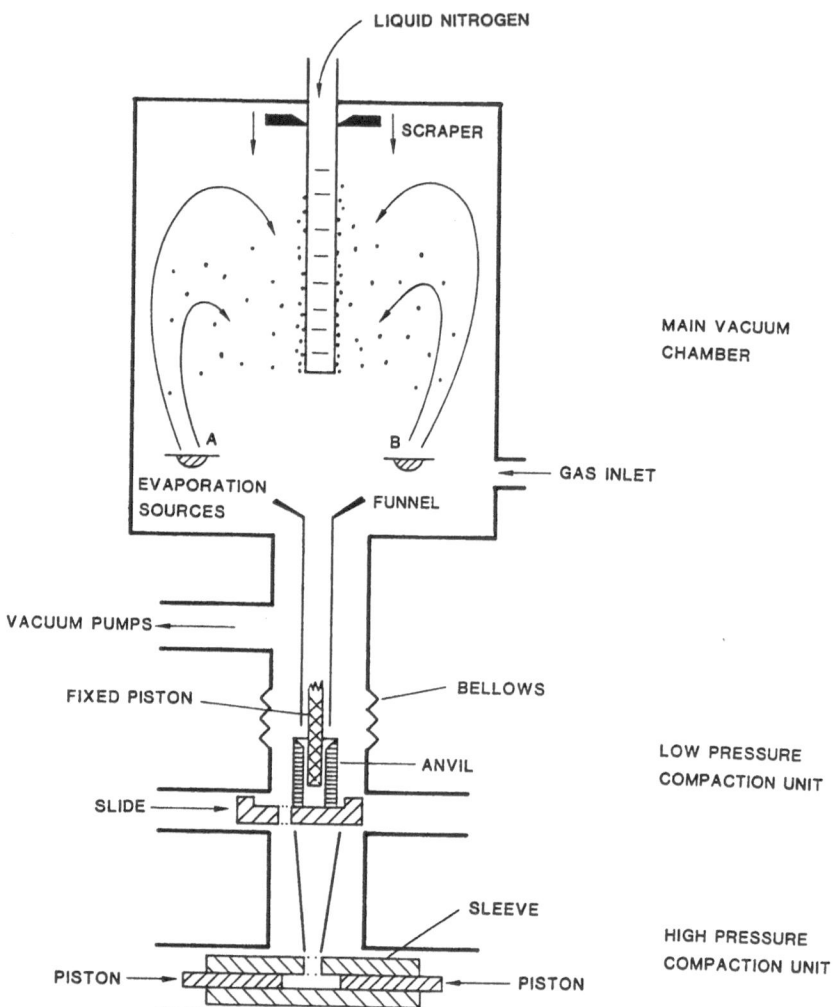

Fig. 4.4. Schematic drawing of a gas-condensation chamber for the synthesis of nanophase materials. Precursor material evaporated from sources A and/or B condenses in the gas and is transported via convection to the liquid-nitrogen filled cold finger. The clusters are then scraped from the cold finger, collected via the funnel, and consolidated first in the low-pressure compaction unit and then in the high-pressure compaction unit, all in vacuum. From [4.22]

see Fig. 4.4, capable of consolidation pressures up to about 1–2 GPa, in which the nanophase samples are formed at room temperature, or at elevated temperatures if needed. The pellets formed in this conventional research apparatus are typically about 9 mm in diameter and 0.1–0.5 mm thick, depending upon the amount of material made (usually a few hundred milligrams) and the experiments to be performed. The sizes of these samples have been more a matter of laboratory convenience than any real limitation of the gas-condensation method itself. All of the fundamental rates involved in the cluster synthesis can be significantly increased above those presently in use in the laboratory, and scaled up production rates leading to samples with greatly increased sizes can be expected. The scraping and consolidation are performed under UHV conditions after removal of the inert or reactive gases from the chamber, in order to maximize the cleanliness of the particle surfaces and the interfaces that are subsequently formed. Also, any possibility of trapping remnants of these gases in the nanophase compact is minimized by consolidation in vacuum. It should be noted in this regard that the total surface area of the nanophase powders produced in a given run is so great that, for a residual gas pressure of less than 10^{-5} Pa in a volume the size of the UHV chamber used, little gas contamination of the cluster surfaces would be expected.

An example of the grain morphology that results from the consolidation of powders such as those shown in Fig. 4.5 is presented in the transmission electron micrograph of Fig. 4.6. It can be seen that the clusters are rather equiaxed when formed and that the grains remain so even after consolidation. The grain size distribution for this as-consolidated TiO_2 is shown in Fig. 4.7. It is quite narrow and has the log-normal shape typical of clusters formed via gas-condensation [4.14]; it is thus essentially identical to that for the clusters from which the nanophase sample was assembled. Indeed, this shape is rather typical for the

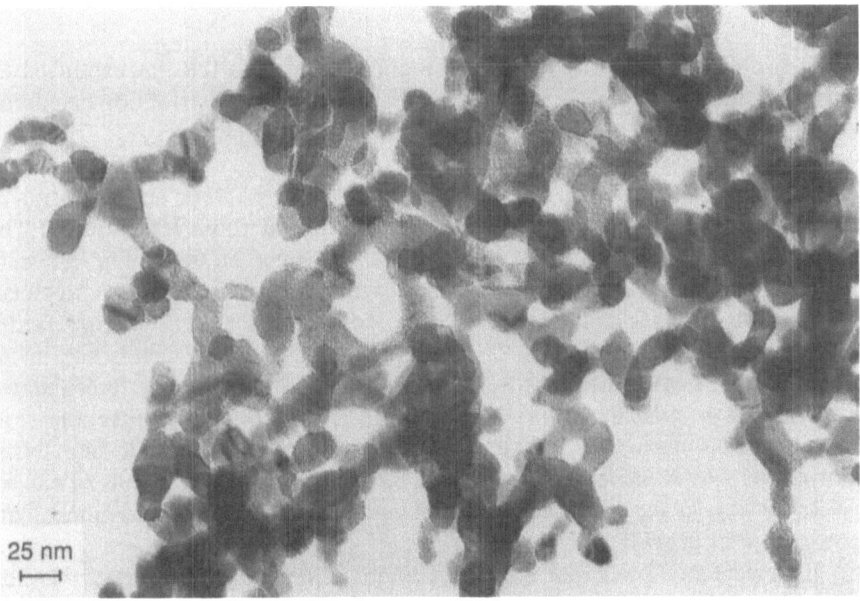

Fig. 4.5. Transmission electron micrograph of as-collected and oxidized TiO_2 (rutile) clusters synthesized in the apparatus shown in Fig. 4.4. From [4.22]

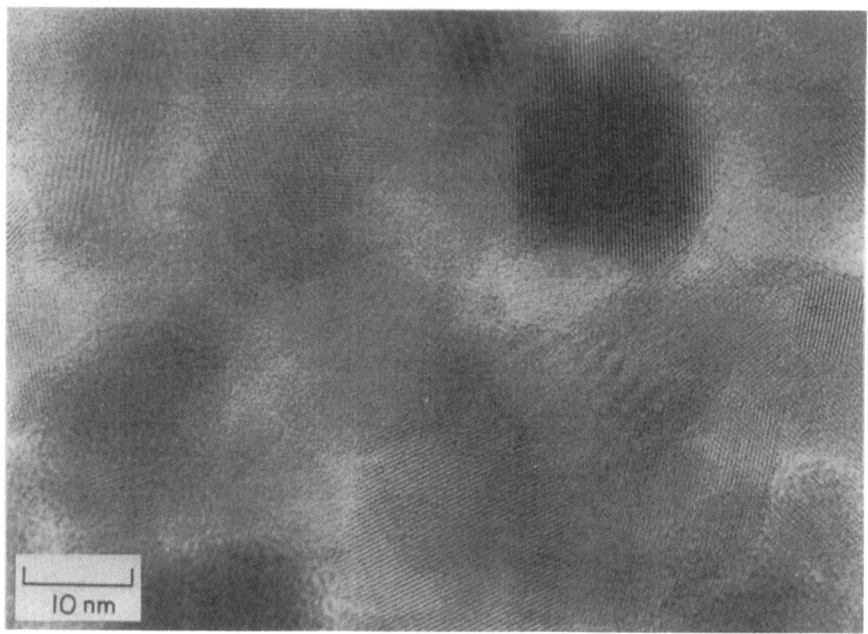

Fig. 4.6. Transmission electron micrograph of nanophase TiO_2 (rutile) after in situ consolidation at room temperature and 1.4 GPa pressure in the apparatus shown in Fig. 4.4, followed by sintering in air for 0.5 h at 500°C. From [4.10]

grain size distribution in any of the nanophase materials thus far produced by the gas-condensation method. The asymmetric tail of the distribution extending to larger sizes is a manifestation of cluster–cluster coalescence.

Since the as-collected gas-condensed clusters are generally aggregated in rather open fractal arrays [4.13, 15], as shown in Fig. 4.5, their consolidation at pressures of 1–2 GPa is easily accomplished, even at room temperature. The difficulties in consolidating the hard equiaxed agglomerates of fine powders resulting from conventional wet chemistry synthesis routes are mostly avoided. The sample densities resulting from cluster consolidation at room temperature have ranged up to about 97% of theoretical for nanophase metals and up to about 75–85% of theoretical for nanophase oxide ceramics. This "green-state" porosity will be considered further below, but it probably represents (at least in part) a manifestation of powder agglomeration leading to void-like flaws. Fortunately, these appear to be capable of being removed by means of cluster consolidation at elevated temperatures and pressures without significant attendant grain growth.

If an elemental precursor is evaporated in an inert gas atmosphere, then the atom clusters formed and collected are the same material, only in a reconstituted form. However, if clusters of a compound, such as a ceramic oxide are desired, the synthesis process can become somewhat more complex. For example, in

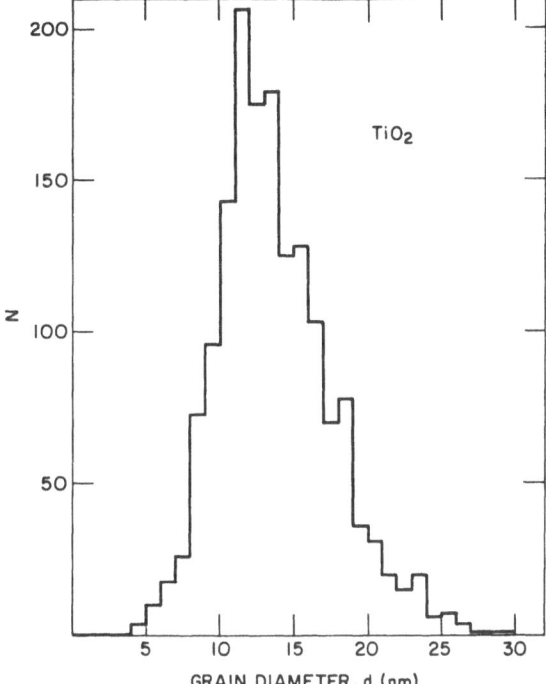

Fig. 4.7. Grain size distribution for a nanophase TiO_2 (rutile) sample compacted to 1.4 GPa at room temperature, as determined by dark-field transmission electron microscopy. From [4.11]

order to produce the nanophase TiO_2 with rutile structure and 12 nm average grain size shown in Figs. 4.5, 6, Ti metal clusters condensed in He were first collected on the cold finger and subsequently oxidized by the introduction of oxygen into the chamber [4.11]. A similar method has been used to produce α-Al_2O_3 [4.20] with an 18 nm average grain size after oxidizing Al clusters in air at 1000°C. If the vapor pressure of a compound is sufficiently large, as in the cases of MgO and ZnO, for example, it is possible to sublime the material directly from the oxide precursor in a He atmosphere containing, in addition, a partial pressure of O_2 to attempt to maintain oxygen stoichiometry during cluster synthesis. Such a method has been used [4.20] to produce such nanophase oxides with average grain sizes down to about 5 nm. Frequently, oxygen stoichiometry is not maintained.

In the case of nanophase TiO_2 cited above [4.11], the oxygen deficiency, while still present, is rather small and easily remedied as a result of the small grain sizes and short diffusion distances involved. Raman spectroscopy has been a useful tool in studying the oxidation state of nanophase TiO_2 owing to the intense and well studied Raman bands [4.23] in both the anatase and rutile forms of this oxide and the observation that these bands were altered in nanophase samples [4.24]. Schematic representations of the two most affected local vibrational modes in anatase and rutile are shown in Fig. 4.8. A series of Raman spectra from two as-consolidated titanium dioxide nanophase samples

(a)

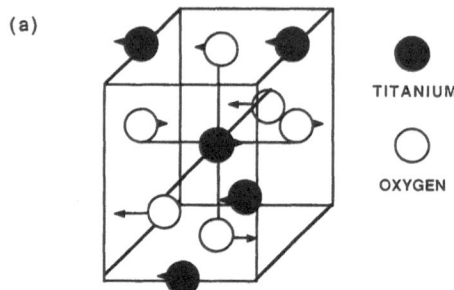

(b)

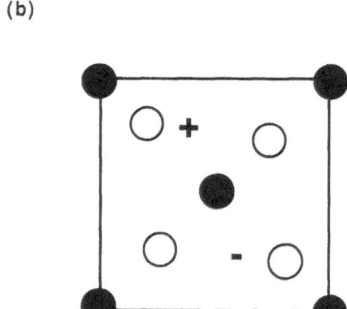

Fig. 4.8. Schematic representations of the atomic displacements of the vibrational mode associated with the (a) $144 \, \text{cm}^{-1}$ Raman line of anatase and (b) $447 \, \text{cm}^{-1}$ Raman line of rutile. From [4.25]

and from one of these samples annealed in air until fully oxidized to TiO_2 is shown in Fig. 4.9. The band broadening observed in the nanophase samples (and also band shifting in both the anatase and rutile phases) was confirmed [4.25] to be the result of an oxygen deficiency which could be subsequently removed in these samples by annealing in air. A subsequent calibration of this deviation from stoichiometry [4.26], shown in Fig. 4.10, indicated that $TiO_{1.89}$ was the actual material produced in the apparatus of Fig. 4.4, but that it could be easily oxidized to fully stoichiometric TiO_2, if desired, without sacrificing its small grain size (12 nm). Also, if intermediate deviations from stoichiometry were sought, in order to select particular properties of this material sensitive to the presence of oxygen deficient defects, they could be readily accessed as well.

Most of the atom clusters assembled into nanophase materials to date have been generated from Joule-heated evaporation sources. However, such sources have limitations that need not be suffered, since a wide variety of other sources are also available. The primary limitations are source-precursor incompatibility, temperature range, uniformity and control, and dissimilar evaporation rates for different constituents in an alloy or compound precursor. Each of these limitations can be avoided by a host of alternative sources that have been developed over the years of ultrafine particle research [4.16], but which are only now beginning to enter the field of nanophase materials synthesis. Among other sources for bringing atom supersaturations into a condensing gas medium that have been successfully used to produce clusters or ultrafine particles are Joule-heated ovens [4.27–29], sputtering [4.30–34], electron-beam heating [4.35–38],

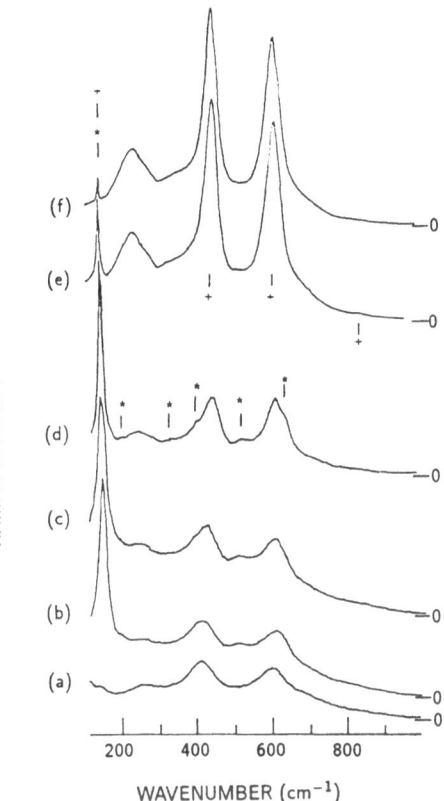

Fig. 4.9. Raman spectra of two areas (a) and (b) of an as-compacted nanophase TiO_2 sample compared with spectra for area (b) after successive annealing in air for 0.5 h at 300°C (c), 500°C (d), 700°C (e), and 1000°C (f) (∗ indicates anatase $k = 0$ phonons; + indicates rutile $k = 0$ phonons). From [4.24]

laser ablation [4.39], and plasma methods [4.40, 41]. It should be clear that this wide variety of evaporation methods will allow for greatly increased flexibility in the use of refractory or reactive precursors for clusters, and will be especially useful as one moves toward synthesizing technological quantities of more complex multicomponent or composite nanophase materials in the future.

Before turning to the structure and properties of nanophase materials, it might be useful to mention that a number of other methods for producing these materials exist. Among the physical methods, the gas-condensation method for the synthesis of nanophase materials appears to have the greatest flexibility and control for engineering new materials. However, other physical methods, such as spark erosion [4.42] and mechanical attrition [4.43–45], can complement this approach. The first of these methods is an alternative to gas-condensation in that it can yield individual clusters which can be subsequently assembled via consolidation, although a wide range of powder sizes is usually produced and a subsequent size separation ("classification") process is then necessary. The second method, however, produces its nanostructures by means of what is essentially a mechanical decomposition of coarser-grained structures via shear bands, so that the individual grains are never isolated clusters and much synthesis and processing flexibility is lost. Nevertheless, grain sizes down into

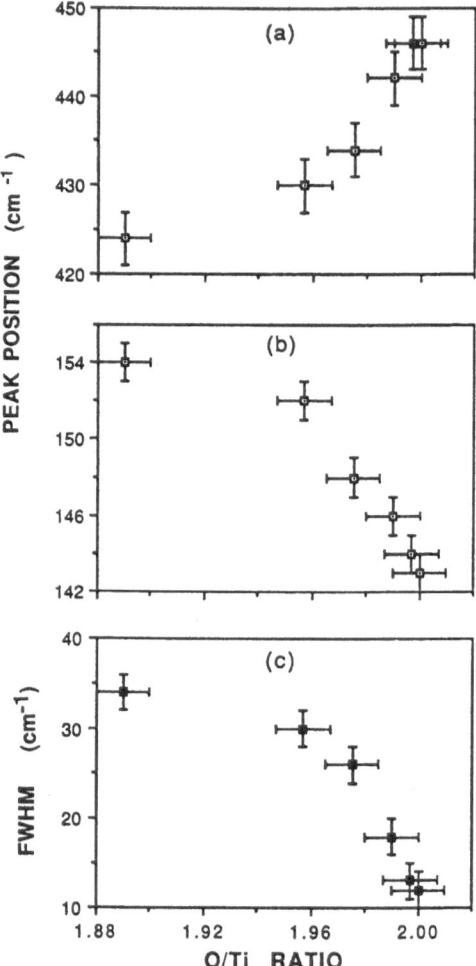

Fig. 4.10. Variation with O/Ti ratio of the peak position of (a) the rutile "447–cm^{-1}" vibrational mode and (b) the anatase "144 cm^{-1}" vibrational mode, as well as (c) this anatase mode's full width at half maximum (FWHM). From [4.26]

the nanometer regime can be accessed, at least in relatively hard materials. There are also various chemical and biological methods for the synthesis of nanophase materials, such as sol–gel synthesis, molecular self-assembly, spray pyrolosis, and biomimetic methods. These will not be covered here as they are essentially outside the scope of this chapter. However, some recent references to the variety of interesting methods available can be found in [4.4, 5, 46–51].

4.3 Structure and Stability

The structures of nanophase materials are dominated by their ultrafine grain sizes and their large number of grain boundaries. In addition, however, other structural features such as pores and their associated free surfaces, grain boundary junctions, and other crystal lattice defects play a significant role in their properties. It has become increasingly clear that all of these aspects of the nature of nanophase materials must be considered in trying to fully understand these new materials [4.52].

4.3.1 Grains and Pores

Our present knowledge of the grain structures of nanophase materials has resulted primarily from direct observations using transmission electron microscopy (TEM) [4.11, 53–55]. A typical high resolution image of a nanophase palladium sample is shown in Fig. 4.11. TEM has shown that the grains in nanophase compacts are rather equiaxed, similar to the atom clusters from which they were formed. This can be seen by a comparison of the micrographs shown in Figs. 4.5, 6. The grains also appear to retain the narrow log-normal size distributions (Fig. 4.7) typical of the clusters formed in the gas-condensation method [4.14], since measurements of these distributions before or after cluster consolidation by dark-field electron microscopy yield rather similar results. On

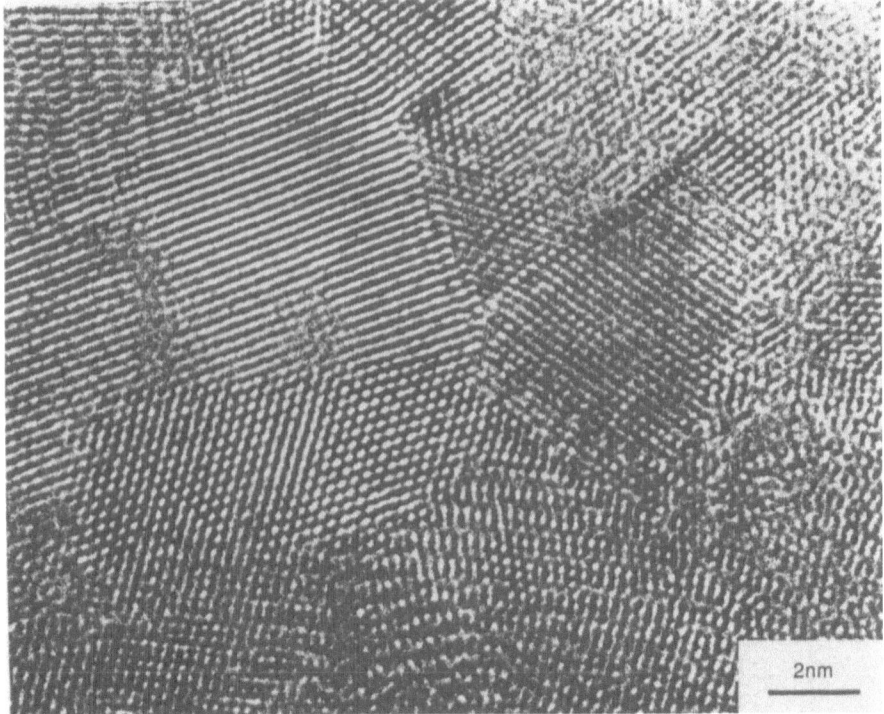

Fig. 4.11. High resolution transmission electron micrograph of a typical area in nanophase palladium. From [4.54]

the other hand, the observations that the densities of nanophase materials consolidated from equiaxed clusters extend well beyond the theoretical limit (78%) for close packing of identical spheres indicate that an extrusion-like deformation of the clusters must result during the consolidation process, filling in (at least in part) the pores among the grains. Observations by electron and X-ray scattering indicate, however, that no apparent preferred orientation or "texture" of the grains results from their uniaxial consolidation and that the grains in the nanophase compact are essentially randomly oriented with respect to one another. Taken together, these observations indicate that cluster extrusion in forming nanophase grains may result from a combination of local deformation and diffusional processes.

All of the nanophase materials consolidated at room temperature to date have invariably posessed a degree of porosity ranging from about 25% to less than 5%, as measured by Archimedes densitometry, with the larger values for ceramics and the smaller ones for metals. Evidence for this porosity was first obtained by positron annihilation spectroscopy (PAS) [4.11, 56, 57] and more recently by precise densitometry [4.58] and porosimetry [4.59, 60] measurements. PAS is primarily sensitive to small pores [4.61], ranging from single vacant lattice sites to larger voids, but can probe these structures enclosed in the bulk of the material. On the other hand, porosimetry measurements using the BET (Brunauer–Emmett–Teller) N_2 adsorption method [4.62] probe only pore structures open to the free surface of the sample, but can yield pore size

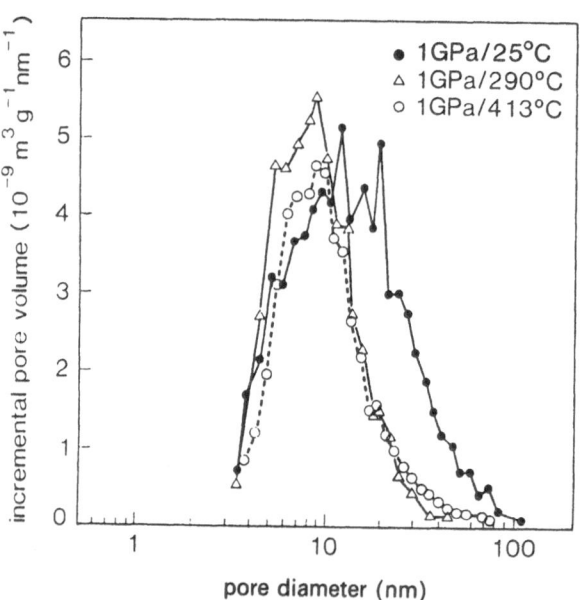

Fig. 4.12. Pore size distributions in nanophase TiO_2 consolidated under 1 GPa pressure at room temperature, 290°C and 413°C, as measured by the BET method. From [4.60]

distributions, which are essentially unavailable from PAS. Densitometry using an Archimedes method, of course, integrates over all densities in the sample, including grains, pores (open or closed), and density decrements at defect sites. These measurements have together shown that the porosity in as-consolidated nanophase metals and ceramics is primarily in the less than 100 nm size regime (although some larger porous flaws have been observed), and usually of comparable sizes to the grains, but that the porosity is to a great extent interconnected and intersects with the specimen surfaces. This is shown for example in Fig. 4.12, where BET porosimetry data [4.60] are presented for nanophase TiO_2 as-consolidated and after being sintered at elevated temperatures under 1 GPa pressure. The ability to probe this porosity via PAS is shown in Fig. 4.13, where positron lifetime data [4.11] from nanophase TiO_2 are presented as a function of sintering, which eventually removes the porosity and hence the positron trapping at voids that leads to the long positron lifetimes observed (τ_2). This will be discussed further in Sec. 4.4.1. However, consolidation at elevated temperatures can remove this porosity without sacrificing the ultrafine grain sizes in these materials.

4.3.2 Grain Boundaries

Owing to their ultrafine grain sizes, nanophase materials have a significant fraction of their atoms in grain boundary environments, where they occupy positions relaxed from their normal lattice sites. For conventional high-angle grain boundaries, these relaxations extend over about two atom planes on either side of the boundary, with the greatest relaxation existing in the first plane [4.63–65]. The significance of these atomic relaxations in nanophase materials can be better appreciated by the simple estimate of the fraction of atoms in grain boundary environments as a function of grain size shown in Fig. 4.14. It can be seen that in the average grain diameter range between 5 and 10 nm, where much of the research on nanophase materials has focussed, grain boundary atom percentages range between about 15 and 50%. Since such a large fraction of their atoms reside in the grain boundaries of nanophase materials, it is clear that the interface structures can play a significant role in affecting the properties of these materials.

A number of earlier investigations on nanocrystalline metals by *Gleiter* and coworkers [4.17], including X-ray diffraction [4.66], Mössbauer spectroscopy [4.67], positron lifetime studies [4.56, 57], and extended X-ray absorption fine structure (EXAFS) measurements [4.68, 69], were interpreted in terms of grain boundary atomic structures that may be random, rather than possessing either the short-range or long-range order normally found in the grain boundaries of conventional coarser-grained polycrystalline materials. This randomness was variously associated [4.17] with either the local structure of individual boundaries (as seen by a local probe such as EXAFS or Mössbauer spectroscopy) or the structural coordination among boundaries (as might be seen by X-ray

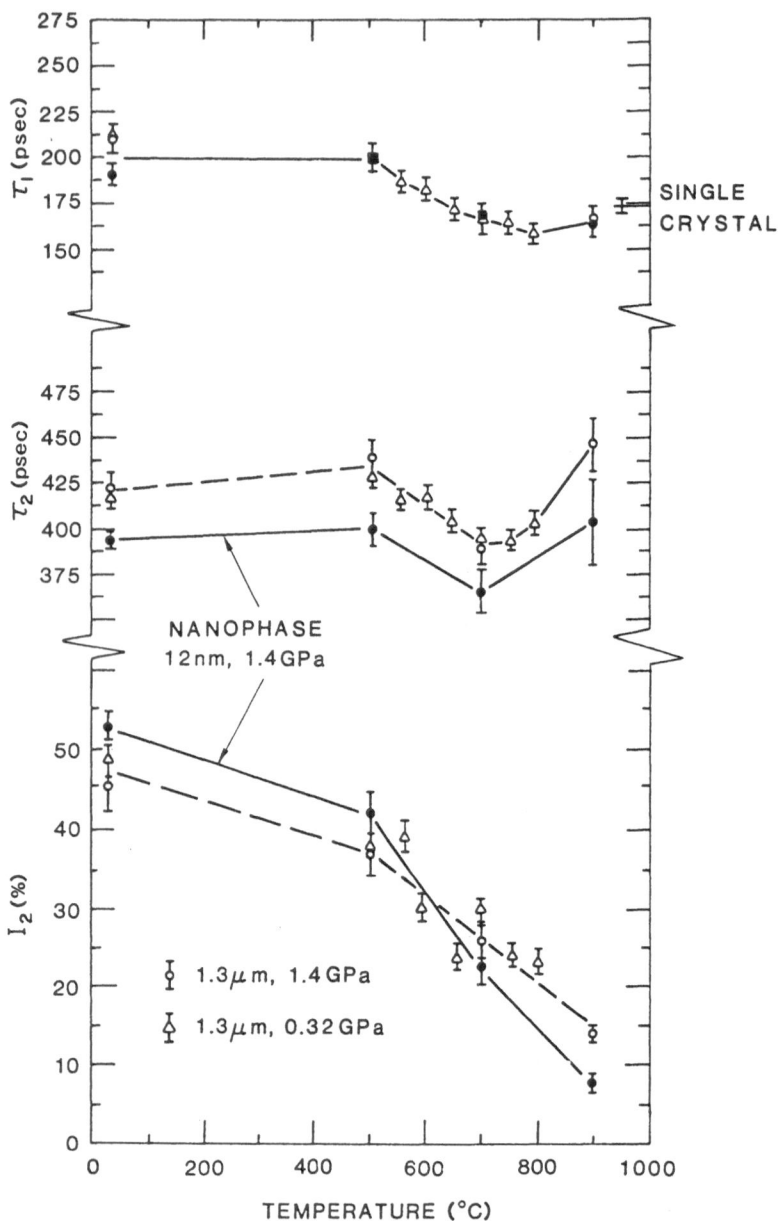

Fig. 4.13. Results of two-component (τ_1, τ_2) lifetime fits to positron annihilation data from three TiO$_2$ samples as a function of sintering temperature. A 12 nm grain size nanophase sample (filled circles) compacted at 1.4 GPa is compared to 1.3 μm grain size samples compacted at 1.4 GPa (open circles) and 0.32 GPa (triangles) from commercial powder. The PAS data were taken at room temperature; no sintering aids were used. From [4.11]

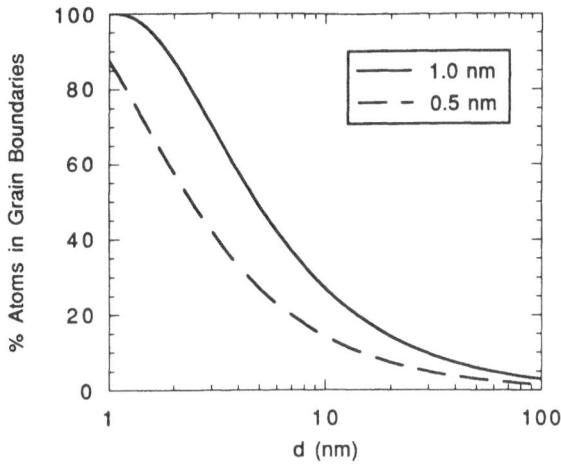

Fig. 4.14. Percentage of atoms in grain boundaries (including grain boundary junctions) of a nanophase material as a function of grain diameter, assuming that the average grain boundary thickness ranges from 0.5 to 1.0 nm (ca. 2 to 4 atomic planes wide). From [4.21]

diffraction). However, recent studies of these grain boundaries by high resolution electron microscopy [4.53, 54] have shown that their structures are rather similar to those of conventional high-angle grain boundaries. An extensive review of these results has recently appeared elsewhere [4.52]; a brief summary is given here.

X-ray diffraction data [4.66] from nanocrystalline Fe with an average grain size of about 6 nm (measured by TEM) and a density of about 83% that of conventional α-Fe, shown in Fig. 4.15, were best fit by the computed scattering from a four atom layer thick grain boundary model in which the atoms of the inner two layers (planes) were randomly displaced by 0.5 nnd and those of the outer two by 0.25 nnd, where nnd is the nearest neighbor distance in bulk bcc α-Fe. Such a relaxed structure was shown to yield a radial distribution function that resembled a structure without short-range order. However, such diffraction behavior has not been recently observed [4.70] on approximately 8 nm grain size and about 80% dense Pd samples, as shown in Fig. 4.16. Whether the differences observed between these two investigations is a matter of the different materials investigated or other differences has not been resolved. Further apparent support for randomly structured nanophase boundaries was drawn from Mössbauer spectroscopy results [4.67] on nanocrystalline Fe samples similar in grain size and density to those used in the X-ray scattering study by *Zhu* et al. [4.66]. Positron lifetime behavior [4.56, 57] in nanocrystalline Fe with 6 nm grain size and densities ranging from about 51 to 80%, which was found to be intermediate between that for uncompacted (but agglomerated) 6 nm powder and that for either a glassy Fe alloy ($Fe_{85}B_{15}$) or polycrystalline bulk Fe, was also taken as support for the hypothesis of an interfacial structure in nanocrystalline metals with a wide distribution of interatomic distances.

EXAFS measurements [4.68, 69] on mechanically powdered (< 5 µm) nanocrystalline samples of Pd and Cu with average grain sizes in the range 10–24 nm were also performed. Results from the Cu are shown in Fig. 4.17; the Pd results

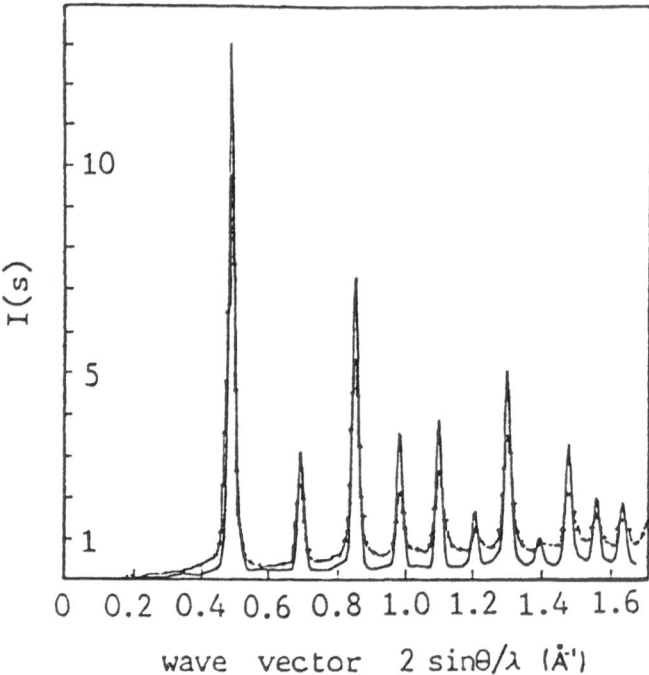

Fig. 4.15. Comparison of measured (-- + --) and computed (———) X-ray interference functions for nanocrystalline Fe. The computational model system assumes a grain boundary structure consisting of four atomic layers in which the atoms on the inner two (the boundary core) are randomly displaced by 0.15 nnd and those in the outer two by 0.07 nnd, where nnd is the equilibrium nearest neighbor distance in α-Fe. From [4.66]

are similar. The amplitudes of the weighted EXAFS oscillations and their Fourier transform were observed to be reduced in the nanocrystalline material relative to those from a coarse grained polycrystalline control sample, and this intensity reduction increased with increasing coordination shell distance. This intensity reduction was attributed to the nanophase grain boundaries and thought to be primarily due to a wide distribution of interatomic bond lengths in these internal interfaces. Recent EXAFS measurements [4.71] performed on both unconsolidated and consolidated nanocrystalline Pd, as well as on a coarse grained Pd reference sample, yielded similar results to the earlier work, but showed that the EXAFS amplitude reduction was even greater in the unconsolidated powder than in the consolidated nanocrystalline sample. Since the

Fig. 4.16. Corrected X-ray intensity data from (a) coarse-grained and (b) nanocrystalline Pd plotted on a logarithmic scale versus scattering vector magnitude, $\tau = 4\pi\sin\theta/\lambda$. Dashed lines indicate the intensity in each case unaccounted for by the Lorentzian-shaped Bragg reflections. Quadratic polynomial representations of the background intensities from (dashed line) coarse-grained and (solid line) nanocrystalline Pd are shown in (c). From [4.70]

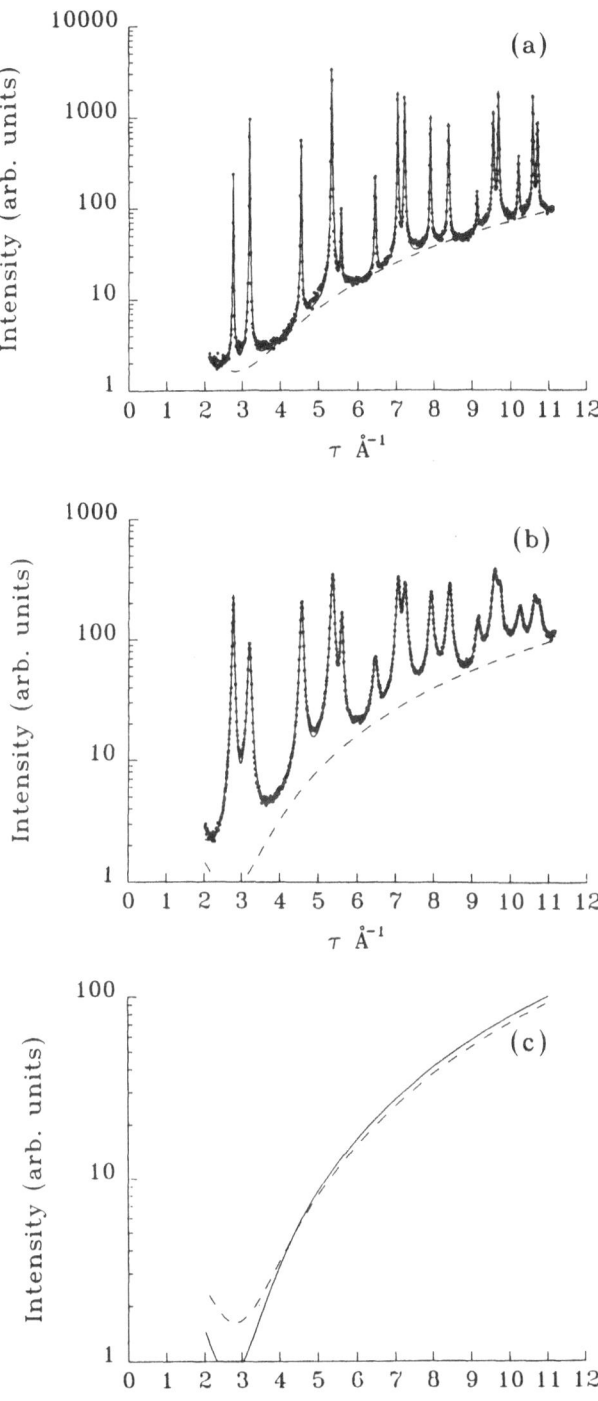

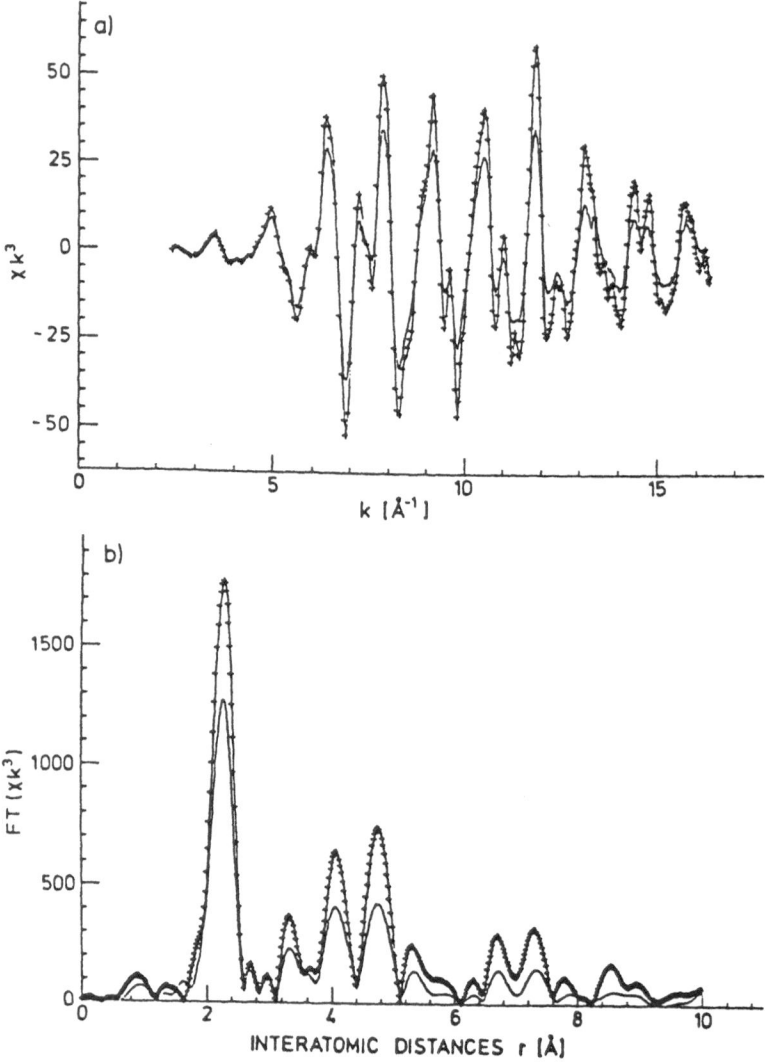

Fig. 4.17. The weighted EXAFS χk^3 (a) and its Fourier transform FTχk^3 (b) with phase shift not included of nanocrystalline Cu with a 10 nm grain size (——) and of coarse-grained polycrystalline Cu (+ + + +) are compared. From [4.68]

consolidated sample, even with its approximately 80% density, had more grain boundary area and less free surface area than the unconsolidated (albeit agglomerated) powder, it seems clear that the attribution of EXAFS amplitude reductions to the grain boundaries alone is inappropriate and that contributions from atoms at free surfaces associated with the significant porosity in the samples investigated must be taken into account.

Investigations of nanophase TiO_2 by Raman spectroscopy [4.24–26] and of nanophase Pd by atomic resolution TEM [4.53, 54] indicate that the grain boundary structures in these materials are rather similar to those in coarse grained conventional materials. These studies show that nanophase grain boundaries contain short-range ordered structural units representative of the bulk material and distortions that are localized to about ± 0.2 nm on either side of the grain boundary plane. Such conclusions are consistent with the results from small-angle neutron scattering (SANS) measurements [4.72–74] and with the expectations for the structures of conventional high-angle grain boundaries from condensed matter theory [4.63–65]. The Raman bands [4.24] were found to be significantly broadened in the as-consolidated (about 75% dense) nanophase TiO_2 samples relative to single-crystal or coarse-grained (air-annealed) rutile, as shown in Fig. 4.9. However, the degree of broadening (shown subsequently to be due to oxygen deficiency [4.25, 26]) was found to be independent of grain size, and hence grain boundary volume fraction, see Fig 4.14, as was the scattered intensity in the broad background. The SANS investigations of TiO_2 and Pd indicated grain boundary thicknesses of 0.5–1.0 nm, but rather low apparent average grain boundary densities (below 70% of bulk density). However, these low apparent densities may partly result from sample porosity, clearly demonstrated by PAS [4.11, 56, 57] and BET [4.59, 60] measurements, which has not been taken fully into account. A clear and complete separation of these effects needs to be accomplished before reliable comparisons between experiment and theory can be made.

The direct imaging of grain boundaries with high resolution electron microscopy (HREM) can avoid the complications that may arise from porosity and other defects in the interpretation of data from less direct methods. Typical grain boundaries in nanophase palladium are shown in Fig. 4.11; a higher magnification view of one such boundary is shown in Fig. 4.18. This HREM study [4.53, 54], which included both experimental observations and complementary image simulations, indicated no manifestations of grain boundary structures with random displacements of the type or extent suggested by earlier X-ray studies on nanophase Fe, Pd, and Cu [4.66, 68, 69]. For example, contrast features at the grain boundary shown in Fig. 4.18 that may be associated with disorder do not appear wider than 0.4 nm on this micrograph, indicating that any significant structural disorder which may be present essentially extends no further than the planes immediately adjacent to the boundary plane. Such localized lattice relaxation features are typical of high-angle grain boundary structures found in coarse-grained metals. HREM investigations of grain boundaries in nanophase Cu [4.75] and Fe alloys [4.76] produced by surface wear and high-energy ball milling, respectively, appear to support this view. The HREM image simulations [4.53, 54], examples of which are shown in Fig. 4.19, indicate that random atomic displacements of average magnitude greater than about 12% of a nearest neighbor distance, if present, could be readily observable by HREM for the assumed contrast conditions, which were identical to those used experimentally.

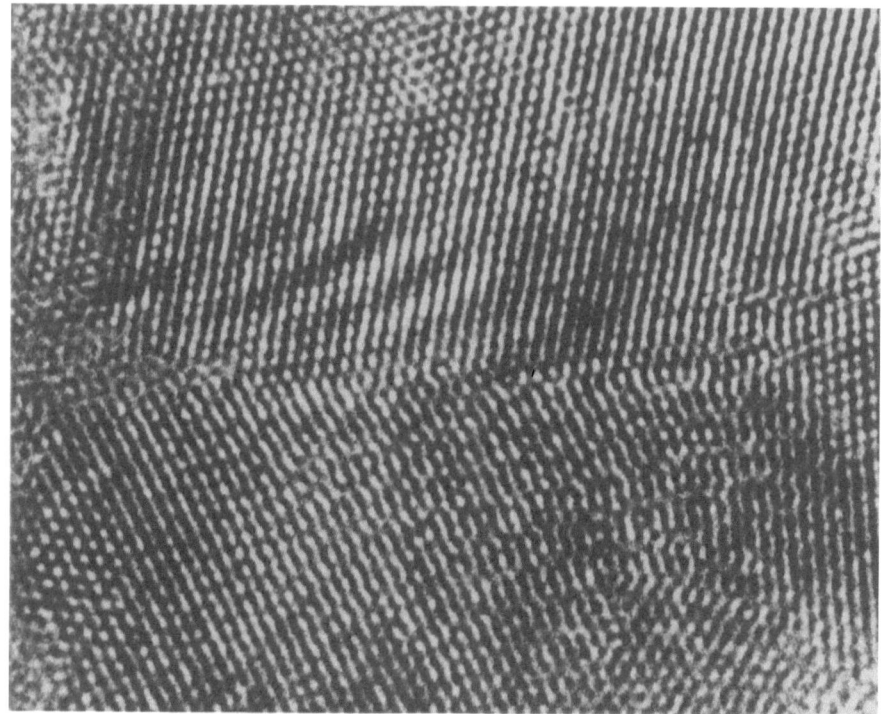

Fig. 4.18. High resolution transmission electron micrograph of a grain boundary in nanophase palladium from an area as shown in Fig. 4.11. The magnification is indicated by the lattice fringe spacings of 0.225 nm for (111) planes. From [4.53, 54]

Since the displacements observed by HREM appear to fall off rapidly for distances much smaller than even the small grain diameters of the nanophase materials thus far investigated, the atomic relaxations must be dominated by the influence of only the closest boundaries, as they are in conventional polycrystals. This implies that the action of thinning the HREM foil, and hence removing the grains above and below the volume under observation, does not in itself significantly affect the structures observed [4.55]. Calculations [4.77] of the possible effects of HREM specimen surfaces on grain boundary structure further support the reliability of such structural observations. Indeed, as shown in Figs. 4.11, 18, the nanophase grain boundaries appear to be rather low energy

Fig. 4.19. Image simulations for a $\Sigma 5$ symmetric $\langle 0\,0\,1 \rangle$ tilt boundary in 7.6 nm thick Pd using microscope parameters and imaging conditions consistent with those used in HREM experimental observations [4.53, 54]: (a) 'perfect' structure with no atomic displacements; (b) randomly disordered structure of the grain boundary with a maximum atomic displacement of 0.5 nnd and average of 0.25 nnd; (c) randomly disordered structure of the grain boundary with a maximum atomic displacement of 0.25 nnd and average of 0.125 nnd, where nnd is the nearest neighbor distance in Pd

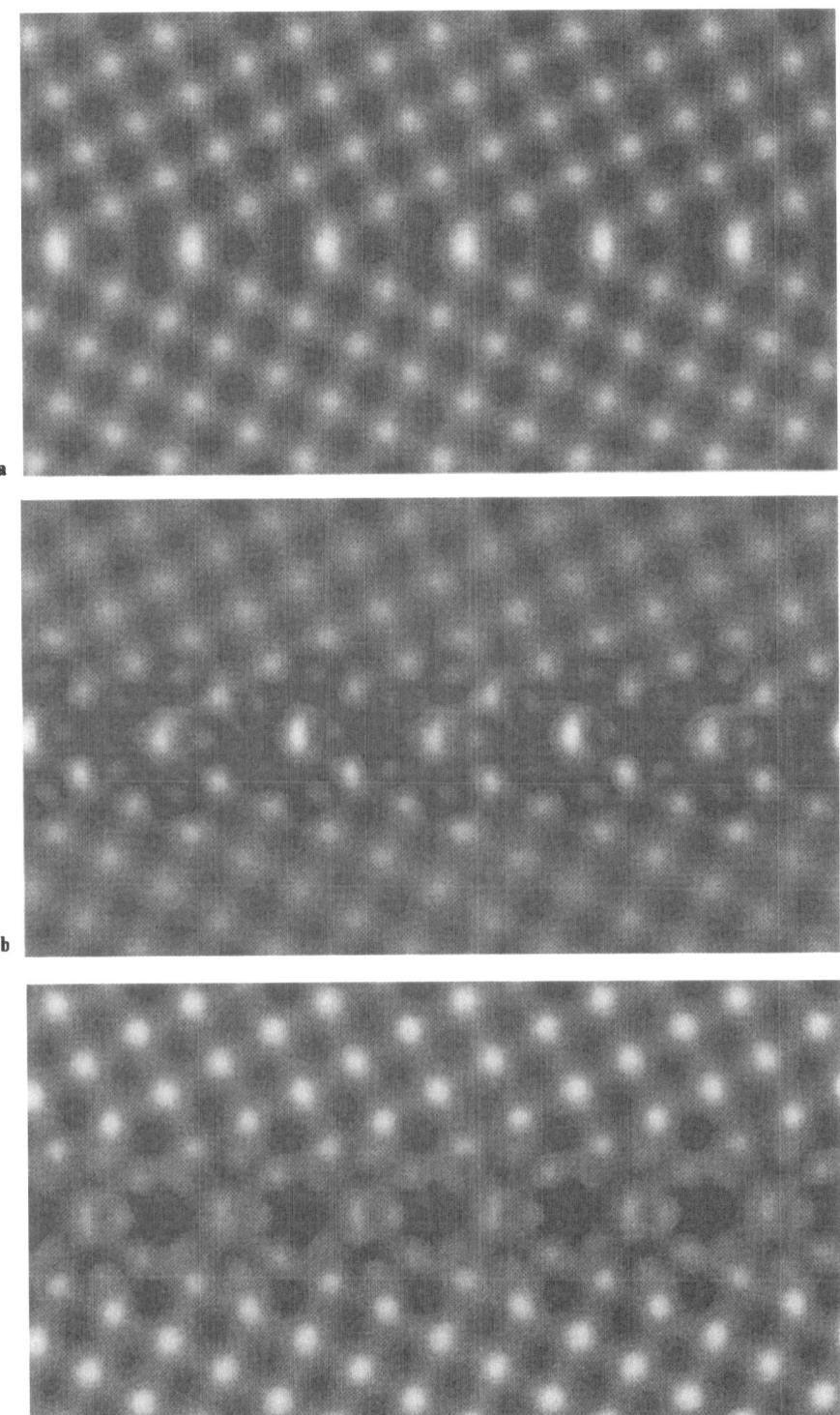

configurations exhibiting flat facets interspersed with steps. Such structures could only arise if sufficient local atomic motion occurred during the cluster consolidation process to allow the system to reach at least a local energy minimum. Such observations suggest at least two conclusions [4.78]: first, that the atoms that constitute the grain boundary volume in nanophase materials have sufficient mobility during cluster consolidation to accommodate themselves into relatively low energy grain boundary configurations; and second, that the local driving forces for grain growth are relatively small, despite the large amount of energy stored in the many grain boundaries in these materials. These conclusions also have an impact upon the inherent grain size stability of nanophase materials, as described in Sect. 4.3.3.

The foregoing discussion suggests [4.78] that nanophase materials should be a valuable resource for studying the average properties of grain boundaries. The high number density of such defects in these materials enhance their influence on macroscopic properties, allowing these effects to be studied by a variety of experimental techniques. Indeed, a number of the effects observed in nanophase materials to date that have been deemed unusual may simply result from so many grain boundaries being available for study in a sample for the first time. For the careful study of grain boundary properties to be successful in the future, however, specimen porosity will need to be removed, via consolidation at elevated temperature and/or pressure, so that its property contributions can be eliminated. By varying their grain size, the effects from interfaces and junctions in nanophase materials could be effectively separated in such future studies.

4.3.3 Grain Size Stability

The ultrafine grain sizes of nanophase materials immediately raise a question regarding their relative stability against spontaneous grain growth. With so many grain boundaries present in these materials, and their concomitantly large stored interface energy, this question of stability against grain growth becomes important to address from both scientific and technological viewpoints. Interestingly, nanophase materials assembled from atom clusters appear to possess an inherent grain size stability. Their grain sizes, as measured by dark-field TEM, remain rather deeply metastable to elevated temperatures. For example, as shown in Fig. 4.20 for nanophase TiO_2 [4.11], the 12 nm initial average grain diameter for the grain size distribution shown in Fig. 4.7 changes very little with annealing to elevated temperatures, until about 40–50% of the absolute melting temperature (T_m) of TiO_2 is reached. This behavior appears to be rather typical for the nanophase oxides already investigated [4.20] and for nanophase metals as well [4.79], as shown in the Arrhenius plot in Fig. 4.21. In the case of the TiO_2, rapid grain growth only develops above the temperature at which the mean bulk diffusion distance $(D_{Ti}t)^{1/2}$ of Ti, the slower moving element in this compound, becomes comparable to the mean grain size, at which temperature any local barriers to grain growth would cease to be significant.

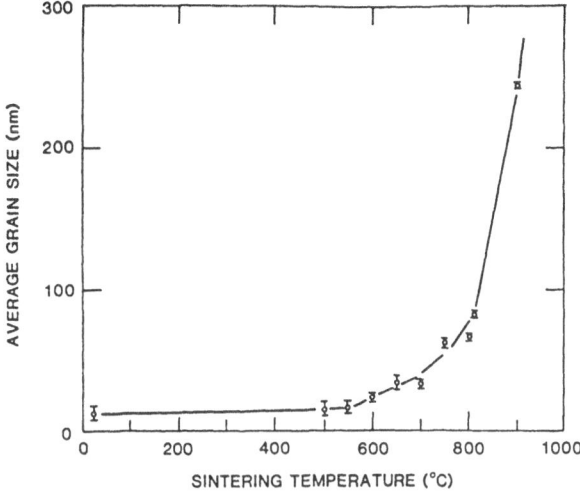

Fig. 4.20. Variation of average grain size with sintering temperature (0.5 h at each) for a nanophase TiO$_2$ (rutile) sample compacted to 1.4 GPa at room temperature, as determined by dark-field transmission electron microscopy. After [4.11]

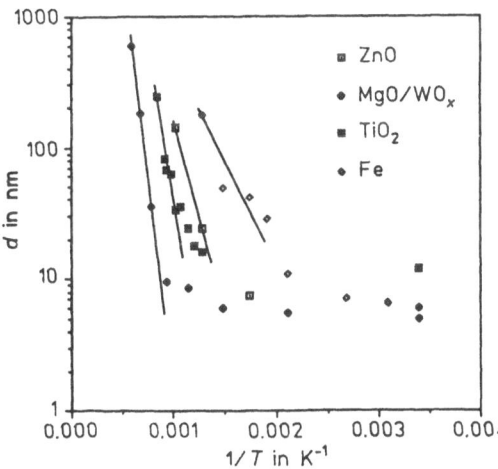

Fig. 4.21. Arrhenius plot of the variation of average grain size, measured by dark-field transmission electron microscopy, with sintering temperature for nanophase Fe [4.79], TiO$_2$ [4.11], MgO/WO$_x$ [4.20], and ZnO [4.20]. The oxide samples were annealed for 0.5 h in air at each temperature; the iron for 10 h in vacuum. From [4.80]

The observations of narrow grain size distributions, equiaxed grain morphologies, and low energy grain boundary structures in nanophase materials suggest that the inherent resistance to grain growth observed for cluster-assembled nanophase materials results primarily from a sort of frustration [4.80]. It appears that the narrow grain size distributions normally observed in these cluster-assembled materials coupled with their relatively flat grain boundary configurations (and also enhanced by their multiplicity of grain boundary junctions) place these nanophase structures in a local minimum in energy from which they are not easily extricated. There are normally no really large grains to grow at the expense of small ones through an Otswald ripening process, and the local grain boundaries, being essentially flat, have no curvature to tell them in which direction to migrate. Their stability thus appears to be analogous to that of a variety of closed-cell foam structures with narrow cell size distributions, which are stable (really, deeply metastable) despite their large stored surface energy. Under such conditions, only at temperatures above which bulk diffusion distances are comparable to or greater than the grain size and communication among grains becomes possible, as in the case of nanophase TiO_2 cited above, will this metastability give way to global energy minimization via rapid grain growth. Such a picture now appears to have some theoretical support [4.81]. It has also been recently suggested [4.82] that the porosity present in nanophase ceramics may assist in their stability against grain growth, as it does in conventional coarser-grained ceramics.

The diffusion controlled grain-growth behavior in nanophase materials is apparent in Fig. 4.21. The data in their high temperature limit fall, within their scatter, along straight lines in this Arrhenius plot. The effective activation energy of this high temperature limiting behavior is about $9\,kT_m$ for all of the materials; it is thus approximately one half that for self-diffusion, or close to the value usually observed for normal grain boundary diffusion in polycrystalline materials. It should be pointed out that exceptions to this frustrated grain growth behavior would be expected if considerably broader grain size distributions were accidentally present in a sample, which would allow a few larger grains to grow at the expense of smaller ones, or if significant grain boundary contamination were present, allowing enhanced stabilization of the small grain sizes to further elevated temperatures. Occasional observations of each of these types of behavior have been made. One could, of course, intentionally stabilize against grain growth by appropriate doping by insoluble elements or composite formation in the grain boundaries. For these cluster assembled materials, such stabilization should be especially easy, since the grain boundaries are available as cluster surfaces prior to consolidation. The ability to retain the ultrafine grain sizes of nanophase materials is important when one considers the fact that it is their grain size and large number of grain boundaries that determine to a large extent their special properties.

4.4 Properties

The unique properties of nanophase materials appear to result from an interplay among their three fundamental features: atomic domains (e.g., grains) spatially confined to less than 100 nm, significant atom fractions associated with interfacial environments (e.g., grain boundaries or free surfaces), and interactions between their constituent domains. In some cases one of these features dominates, in other cases another. Research on a variety of chemical, mechanical, and physical properties is beginning to yield a glimmer of understanding of just how this interplay manifests itself in the properties of these new materials.

4.4.1 Chemical Properties

Nanophase materials exhibit properties that are different and often considerably improved in comparison with those of conventional coarse-grained structures. Because of their small sizes and radii of curvature, coupled with their surface cleanliness, the constituent clusters of nanophase materials can react with one another rather aggressively, even at relatively low temperatures. For example, nanophase TiO_2 (rutile) exhibits significant improvements in both sinterability and resulting mechanical properties relative to conventionally synthesized coarser-grained rutile [4.11, 59, 83, 84]. Nanophase TiO_2 with a 12 nm initial mean grain diameter has been shown [4.11] to sinter under ambient pressures at 400–600°C lower temperatures than conventional coarse-grained rutile, and without the need for any compacting or sintering aid, such as polyvinyl alcohol, which is usually required. This behavior is shown in Fig. 4.22. Furthermore, it has been recently demonstrated [4.59] that sintering the same nanophase material under pressure (1 GPa), or with appropriate dopants such as Y, can

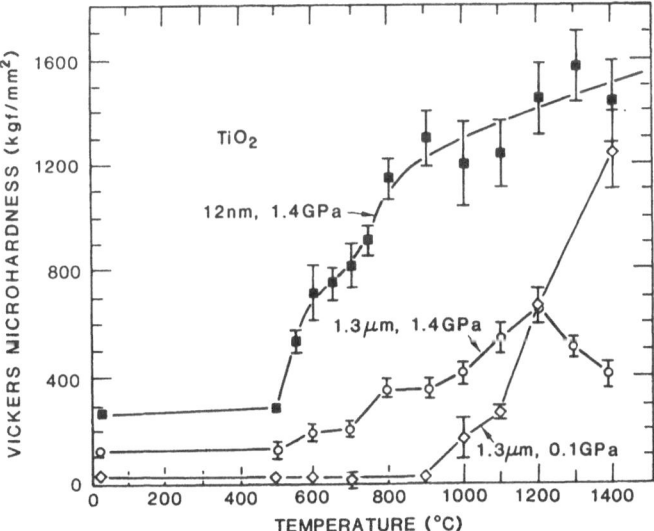

Fig. 4.22. Vickers microhardness of TiO_2 (rutile) measured at room temperature as a function of 0.5 h sintering at successively increased temperatures. Results for a nanophase sample (filled squares) with an initial average grain size of 12 nm consolidated at 1.4 GPa are compared with those for coarser-grained samples with 1.3 μm initial average grain size sintered with (diamonds) or without (circles) the aid of polyvinyl alcohol from commercial powder consolidated at 0.1 GPa and 1.4 GPa, respectively. After [4.11]

further reduce the sintering temperatures, while suppressing grain growth as well, thus allowing for the unique possibility to sinter nanophase ceramics to full density while retaining their ultrafine grain size. The resulting fracture characteristics [4.82, 83, 85] developed for sintered nanophase TiO_2 are as good, or in some aspects improved, relative to those for conventional coarser-grained rutile.

Sintering behavior has also been followed by PAS and SANS measurements. We have already discussed in Sect. 4.3.1 how PAS can be a useful tool in the study of the ultrafine-scale porosity inherent in as-consolidated nanophase compacts [4.11, 56, 57]; it can as well probe such porosity as a function of sintering temperature, to observe densification via the removal of voids. An example of PAS lifetime results [4.11] used to follow the sintering behavior of nanophase TiO_2 was already shown in Fig. 4.13. The intensity I_2 of the lifetime (τ_2) signal corresponding to positron annihilation from void-trapped states in the nanophase sample is seen to decrease rapidly during sintering above 500°C as a result of the densification of this ultrafine-grained ceramic, even though rapid grain growth does not set in until above 800°C, see Fig. 4.20. Furthermore, the variation of τ_2 with sintering indicates that there is a redistribution of void sizes accompanying this densification. Similar behavior is also observed for the coarser-grained samples investigated, but as expected, the densification proceeds more slowly in these latter samples and the average pore sizes are larger according to the larger values of τ_2 [4.61]. The redistribution of pore sizes can also be monitored by means of BET measurements (if the pores are still open to the sample surfaces so that N_2 can enter) as demonstrated by [4.59, 60]. Nanophase TiO_2 in its as-consolidated state and as a function of sintering in air was also followed by SANS [4.72]. While SANS contrast can yield information regarding the nature of the intergrain nanophase boundaries, particularly their average density via-à-vis the grain density [4.72–74], it can also yield information about the presence of voids (or porosity) and their removal during the sintering process. However, the clear separation of void and grain boundary contributions to small angle neutron scattering is still an open question, as discussed in Sect. 4.3.2, even though an attempt at using pressure-assisted sintering to remove the voids has recently been made [4.60].

Atomic diffusion in nanophase materials, which can have a significant bearing on their mechanical properties, such as creep and superplasticity, and physical properties as well, has been found to be very rapid. Measurements of self-diffusion and impurity diffusion [4.83, 86–90] in as-consolidated nanophase metals (Cu, Pd) and ceramics (TiO_2) indicate that atomic transport can be orders of magnitude faster in these materials than in coarser-grained polycrystalline samples, exhibiting 'surface-diffusion-like' behavior. However, the very rapid diffusion in as-consolidated nanophase materials appears to be intrinsically coupled with the porous nature of the interfaces in these materials. It has been recently shown in at least one case (Hf in TiO_2) that the rapid 'surface-like' diffusivities can be suppressed back to conventional values by sintering samples to full density [4.83], as demonstrated by a comparison between the Hf diffusion observations shown in Figs. 4.23(a), (b) before and after

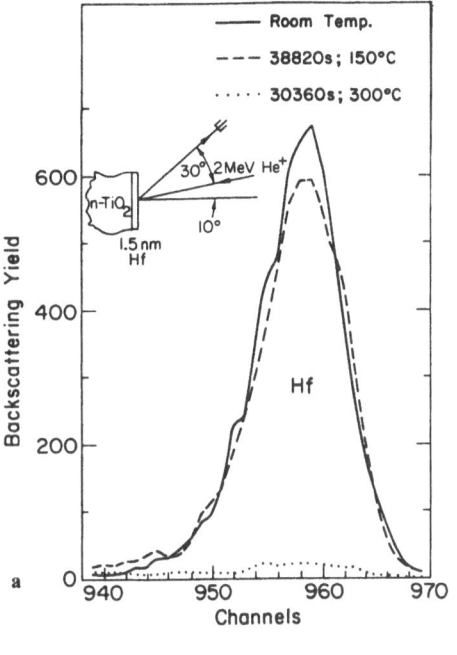

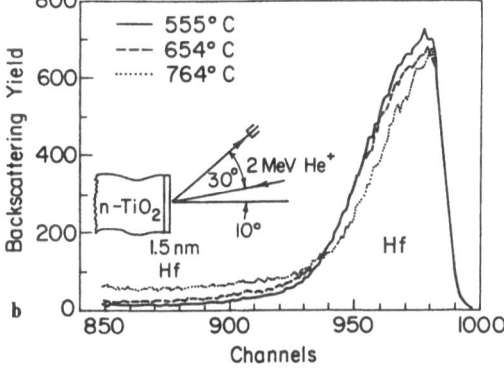

Fig. **4.23.** Diffusion profiles of Hf in nanophase TiO_2 measured by Rutherford backscattering after (a) sintering in air at atmospheric pressure at 100°C or (b) pressure-assisted sintering in air at 1 GPa at 550°C, with subsequent Hf deposition on the sample surface. From [4.83, 88]

pressure-assisted sintering of nanophase TiO_2, respectively. Nonetheless, there exist considerable possibilities for efficiently doping nanophase materials at relatively low temperatures via the rapid diffusion available along their ubiquitous grain-boundary networks and interconnected porosity, with only short diffusion paths remaining into their grain interiors, to synthesize materials with tailored chemical, mechanical, or physical properties.

The chemical reactivity of nanophase materials compared to conventional materials can be rather striking. Recently completed measurements [4.91] of the decomposition of H_2S over lightly consolidated nanophase TiO_2 at 500°C demonstrate this rather well. Figure 4.24 shows the activity for S removal from

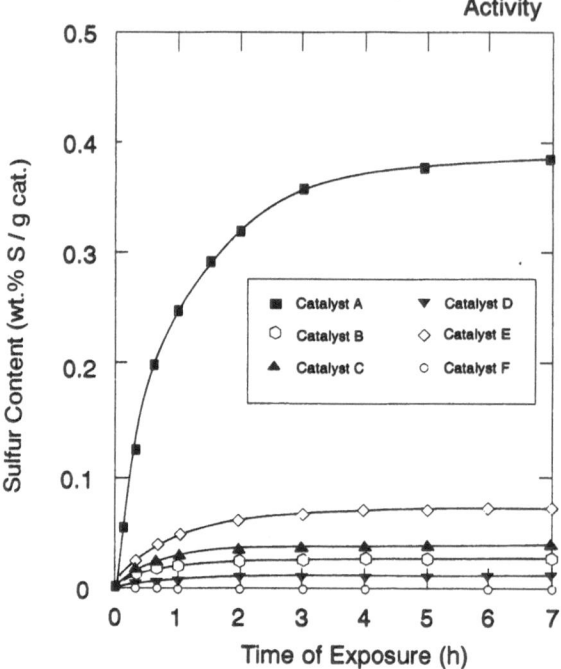

Fig. 4.24. Activity of nanophase TiO_2 for H_2S decomposition as a function of exposure time at 500°C compared with that from several commercially available TiO_2 materials and a reference (A: 76 m²/g nanophase rutile; B: 61 m²/g anatase; C: 2.4 m²/g rutile; D: 30 m²/g anatase; E: 20 m²/g rutile; F: reference alumina). From [4.91]

H_2S via decomposition for nanophase TiO_2 (rutile) compared with that for a number of other commercially available forms of TiO_2 having either the rutile or anatase structure. It can be seen that the nanophase sample was far more reactive than any of the other samples tested, both initially and after extended exposure to the H_2S. This greatly enhanced activity was shown to result from a combination of unique features of the nanophase material, its high specific surface area combined with its rutile structure and its oxygen deficient composition, see Sect. 4.2.

4.4.2 Mechanical Properties

Owing to their ultrafine grain sizes, nanophase ceramics are easily formed, as has been clearly evident from the sample compaction process [4.10, 11, 18] and from demonstrations via deformation [4.92] as well. However, the degree to which nanophase ceramics are truly ductile is only beginning to be understood. Nanoindenter measurements on nanophase TiO_2 [4.84] and ZnO [4.93] have recently demonstrated that a dramatic increase of strain rate sensitivity occurs with decreasing grain size, as shown in Fig. 4.25, that is remarkably similar for these two materials, even though they had as-consolidated densities of about 75% and 85%, respectively. Since this strong grain-size dependence was found for sets of samples in which the porosity was changing very little, it appears to be

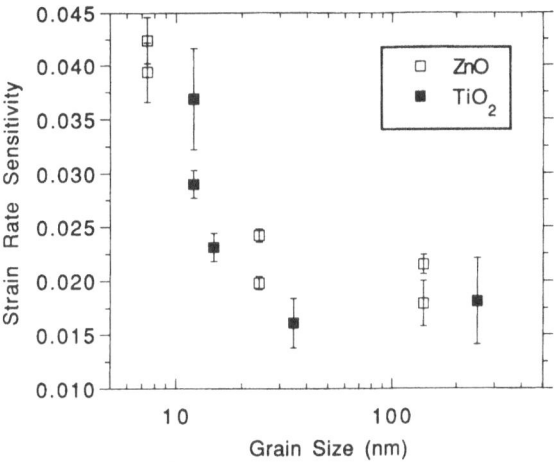

Fig. 4.25. Strain rate sensitivity of nanophase TiO_2 [4.84] and ZnO [4.93] as a function of grain size. The strain rate sensitivity was measured by a nano-indentation method [4.84] and the grain size was determined by dark-field transmission electron microscopy

an intrinsic property of these ultrafine-grained ceramics. The strain rate sensitivity values at the smallest grain sizes yet investigated (12 nm in nanophase TiO_2 and 7 nm in ZnO) indicate not only ductile behavior of these nanophase ceramics at room temperature, but also a significant potential for increased ductility at even smaller grain sizes and at elevated temperatures. The maximum strain rate sensitivities measured in these studies, about 0.04, are approximately one-quarter that for Pb at room temperature, for example. However, no superplasticity has yet been observed in nanophase materials at room temperature, which would yield m values about an order of magnitude higher than the maximum observed. Nevertheless, it already seems clear that in the future, at smaller grain sizes and/or at elevated temperatures, superplasticity of these materials will indeed be observed.

The possibilities for plastic forming nanophase ceramics to near net shape appear to be well on their way to realization. *Karch* and *Birringer* [4.94] have recently demonstrated that nanophase TiO_2 could be readily formed to a desired shape with excellent detail below 900°C, and the fracture toughness was found to increase by a factor of two as well. The ability to extensively deform nanophase TiO_2 at elevated temperatures (ca. 800°C) without cracking or fracture has been demonstrated by *Hahn* and coworkers [4.95, 96], as shown rather dramatically in Fig. 4.26. While these latter demonstrations have been accompanied by significant grain growth in the samples at the elevated temperatures employed, it can be expected that lower temperature studies (below 0.4–0.5 T_m) in the future will also allow for near net shape forming of nanophase ceramics with both their ultrafine grain sizes and their attendant properties retained.

The enhanced strain rate sensitivity at room temperature found in the nanophase ceramics TiO_2 and ZnO [4.84, 93] appears to result from increased grain boundary sliding in this material, aided by the presence of porosity, ultrafine grain size, and probably rapid short-range diffusion as well. This

Fig. 4.26. Nanophase TiO$_2$ sample before and after compression at 810°C for 15 h. The total true strains were as high as 0.6, which represents deformation to a final thickness of less than 2 mm from an initial length of 3.5 mm at about 0.5 of its melting temperature (1830°C). The grain size increased from 40–50 nm to about 1 μm. The small rule divisions are millimeters. From [4.95]

behavior is therefore dominated by the presence of the numerous interfaces in these materials and the very short diffusion distances involved in effecting the necessary atomic healing of incipient cracks for grain boundary sliding to progress at the strain rates utilized without fracturing the sample. Extrapolating from this apparently generic behavior, one can expect that grain boundary sliding mechanisms, accompanied by short-range diffusion assisted healing events, would be expected to increasingly dominate the deformation of a wide range of nanophase materials. Enhanced forming and even superplasticity in a wide range of nanophase materials, including intermetallic compounds, ceramics, and semiconductors might become a reality. Consequently, increased opportunities for high deformation or superplastic near net shape forming of a very wide range of even conventionally rather brittle and difficult to form materials could result.

The predominant mechanical property change resulting from reducing the grain sizes of nanophase metals, in contrast to the behavior found for nanophase ceramics, is the significant increase in their strength. While the microhardness of as-consolidated nanophase oxides is reduced relative to their fully-dense counterparts (Fig. 4.22), owing to significant porosity in addition to their ultrafine grain sizes, the case for nanophase metals is quite different. Figures 4.27, 28 show recent microhardness and stress–strain results for nanophase Pd and Cu compared with similar results for their coarser-grained counterparts

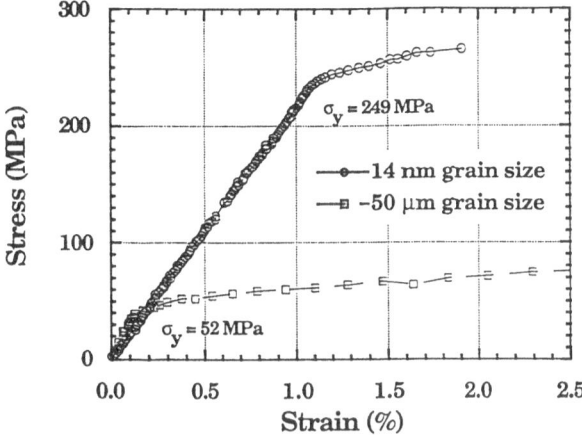

Fig. 4.27. Stress–strain curve for a nanophase (14 nm) Pd sample compared with that for a coarse-grained (50 μm) Pd sample. The strain rate $\dot{\varepsilon} \approx 2 \times 10^{-5}$ s^{-1}. After [4.98]

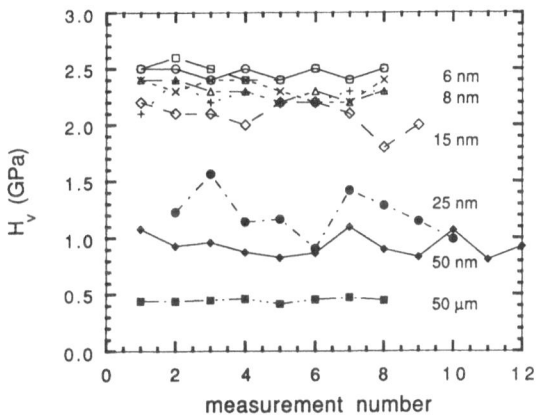

Fig. 4.28. Vickers microhardness measurements at a number of positions across several nanophase Cu samples ranging in grain size from 6 to 50 nm, compared with similar measurements from an annealed conventional 50 μm grain size Cu sample. After [4.58]

[4.58, 97, 98]. In their as-consolidated state, nanophase Pd samples with 5–10 nm grain sizes have been observed to exhibit up to about a 500% increase in hardness over coarser-grained (ca. 100 μm) samples [4.97], with concomitant increases in yield stress σ_y, as shown in Fig. 4.27. Similar results have been observed in nanophase Cu as well, as shown in Fig. 4.28. The common strengthening behavior found in nanophase Pd and Cu indicates that this response is generic to nanophase metals, at least those with a fcc structure. The likelihood that such mechanical behavior is more broadly generic to nanophase metals in general is enhanced by the observations that nanophase metals and alloys produced via mechanical attrition also exhibit significantly enhanced strength. For example, *Koch* and coworkers [4.99, 100] have found hardness increases of factors of 4–5 in nanophase Fe and a factor of about 1.2 in nanophase Nb$_3$Sn when the grain size drops from 100 nm to 6 nm.

In apparent contrast to these observations of enhanced strength with decreasing grain size, *Chokshi* et al. [4.101] have reported a softening with decreasing grain size in the nanometer regime for cluster-assembled Cu and Pd samples, which was rationalized in terms of their expectation [4.92, 102] of room temperature diffusional creep in these ultrafine-grained metals. A similar apparent softening was also reported [4.103] for TiAl. The rapid atomic diffusion observed in nanophase materials, see Sect. 4.4.1, along with their nanometer grain sizes, has suggested that a large creep enhancement might result in nanophase materials, even at room temperature [4.92, 102]. However, recent constant-stress creep measurements on nanophase Pd and Cu [4.58, 98] show that the observed creep rates at room temperature are at least three orders of magnitude smaller than predicted on the basis of a Coble creep model, in which the creep rate varies as D_b/d^3, where D_b is the grain boundary diffusivity and d is the mean grain size. Such creep resistance will need to be explored further at elevated temperatures in these and other nanophase materials. However, it appears that this apparent softening may only be a product of the manner in which the grain sizes were varied in these samples via annealing (sintering) leading to grain growth [4.104].

The increased strength observed in ultrafine-grained nanophase metals, although apparently analogous to conventional Hall-Petch strengthening [4.105, 106] observed with decreasing grain size in coarser-grained metals, must result from fundamentally different mechanisms. The grain sizes in the nanophase metals considered here are smaller than the necessary critical bowing lengths for Frank-Read dislocation sources to operate at the stresses involved and smaller also than the normal spacings between dislocations in a pile-up. It is therefore clear that an adequate description of the mechanisms responsible for the increased strength observed in nanophase metals will clearly need to accommodate to the ultrafine grain-size scale in these materials. As this scale is reduced, and conventional dislocation generation and migration become increasingly difficult, it is apparent that the energetic hierarchy of microscopic deformation mechanisms will become successively accessed. Thus, easier mechanisms (such as dislocation generation from Frank-Read sources) will become frozen out at sufficiently small grain sizes and more costly mechanisms will become necessary to effect deformation. Hence, grain confinement appears to be the dominant cause for the increased strength of nanophase metals. It can be generally expected that as nanophase grain sizes decrease and fall below the critical length scale for a given mechanism to operate, the associated property will be significantly changed. Clearly, much work in this area remains.

4.4.3 Physical Properties

Rather limited research has been carried out so far on the physical properties of cluster-assembled nanophase materials. However there appear to be interesting prospects, based upon what little has been accomplished and on the expectations for confined systems of atoms in which the sizes of constituent domains fall below the critical length scales pertinent to a given property. Two examples of the physical property changes that can occur in nanophase samples will be presented here. Many more can be expected in the near future, as research begins to focus on these aspects of nanophase materials.

Nanophase TiO_2 with a 12 nm grain size was doped at about the 1 atomic % level with Pt diffused in from the surface, as demonstrated by Rutherford backscattering measurements [4.20]. After annealing in air for 4 h at about 500°C, the ac conductivity of the sample was measured as a function of temperature [4.107]. The strongly nonlinear, and reversible, electrical response shown in Fig. 4.29, caused presumably by the Pt doping into the band gap of this wide band gap (3.2 eV) semiconductor, suggests that the rather easy impurity doping of nanophase electroceramics via rapid diffusion down their many grain boundaries and interconnected pores may lead to a wide range of interesting device applications in the future. The enhanced low temperature sinterability of nanophase ceramics, without the need for any possibly contaminating additives, should help with device compatibility problems frequently encountered in such applications. However, much work remains to be done in this area.

The ability to control the porosity in the synthesis of nanophase ceramics will apparently lead to some rather interesting optical applications in the near future. It has been recently demonstrated [4 108] that the oxidation step in the synthesis of cluster-consolidated nanophase Y_2O_3 can be so controlled that the resulting "green-state" porosity of 25–35% has a size distribution sufficiently small and narrow, as shown in Fig. 4.30, that the as-consolidated material is effectively transparent. Samples of this material are shown in the inset of Fig. 4.30 along with a similar opaque sample containing a larger population of pores. The ability to thus reduce the porosity in such material to well below the wavelengths of visible or other radiation, coupled with the possibilities for

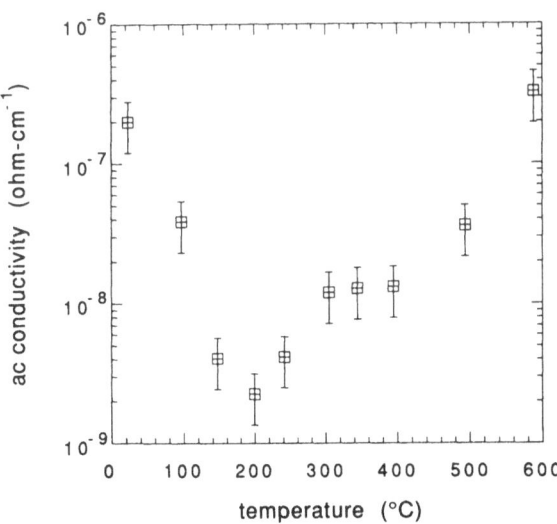

Fig. 4.29. The ac conductivity of Pt-doped nanophase TiO_2 as a function of temperature. The sample was pre-annealed in air at about 500°C for 4 h prior to the conductivity measurements; the electrical response is reversible with temperature. From [4.107]

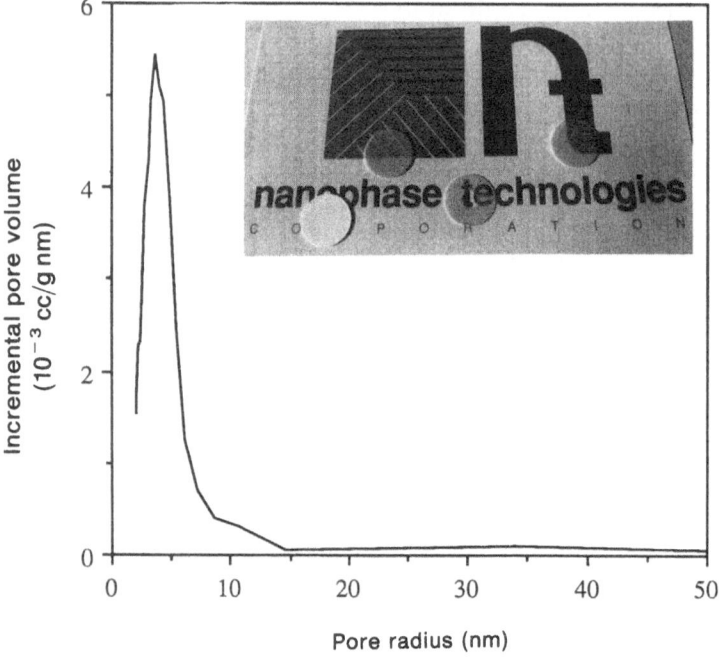

Fig. 4.30. Pore size distribution in nanophase Y_2O_3 compacted at room temperature under 500 Pa pressure to a density of between 65–75% of the theoretical value. A similar set of samples is shown in the inset along with an opaque sample with larger pores. After [4.108]

efficient doping of these materials cited previously, should lead to a variety of interesting optical properties and related applications.

4.5 Future Directions

It is very clear that in such a broad field as described in this chapter, we are just beginning to scratch the surface of the tremendous opportunities for synthesizing nanophase materials via the assembly of atom clusters. The prognosis for the field has been recently considered [4.21] and will for the most part be recapitulated here. Based upon the limited knowledge that has already been accumulated, the future appears to hold great promise for nanophase materials. The property changes that are found to occur when grain sizes are scaled down into the nanometer regime and interfaces take up significant volume fractions of the material can make significant differences in the range of use of a variety of materials. The cluster sizes accessed to date (down to about 5 nm) indicate that the high reactivities and short diffusion distances available in cluster-assembled materials can have profound effects upon both the processing characteristics of these materials and their properties. These characteristics should be further enhanced as even smaller and more uniformly sized clusters become available in sufficiently large numbers to effect their assembly into usable and commercially viable materials.

The enhanced diffusivities along their grain boundary networks, with only few atomic jumps separating grain interiors from grain boundaries, should enable efficient impurity doping of these materials. Nanophase insulators and semiconductors, for example, could be easily doped with impurities at relatively low temperatures allowing efficient introduction of impurity levels into their band gaps and control over their electrical and optical properties. Moreover, the ability to produce via cluster assembly fully dense ultrafine-grained nanophase ceramics with controlled or flaw-free microstructures that are readily formable and exhibit ductility should have a significant technological impact in a wide variety of applications. For example, near net-shape forming of nanophase ceramic parts with complex and ultrafine detail would seem to be possible. Subsequent controlled grain growth could then be used to alter the grain-size dependent properties of these ceramics.

Research on cluster-assembled nanophase materials, although currently being carried out in only a few laboratories, appears now to be rapidly expanding [4.48, 51]. Much work still remains to be done. Further research on the synthesis of a broader range of nanophase materials, encompassing metals, alloys, ceramics, semiconductors, and composites will have to be carried out in order to arrive at a fuller appreciation of just how broad an impact nanophase materials will have on materials technology. Investigations of their structure that will need to accompany such research will begin to elucidate the interplay between the effects of spatial confinement and large numbers of interfaces on the chemical, mechanical, and physical properties of these new materials. A knowledge of the variation of such properties with the detailed structures and the synthesis and processing parameters of a variety of nanophase materials should eventually lead to an understanding of these new materials and consequently to the realization of their full impact on materials science and technology.

Acknowledgements. This work was supported by the U.S. Department of Energy, BES-Materials Sciences, under Contract W-31-109-Eng-38. The author wishes to thank his many collaborators at Argonne National Laboratory and elsewhere, without whose efforts and contributions this work would not have been possible.

References

4.1   M. Blander, J.L. Katz: Geochim. Cosmochim. Acta **31**, 1025 (1967)
4.2   M. Blander, M. Abdel-Gawad: Geochim. Cosmochim. Acta **33**, 701 (1969)
4.3   L. Grossman: Geochim. Cosmochim. Acta **36**, 597 (1972)
4.4   B.H. Kear, L.E. Cross, J.E. Keem, R.W. Siegel, F. Spaepen, K.C. Taylor, E.L. Thomas, K.-N. Tu: *Research Opportunities for Materials with Ultrafine Microstructures* (National Academy, Washington, DC 1989) Vol. NMAB-454

4.5   R.P. Andres, R.S. Averback, W.L. Brown, L.E. Brus, W.A. Goddard, III, A. Kaldor, S.G. Louie, M. Moskovits, P.S. Peercy, S.J. Riley, R.W. Siegel, F. Spaepen, Y. Wang: J. Mater. Res. **4**, 704 (1989)
4.6   R.W. Siegel: In *Materials Science and Technology*, Vol. 15, Processing of Metals and Alloys, ed. by R.W. Cahn (VCH, Weinheim 1991) p. 583
4.7   H. Gleiter: In *Deformation of Polycrystals: Mechanisms and Microstructures*, ed. by N. Hansen, A. Horsewell, T. Leffers, H. Lilholt (Risø National Laboratory, Roskilde 1981) p. 15
4.8   R. Birringer, H. Gleiter, H.-P. Klein, P. Marquardt: Phys. Lett. **102A**, 365 (1984)
4.9   R. Birringer, U. Herr, H. Gleiter: Suppl. Trans. Jpn. Inst. Met. **27**, 43 (1986)
4.10   R.W. Siegel, H. Hahn: In *Current Trends in the Physics of Materials*, ed. by M. Yussouff (World Scientific, Singapore 1987) p. 403
4.11   R.W. Siegel, S. Ramasamy, H. Hahn, Z. Li, T. Lu, R. Gronsky: J. Mater. Res. **3**, 1367 (1988)
4.12   A.H. Pfund: Rev. Sci. Inst. **1**, 397 (1930)
4.13   K. Kimoto, Y. Kamiya, M. Nonoyama, R. Uyeda: Jpn. J. Appl. Phys. **2**, 702 (1963)
4.14   C.G. Granqvist, R.A. Buhrman: J. Appl. Phys. **47**, 2200 (1976)
4.15   A.R. Thölén: Acta Metall. **27**, 1765 (1979)
4.16   R. Uyeda: Prog. Mater. Sci. **35**, 1 (1991)
4.17   R. Birringer, H. Gleiter: In *Encyclopedia of Materials Science and Engineering*, Suppl. Vol. 1, ed. by R.W. Cahn (Pergamon, Oxford 1988) p. 339
4.18   H. Hahn, J.A. Eastman, R.W. Siegel: In *Ceramic Transactions, Ceramic Powder Science*, Vol. 1, Part B, ed. by G.L. Messing, E.R. Fuller, Jr., H. Hausner (American Ceramic Society, Westerville 1988) p. 1115
4.19   H. Gleiter: Prog. Mater. Sci. **33**, 223 (1989)
4.20   J.A. Eastman, Y.X. Liao, A. Narayanasamy, R.W. Siegel: Mater. Res. Soc. Symp. Proc. **155**, 255 (1989)
4.21   R.W. Siegel: Ann. Rev. Mater. Sci. **21**, 559 (1991)
4.22   R.W. Siegel, J.A. Eastman: Mater. Res. Soc. Symp. Proc. **132**, 3 (1989)
4.23   V.A. Maroni: J. Phys. Chem. Solids **47**, 307 (1988)
4.24   C.A. Melendres, A. Narayanasamy, V.A. Maroni, R.W. Siegel: J. Mater. Res. **4**, 1246 (1989)
4.25   J.C. Parker, R.W. Siegel: J. Mater. Res. **5**, 1246 (1990)
4.26   J.C. Parker, R.W. Siegel: Appl. Phys. Lett. **57**, 943 (1990)
4.27   K. Sattler, J. Mühlbach, E. Recknagel: Phys. Rev. Lett. **45**, 821 (1980)
4.28   K. Sattler: Ultramicroscopy **20**, 55 (1986)
4.29   K. Sattler: Z. Phys. D **3**, 223 (1986)
4.30   H. Oya, T. Ichihashi, N. Wada: Jpn. J. Appl. Phys. **21**, 554 (1982)
4.31   S. Yatsuya, K. Yamauchi, T. Kamakura, A. Yanagada, H. Wakaiyama, K. Mihama: Surf. Sci. **156**, 1011 (1985)
4.32   S. Yatsuya, T. Kamakura, K. Yamauchi, K. Mihama: Jpn. J. Appl. Phys., P. 2, **25**, L42 (1986)
4.33   H. Hahn, R.S. Averback: J. Appl. Phys. **67**, 1113 (1990)
4.34   G.M. Chow, R.L. Holtz, A. Pattnaik, A.S. Edelstein, T.E. Schlesinger, R.C. Cammarata: Appl. Phys. Lett. **56**, 1853 (1990)
4.35   S. Iwama, E. Shichi, T. Sahashi: Jpn. J. Appl. Phys. **12**, 1531 (1973)
4.36   S. Iwama, K. Hayakawa, T. Arizumi: J. Cryst. Growth **56**, 265 (1982); J. Cryst. Growth **66**, 189 (1984)
4.37   S. Iwama, K. Hayakawa: Surf. Sci. **156**, 85 (1985)
4.38   B. Günther, A. Kumpmann: Nanostructured Mater. **1**, 27 (1992)
4.39   A. Matsunawa, S. Katayama: In *Laser Welding, Machining and Materials Processing*, Proc. ICALEO '85, ed. by C. Albright (IFS, 1985) p. 205
4.40   K. Baba, N. Shohata, M. Yonezawa: Appl. Phys. Lett. **54**, 2309 (1989)
4.41   M. Uda: Nanostructured Mater. **1**, 101 (1992)
4.42   A.E. Berkowitz, J.L. Walter: J. Mater. Res. **2**, 277 (1987)
4.43   E. Hellstern, H.J. Fecht, Z. Fu, W.L. Johnson: J. Appl. Phys. **65**, 305 (1989)
4.44   C.C. Koch, J.S.C. Jang, S.S. Gross: J. Mater. Res. **4**, 557 (1989)
4.45   H.J. Fecht, E. Hellstern, Z. Fu, W.L. Johnson: Adv. Powder Metall. **1-2**, 111 (1989)

4.46   I.A. Aksay, G.L. McVay, D.R. Ulrich (eds.): *Processing Science of Advanced Ceramics*, Mater. Res. Soc. Symp. Proc. **155** (1989)

4.47   P.C. Rieke, P.D. Calvert, M. Alper (eds.): *Materials Synthesis Utilizing Biological Processes*, Mater. Res. Soc. Symp. Proc. **174** (1990)

4.48   R.S. Averback, D.L. Nelson, J. Bernholc (eds.): *Clusters and Cluster-Assembled Materials*, Mater. Res. Soc. Symp. Proc. **206** (1991)

4.49   G.M. Whitesides, J.P. Mathias, C.T. Seto: Science **254**, 1312 (1991)

4.50   A.H. Heuer, D.J. Fink, V.J. Laraia, J.L. Arias, P.D. Calvert, K. Kendall, G.L. Messing, J. Blackwell, P.C. Rieke, D.H. Thompson, A.P. Wheeler, A. Veis, A.I. Caplan: Science **255**, 1098 (1992)

4.51   B.H. Kear, R.W. Siegel, T. Tsakalakos (eds.): Nanostructured Mater. **1** (1992)

4.52   R.W. Siegel: In *Materials Interfaces: Atomic-Level Structure and Properties*, ed. by D. Wolf, S. Yip (Chapman and Hall, London 1992) p. 431

4.53   G.J. Thomas, R.W. Siegel, J.A. Eastman: Mater. Res. Soc. Symp. Proc. **153**, 13 (1989)

4.54   G.J. Thomas, R.W. Siegel, J.A. Eastman: Scripta Metall. et Mater. **24**, 201 (1990)

4.55   W. Wunderlich, Y. Ishida, R. Maurer: Scripta Metall. et Mater. **24**, 403 (1990)

4.56   H.E. Schaefer, R. Würschum, M. Scheytt, R. Birringer, H. Gleiter: Mater. Sci. Forum **15-18**, 955 (1987)

4.57   H.E. Schaefer, R. Würschum, R. Birringer, H. Gleiter: Phys. Rev. B **38**, 9545 (1988)

4.58   G.W. Nieman, J.R. Weertman, R.W. Siegel: J. Mater. Res, **6**, 1012 (1991)

4.59   H. Hahn, J. Logas, R.S. Averback: J. Mater. Res. **5**, 609 (1990)

4.60   W. Wagner, R.S. Averback, H. Hahn, W. Petry, A. Wiedenmann: J. Mater. Res. **6**, 2193 (1991)

4.61   R.W. Siegel: Ann. Rev. Mater. Sci. **10**, 393 (1980)

4.62   S.J. Gregg, K.S.W. Sing: *Adsorption, Surface Area and Porosity* (Academic, New York 1982)

4.63   D. Wolf, J.F. Lutsko: Phys. Rev. Lett. **60**, 1170 (1988)

4.64   D. Wolf, J.F. Lutsko: J. Mater. Res. **4**, 1467 (1989)

4.65   S.R. Phillpot, D. Wolf, S. Yip: MRS Bulletin **XV**(10), 38 (1990)

4.66   X. Zhu, R. Birringer, U. Herr, H. Gleiter: Phys. Rev. B **35**, 9085 (1987)

4.67   U. Herr, J. Jing, R. Birringer, U. Gonser, H. Gleiter: Appl. Phys. Lett. **50**, 472 (1987)

4.68   T. Haubold, R. Birringer, B. Lengeler, H. Gleiter: J. Less-Common Metals **145**, 557 (1988)

4.69   T. Haubold, R. Birringer, B. Lengeler, H. Gleiter: Phys. Lett. A **135**, 461 (1989)

4.70   M.R. Fitzsimmons, J.A. Eastman, M. Müller-Stach, G. Wallner: Phys. Rev. B **44**, 2452 (1991)

4.71   J.A. Eastman, M.R. Fitzsimmons, M. Müller-Stach, G. Wallner, W.T. Elam: Nanostructured Mater. **1**, 47 (1992)

4.72   J.E. Epperson, R.W. Siegel, J.W. White, T.E. Klippert, A. Narayanasamy, J.A. Eastman, F. Trouw: Mater. Res. Soc. Symp. Proc. **132**, 15 (1989)

4.73   J.E. Epperson, R.W. Siegel, J.W. White, J.A. Eastman, Y.X. Liao, A. Narayanasamy: Mater. Res. Soc. Symp. Proc. **166**, 87 (1990)

4.74   E. Jorra, H. Franz, J. Peisl, G. Wallner, W. Petry, R. Birringer, H. Gleiter, T. Haubold: Phil. Mag. B **60**, 159 (1989)

4.75   S.K. Ganapathi, D.A. Rigney: Scripta Metall. et Mater. **24**, 1675 (1990)

4.76   M.L. Trudeau, A. Van Neste, R. Schultz: Mater. Res. Soc. Symp. Proc. **206**, 487 (1991)

4.77   M.J. Mills, M.S. Daw: Mater. Res. Soc. Symp. Proc. **183**, 15 (1990)

4.78   R.W. Siegel, G.J. Thomas: Ultramicroscopy **40**, 376 (1992)

4.79   E. Hort, Diploma Thesis, Universität des Saarlandes, Saarbrücken (1986); H. Gleiter: private communication

4.80   R.W. Siegel: Mater. Res. Soc. Symp. Proc. **196**, 59 (1990)

4.81   N. Rivier: In *Physics and Chemistry of Finite Systems: From Clusters to Crystals*, Vol. 1, ed. by P. Jena, S.N. Khanna, B.K. Rao (Kluwer, Boston 1992) p. 189

4.82   H.J. Höfler, R.S. Averback: Scripta Metall. et Mater. **24**, 2401 (1990)

4.83   R.S. Averback, H. Hahn, H.J. Höfler, J.L. Logas, T.C. Chen: Mater. Res. Soc. Symp. Proc. **153**, 3 (1989)

4.84   M.J. Mayo, R.W. Siegel, A. Narayanasamy, W.D. Nix: J. Mater. Res. **5**, 1073 (1990)

4.85   Z. Li, S. Ramasamy, H. Hahn, R.W. Siegel: Mater. Lett. **6**, 195 (1988)

4.86   J. Horváth, R. Birringer, H. Gleiter: Solid State Commun. **62**, 319 (1987)

4.87   J. Horváth: Defect and Diffusion Forum **66-69**, 207 (1989)
4.88   H. Hahn, H. Höfler, R.S. Averback: Defect and Diffusion Forum **66-69**, 549 (1989)
4.89   S. Schumacher, R. Birringer, R. Straub, H. Gleiter: Acta Metall. **37**, 2485 (1989)
4.90   T. Mütschele, R. Kirchheim: Scripta Metall. **21**, 135 (1987); Scripta Metall. **21**, 1101 (1987)
4.91   D.D. Beck, R.W. Siegel: J. Mater. Res. **7**, 2840 (1992)
4.92   J. Karch, R. Birringer, H. Gleiter: Nature **330**, 556 (1987)
4.93   M.J. Mayo, R.W. Siegel, Y.X. Liao, W.D. Nix: J. Mater. Res. **7**, 973 (1992)
4.94   J. Karch, R. Birringer: Ceramics International **16**, 291 (1990)
4.95   H. Hahn, J. Logas, H.J. Höfler, P. Kurath, R.S. Averback: Mater. Res. Soc. Symp. Proc. **196**, 71 (1990)
4.96   M. Guermazi, H.J. Höfler, H. Hahn, R.S. Averback: J. Amer. Cer. Soc. **74**, 2672 (1991)
4.97   G.W. Nieman, J.R. Weertman, R.W. Siegel: Scripta Metall. **23**, 2013 (1989)
4.98   G.W. Nieman, J.R. Weertman, R.W. Siegel: Scripta Metall. et Mater. **24**, 145 (1990)
4.99   J.S.C. Jang, C.C. Koch: Scripta Metall. et Mater. **24**, 1599 (1990)
4.100 C.C. Koch, Y.S. Cho: Nanostructured Mater. **1**, 207 (1992)
4.101 A.H. Chokshi, A. Rosen, J. Karch, H. Gleiter: Scripta Metall. **23**, 1679 (1989)
4.102 R. Birringer, H. Hahn, H. Höfler, J. Karch, H. Gleiter: Defect and Diffusion Forum **59**, 17 (1988)
4.103 H. Chang, H.J. Höfler, C.J. Altstetter, R.S. Averback: Scripta Metall. et Mater. **25**, 1161 (1991)
4.104 G.E. Fougere, J.R. Weertman, R.W. Siegel, S. Kim: Scripta Metall. et Mater. **26**, 1879 (1992)
4.105 E.O. Hall: Proc. Phys. Soc. London B **64**, 747 (1951)
4.106 N.J. Petch: J. Iron Steel Inst. **174**, 25 (1953)
4.107 A. Narayanasamy, J.A. Eastman, R.W. Siegel: unpublished results
4.108 G. Skandan, H. Hahn, J.C. Parker: Scripta Metall. et Mater. **25**, 2389 (1991)
4.109 G.C. Hadjipanayis, R.W. Siegel (eds): *Nanophase Materials: Synthesis-Properties-Applications* (Kluwer, Dordrecht 1994)
4.110 A.S. Edelstein, R.C. Cammarata (eds): *Nanomaterials: Synthesis, Properties and Uses* (IOP, Bristol 1996)
4.111 K.E. Gonsalves, G.-M. Chow, T.D. Xiao, R.C. Cammarata (eds): MRS Symp. **351** (Mater. Res. Soc., Pittsburgh, PA 1994)
4.112 J.C. Parker: In [Ref. 4.111, p. 573]
4.113 R.W. Siegel: In [Ref. 4.111, p. 201]
4.114 J.T. Lee, J.-H. Hwang, J.J. Mashek, T.O. Mason, A.E. Miller, R.W. Siegel: J. Mater. Res. **10**, 2295 (1995)
4.115 R.N. Viswanath, S. Ramasamy, R. Ramamoorthy, P. Jayavel, T. Nagarajan: Nanostruct. Mater. **6**, 993 (1995)
4.116 S.M. Thompson: In *Fundamental Properties of Nanostructured Materials*, ed. by D. Fiorani, G. Sberveglieri (World Scientific, Singapore 1994) p. 255
4.117 S.M. Thompson, J.F. Gregg: In *Fundamental Properties of Nanostructured Materials*, ed. by D. Fiorani, G. Sberveglieri (World Scientific, Singapore 1994) p. 265
4.118 R.W. Siegel, G.E. Fougere: M.R. Soc. Symp. **362**, 219 (Mater. Res. Soc., Pittsburgh, PA 1995)
4.119 M.L. Steigerwald, L.E. Brus: Ann. Rev. Mater. Sci. **19**, 471 (1993)

5 Intercalation Compounds of Transition-Metal Dichalcogenides

K. Motizuki and N. Suzuki

Layered compounds such as graphite and transition-metal dichalcogenide are interesting materials in their own right, and at the same time they are very important as mother materials for intercalation compounds. Intercalation means insertion of guest atoms, ions or molecules between layers of layered compounds, and the resultant substances produced by the intercalation process are called "intercalation compounds" or "intercalates". The central purpose of intercalation lies in synthesizing materials with new functions different from those of the mother materials. Graphite intercalation compounds have been investigated for a long time in comparison with intercalation in transition-metal dichalcogenides. In this chapter, however, we focus our attention to the latter. For the graphite intercalation compounds the reader should consult [5.1].

5.1 Background

Various atoms and organic and inorganic molecules can be intercalated into van der Waals gap sites of layered transition-metal dichalcogenides TX_2 (T = group IV, V, VI transition metal, X = S, Se, Te) [5.2–5] and many efforts have been made to obtain materials with new functions. Intercalation of alkali metals has been studied intensively for the purpose of synthesizing new superconductors [5.6] or new materials for battery electrodes [5.7]. On the other hand, intercalation of 3d transition metals has also been studied actively, and quite dramatic changes in the physical properties of the host transition-metal dichalcogenides have been observed depending on the intercalate species. Analyses of these properties, however, have been done mostly with use of the so-called rigid-band model which assumes that the intercalant transition metals act as electron donors to the host material without changing its original band structures. Further, the 3d states of the intercalant transition metal ions have been treated as localized states.

Recently, experimental studies, including electrical, optical, and magnetic measurements, have been made intensively on intercalation of 3d transition metals into $1T–TiS_2$ [5.8]. The experimental results have revealed itinerant character of the intercalation 3d electrons and they indicate also the inapplicability of the rigid-band model. Further, measurements of the photoemission suggest strong hybridization of the M 3d states with the Ti 3d and S 3p states of the mother crystal. Intercalation of transition-metals into 2H–type TX_2 (T = Nb, Ta; X = S, Se) has been studied also intensively [5.3] and a large variety of magnetic orderings such as ferro, antiferro, and helix have been found.

In general, knowledge of electronic structures of a given compound holds an important key to making clear its physical properties. As for the host transition-metal dichalcogenides TX_2 band calculations have been performed by various groups [5.9, 10] and the structural phase transitions or the charge-density-wave transitions observed in TX_2 have been studied microscopically on the basis of the realistic electronic band structures [5.11]. On the other hand, for the electronic structures of the intercalation compounds of TX_2 there has been few theoretical studies. Very recently, however, intensive studies of the electronic band structures of the transition-metal intercalation compounds M_xTX_2 (M = 3d transition metal) have been made by Motizuki's group in Japan and by Haas' group in the Netherlands. In particular, Motizuki's group has performed systematic band calculations for M_xTiS_2 by a self-consistent augmented-plane-wave (APW) method [5.12–18]. *Motizuki* and co-workers have clarified what changes the electronic state of the mother material TiS_2 undergoes when transition-metals are intercalated. They also clarified how different the electronic structures of M_xTiS_2 are for different transition metals, M. Furthermore, the bond orders have been calculated for TiS_2 and M_xTiS_2 and an important insight in the bonding nature of the intercalation compounds M_xTiS_2 has been obtained. Clarifying the bonding nature of the mother crystal and its intercalation compounds is of great importance to obtain both a new guide for synthesizing new materials and a better understanding of the physical properties of the intercalation compounds.

In this review article we focus our attention to the theoretical aspect of the electronic structures of the transition-metal intercalation compounds, particularly of M_xTiS_2. The arrangement of this article is as follows: In Sect. 5.2 the principle of the APW method and the approximations used in it will be briefly outlined first. Then, the results of systematic band calculations for the nonmagnetic states of $M_{1/3}TiS_2$ (M = Ti, V, Cr, Mn, Fe, Co, Ni) are summarized in Sect. 5.2.1 and the results of the band calculations for the ferromagnetic states of $M_{1/3}TiS_2$ (M = Cr, Fe, Co) are given in Sect. 5.2.2. Comparison of the theoretical results with some experimental results is made in Sect. 5.2.3. In Sect. 5.3 bonding nature in the mother crystal $1T–TiS_2$ and the intercalate $Fe_{1/3}TiS_2$ is discussed on the basis of the bond orders calculated in the framework of the APW band calculation. In Sect. 5.4 the band structures of silver intercalation compound $Ag_{1/3}TiS_2$ are presented and we discuss the differences between the bonding nature of M_xTiS_2 and that of Ag_xTiS_2. In Sect. 5.5 the band structures for the nonmagnetic and ferromagnetic states of $Mn_{1/4}TaS_2$ are given and the results are discussed in connection with the magnetic properties of this compound. Finally, Sect. 5.6 is devoted to discussion.

5.2 Electronic Band Structures of 3d Transition-Metal Intercalated Compounds of 1T–Type TiS$_2$

The mother crystal 1T–TiS$_2$ has the CdI$_2$–type structure, which consists of a sequence of metal layers sandwiched between two non-metal layers. The space group is D$_{3d}^3$. The crystal structure and the first Brillouin zone (BZ) of 1T–TiS$_2$ are shown in Fig. 5.1 and Fig. 5.2, respectively. The unit cell contains one Ti and two S ions and each Ti ion is surrounded by six S ions coordinated almost octahedrally.

In the intercalation compounds M$_x$TiS$_2$, intercalant M atoms occupy the octahedral interstitial sites between the neighbouring sulphur layers (van der Waals gap layers) and form regular array for particular x values such as $x = 1/4, 1/3, 1$. In the case of $x = 1/3$ the intercalant atoms form a $\sqrt{3}a_0 \times \sqrt{3}a_0$ triangular lattice in the van der Waals gap layer [5.19] where a_0 denotes the nearest neighbour Ti–Ti distance. Such triangular lattices of M atoms stack along the c axis as ABCABC..., and the primitive unit cell is rhombohedral as

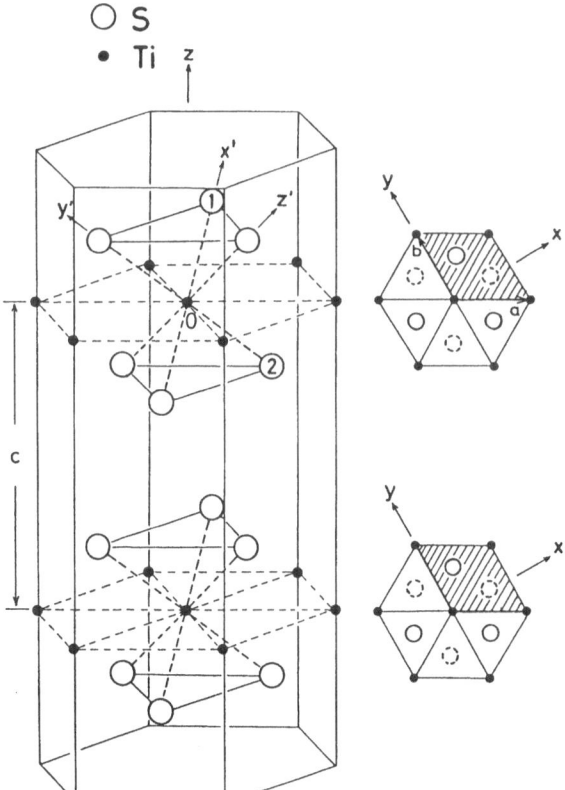

Fig. 5.1. The crystal structure of 1T–TiS$_2$. Ti atom is indicated by a closed circle, and S atoms are indicated by open circles. The unit cell contains one Ti and two S ions. Local axes, x', y' and z', are approximately directed from a Ti ion toward the neighboring S ions

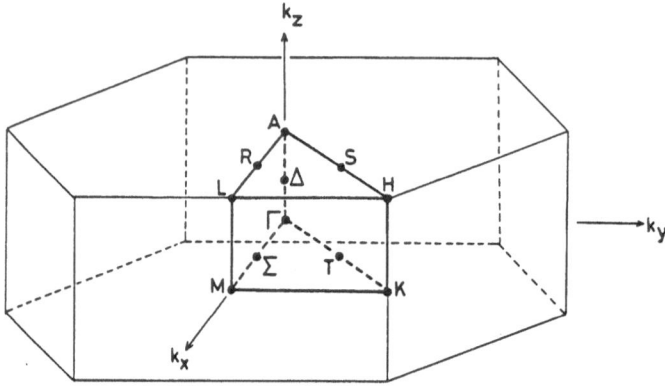

Fig. 5.2. The first Brillouin zone of TiS_2

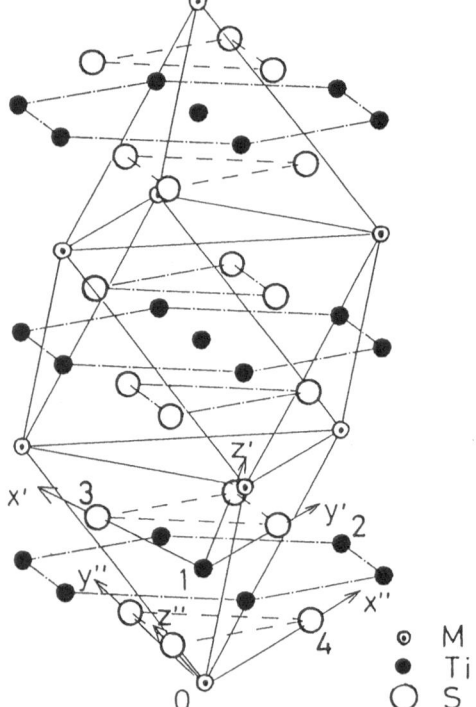

Fig. 5.3. The crystal structure of $M_{1/3}TiS_2$. The unit cell contains one M, three Ti, and six S ions. The local coordinate systems (x', y', z') and (x'', y'', z'') are used in Sect. 5.3

shown in Fig. 5.3 and contains one M ion, three Ti ions and six S ions. The space group is C_{3i}^1. For $x = 1/4$ the intercalant atoms form a $2a_0 \times 2a_0$ triangular lattice in the van der Waals gap layer and for $x = 1$ M atoms occupy all the available octahedral sites. The BZ of $M_{1/3}TiS_2$ has a shape of 14-faced polyhedron as shown in Fig. 5.4(a). But we can construct a hexagonal prismatic BZ

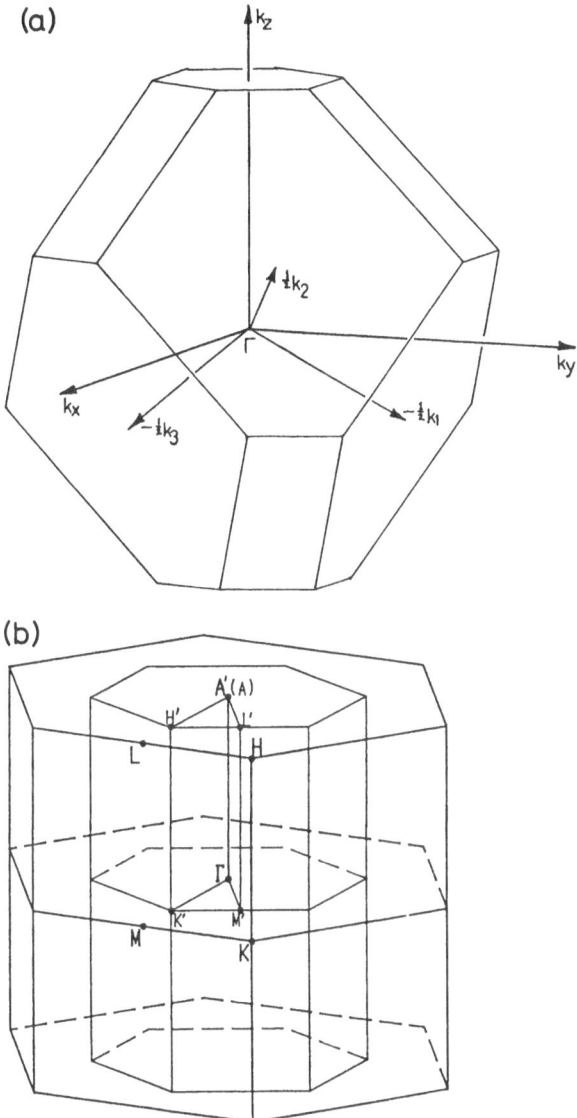

Fig. 5.4. The first Brillouin zone (BZ) of $M_{1/3}TiS_2$: (a) the original 14-faced polyhedral BZ and (b) the deformed hexagonal prismatic BZ. The connection between original BZ of $M_{1/3}TiS_2$ and equivalent hexagonal prismatic BZ is also shown

as shown in Fig. 5.4(b) by deforming the original BZ, and we use this hexagonal prismatic BZ for $M_{1/3}TiS_2$.

For later use we here describe briefly the essence of the self-consistent APW method. The detailed description of this method should be referred to *Loucks* [5.20] and *Mattheiss* et al. [5.21]. The APW method usually assumes that the crystal potential $v(r)$ can be approximated by a muffin-tin (MT) potential which assumes a spherical potential in MT spheres around each atomic site and a constant potential outside the MT spheres. Usually, the constant potential is chosen as the origin of energy and then the MT potential is expressed as

$$v(r) = \begin{bmatrix} V(|r - R_\nu|) & \text{(inside } \nu\text{th MT sphere)}, \\ 0 & \text{(outside MT spheres)}, \end{bmatrix} \tag{5.1}$$

where the subscript ν labels different (or inequivalent) atoms in a unit cell of the crystal and R_ν denotes the νth atomic position. Inside each MT sphere the potential $V(|r - R_\nu|)$ is assumed to be spherically symmetric. The radius of MT sphere, S_ν, is determined so that the spheres do not overlap each other.

Assuming that the potential $v(r)$ has the form of MT type, the solution of one-electron Schrödinger equation can be represented by a linear combination of APW functions, $\chi(k_i, r)$, as follows:

$$\psi_k(r) = \sum_i c_i(k)\chi(k_i, r) , \tag{5.2}$$

where $k_i = k + G_i$ with G_i being a reciprocal lattice vector. The expansion coefficients $c_i(k)$ are determined variationally. The APW secular equation is obtained through this variational procedure. The solution of this secular equation gives eigenvalues $E(k)$ and eigenvectors $c_i(k)$.

Reflecting the MT potential, the APW $\chi(k_i, r)$ is represented by a single plane wave outside the MT spheres and by an expansion in spherical harmonics $Y_{lm}(r)$ inside the MT spheres as follows:

$$\chi(k_i, r) = \begin{cases} e^{ik_i \cdot r} & \text{(outside MT spheres)} , \\ \displaystyle\sum_{lm} A_{lm}^\nu(k_i) Y_{lm}(r - R_\nu)u_l(|r - R_\nu|) & \text{(inside } \nu\text{th MT sphere)} , \end{cases} \tag{5.3}$$

The functions $u_l(|r - R_\nu|)$ are solutions to the radial Schrödinger equation given by

$$-\frac{1}{\rho^2}\frac{d}{d\rho}\left(\rho^2\frac{du_l}{d\rho}\right) + \left[\frac{l(l + 1)}{\rho^2} + V(\rho)\right]u_l = Eu_l , \tag{5.4}$$

where $\rho = |r - R_\nu|$ and $V(\rho)$ is the spherically symmetric potential introduced in (5.1).

The demand that the APW wave functions should be continuous on the MT spheres determines the expansion coefficients inside the νth MT sphere in (5.3) as follows:

$$A_{lm}^\nu(k_i) = 4\pi e^{ik \cdot R_\nu} i^l Y_{lm}(\hat{k}_i)j_l(k_i \cdot S_\nu)/u_l(S_\nu) , \tag{5.5}$$

where $\hat{k}_i \equiv k_i/k_i$ and j_l is a spherical Bessel function of order l.

The MT potential represented by (5.1) consists of the Coulomb and exchange-correlation parts and is related with the electronic charge densities $\rho(r)$

of the crystal. The Coulomb potentials inside and outside the MT spheres are obtained by integrating the Poisson's equation, including both the nuclear charge Z_v and the charge density $\rho(r)$. The Coulomb part of the constant potential outside the MT spheres can be approximated by taking an average over a spherical region having the same volume as that of the outside region.

The exchange-correlation part of the potential are approximated by an analytic formula expressed in terms of the charge density. Various types of the analytic formula have been used for calculating electronic structures of crystals, such as the Slater type [5.22], the Kohn–Sham type [5.23] and so on. These analytic formulas have been obtained by calculating the exchange-correlation potential of the free electron gas with use of the local-density approximation (LDA) or the local-spin-density approximation (LSDA). Motizuki's group has used the Gunnarsson and Lundqvist type formula [5.24] given by

$$v^{xc}_{\pm}(r) = v^x_p(r_s)[\beta(r_s) \pm \tfrac{1}{3}\delta(r_s)\zeta_s/(1 \pm \gamma\zeta_s)] \text{ [Ryd]} ,$$

$$\beta(r_s) = 1 + 0.0545r_s \ln(1 + 11.4/r_s) ,$$

$$\delta(r_s) = 1 - 0.036r_s + 1.36r_s/(1 + 10r_s) , \tag{5.6}$$

where $\gamma = 0.297$, $v^x_p(r_s) = -2/(\pi\alpha r_s)$ with $\alpha = (4/9\pi)^{1/3} = 0.521$, $r_s = (3\pi\rho/4)^{1/3}$ and $\zeta_s = (\rho_+ - \rho_-)/\rho$. $\rho_+(r)(\rho_-(r))$ is the charge density for up (down) spin electrons. This formula can be used also for the non-magnetic band by setting ζ_s to zero.

Once the electronic charge density $\rho(r)$ is obtained, this charge density can in turn be used to calculate a new potential $v(r)$, and the procedure is repeated until the input and output charge densities agree with each other within a certain accuracy. In solving the Schrödinger equation in the MT spheres the relativistic effects are included except the spin–orbit interaction, i.e., all other relativistic effects such as the mass–velocity and Darwin terms [5.25] are retained.

Actual procedure of the APW band calculations is as follows: The starting charge density has been constructed by superposing the self-consistent charge densities of the neutral atoms. The angular momentum expansion in (5.3) is usually truncated at $l_{max} = 6$–8 and the wave vector $|k + G_i|$ is cut off at an appropriate maximum wave vector, $|k + G|_{max}$. The values of $|k + G|_{max}$ restrict the numbers of APW's $\chi(k_i, r)$. The charge density of the crystal in each step of the self-consistent iteration procedure is determined by the special point method [5.26, 27]. The density of states is calculated by means of the linear energy tetrahedron method [5.28, 29]. The accuracy of the eigenvalues obtained by Motizuki's group is within 0.01 Ryd.

5.2.1 Nonmagnetic States

Systematic band calculations have been performed by Motizuki's group for the nonmagnetic state of TiS_2, $M_{1/3}TiS_2$ (M = Ti, V, Cr, Mn, Fe, Co, Ni), and

MTiS$_2$ (M = Fe, Cr). The values of lattice constants, l_{max} and $|k + G|_{max}$ used in the calculations are tabulated in Table 5.1.

Figures 5.5, 6 show the dispersion curves and the density of states of TiS$_2$, respectively. The lowest two bands consist of S 3s state. Eleven bands above the gap are the p–d mixing bands. As seen from Fig. 5.6 the density of states for the

Table 5.1. Hexagonal lattice constants, l_{max} and $|k + G|_{max}$ used in the APW band calculations for TiS$_2$, M$_{1/3}$TiS$_2$ (M = Ti, V, Cr, Mn, Fe, Co, Ni), and MTiS$_2$ (M = Cr, Fe)

	a [Å]	c [Å]	l_{max}	$|k + G|$ ($2\pi/a$)
TiS$_2$	3.4100	5.700	7	2.7
Ti$_{1/3}$TiS$_2$	5.9115	17.151	8	3.8
V$_{1/3}$TiS$_2$	5.9115	17.094	8	3.8
Cr$_{1/3}$TiS$_2$	5.9288	17.133	8	3.8
Mn$_{1/3}$TiS$_2$	5.9253	17.890	8	3.7
Fe$_{1/3}$TiS$_2$	5.9290	17.124	6	3.9
Co$_{1/3}$TiS$_2$	5.8870	16.887	8	3.8
Ni$_{1/3}$TiS$_2$	5.8820	16.872	8	3.9
CrTiS$_2$	3.4180	5.809	7	2.7
FeTiS$_2$	3.4280	5.809	7	2.7

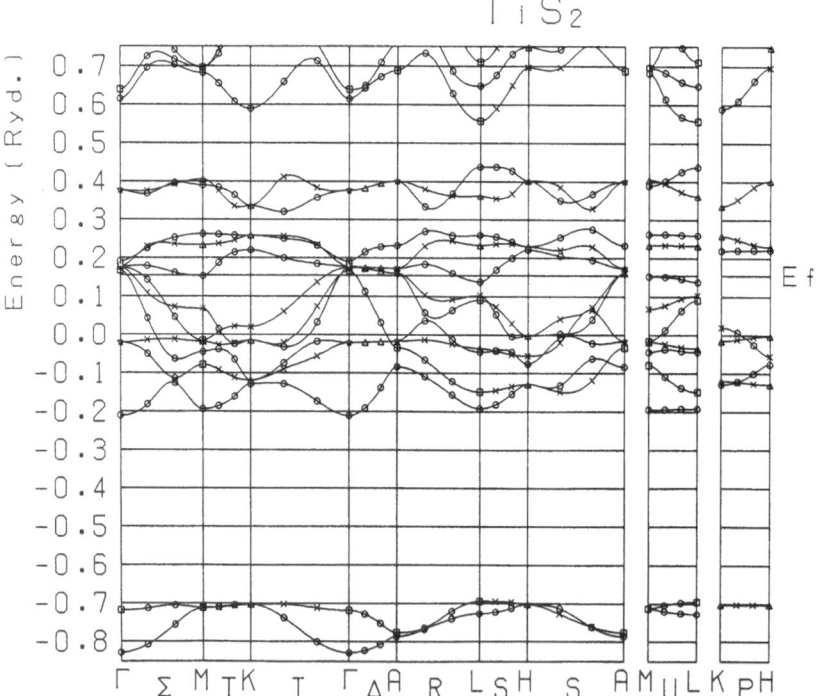

Fig. 5.5. The energy dispersion curves of 1T–TiS$_2$

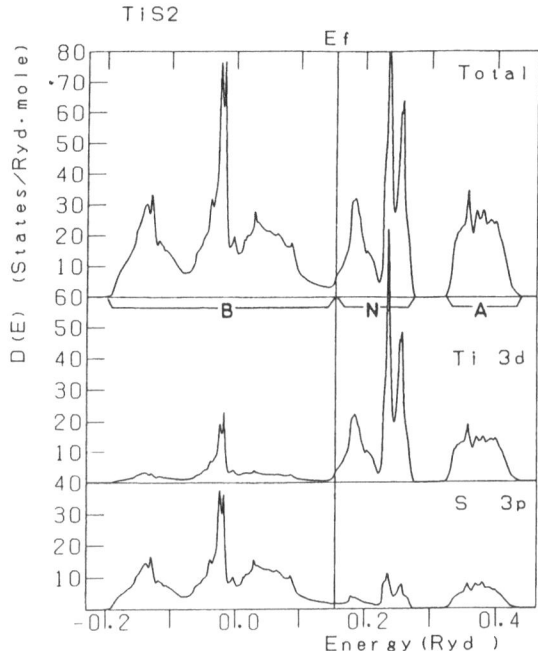

Fig. 5.6. The density of states of the p–d mixing bands of TiS$_2$. The partial density of states arising from the Ti d and S p states within each MT sphere are also shown

p–d mixing bands of TiS$_2$ consists of three parts, (B), (N), and (A):

(B) the lower-energy range between −0.19 and 0.15 Ryd;

(N) the intermediate-energy range between 0.15 and 0.28 Ryd;

(A) the higher-energy range between 0.33 and 0.43 Ryd.

Since Ti ion is surrounded octahedrally by six S ions, out of five 3d orbitals of Ti the two e$_g$ (dγ) orbitals are suited geometrically to covalent bonding, i.e., from e$_g$, 4s, and 4p orbitals we can construct six hybridized orbitals that are directed toward the neighboring six S ions. These hybridized orbitals of the Ti ions and the 3p orbitals of the S ion form the bonding and antibonding orbitals [5.30]. The remaining three 3d orbitals, i.e., t$_{2g}$ (dε) orbitals, of the Ti ion form the basis of nonbonding bands. Therefore, the calculated results of the density of states can be interpreted as follows: (B) and (A) arise from the bonding and antibonding bands of S p and Ti d mixing, and (N) arises mainly from the nonbonding Ti d bands. The bonding nature in TiS$_2$ is discussed in detail in Sect. 5.3.

The density of states in Fig. 5.6 agrees quite well with that obtained by *Umrigar* et al. [5.31] on the basis of a self-consistent linearized APW method. The gross features of the density of states are also in agreement with those obtained by *Zunger* and *Freeman* [5.32] except that there is a small gap between (B) and (N) in their density of states. Experimentally, stoichiometric TiS$_2$ is a semiconductor with a small gap 0.2–0.3 eV [5.33, 34], which is consistent with the result of *Zunger* and *Freeman*. This discrepancy between our result and that of *Zunger* and *Freeman* may be ascribed to our muffin-tin approximation or to the local density approximation itself.

When transition-metal ions are intercalated, 3d orbitals of the intercalants form new hybridized bands, which intervene between the bonding band (B) and the nonbonding band (N) of TiS_2. As an example, the dispersion curves of $Fe_{1/3}TiS_2$ are shown in Fig. 5.7. The lowest six bands consist of S 3s state. 38 bands above the gap are the p–d mixing bands: in order of increasing energy, 18 bands consist of mainly S 3p states, 14 bands consist of mainly Fe 3d and Ti 3d states, and 6 bands consist of mainly Ti 3d states.

The density of states calculated for the p–d mixing bands of $Fe_{1/3}TiS_2$ is shown in Fig. 5.8, where the partial densities of states within the MT spheres are also shown. By comparing the density of states of TiS_2 and that of $Fe_{1/3}TiS_2$ we may divide the energy range of the p–d mixing bands of $Fe_{1/3}TiS_2$ into four energy ranges (i), (ii), (iii), and (iv) as follows (in unit of Ryd):

(i)	(ii)	(iii)	(iv)
-0.17–0.19	0.19–0.27	0.27–0.36	0.39–0.49.

Part (i) consists mainly of S p orbitals and hybridization with Ti d orbitals is fairly large. In part (ii) main contribution arises from the intercalant Fe d states, but nonnegligible amounts of Ti d and S p components are seen. In part (iii) the

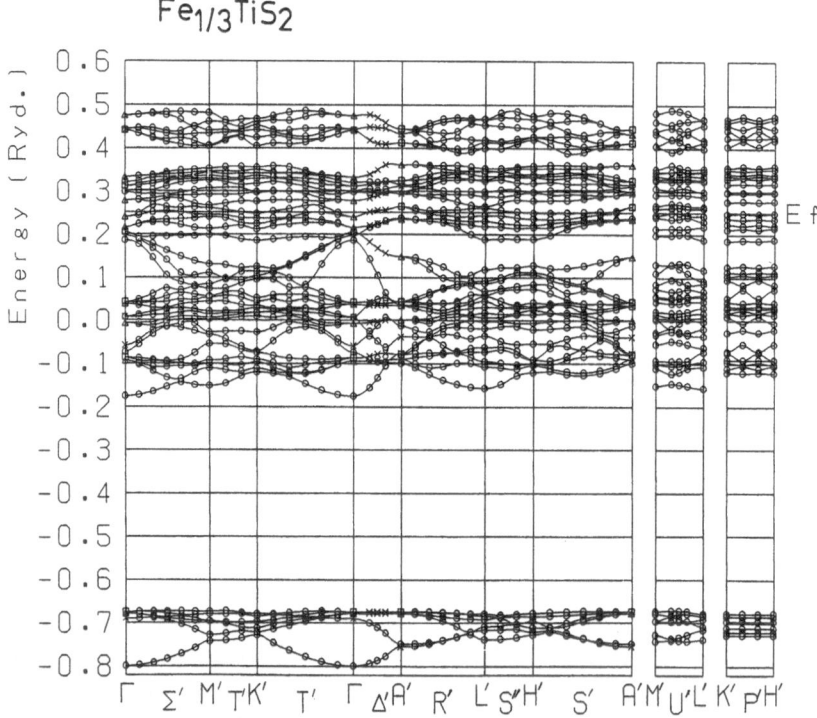

Fig. 5.7. The energy dispersion curves of nonmagnetic $Fe_{1/3}TiS_2$

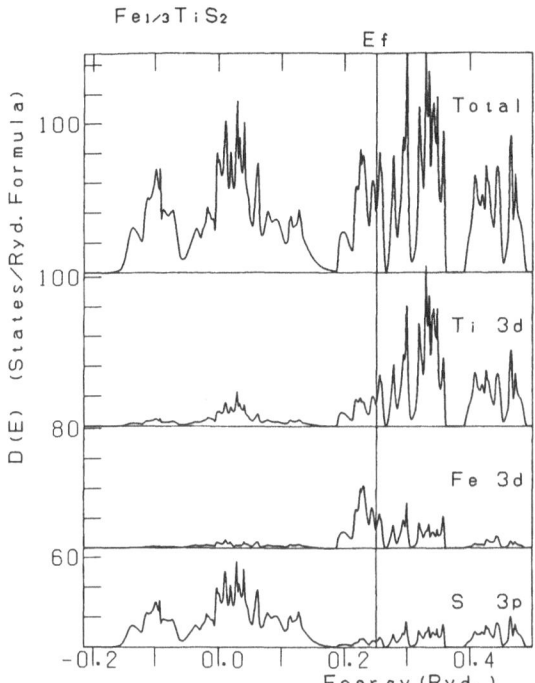

Fig. 5.8. The density of states of nonmagnetic $Fe_{1/3}TiS_2$. The contributions from d orbitals of Fe, d orbitals of Ti, and p orbitals of S, respectively, inside each MT sphere, are also shown

contributions from Ti d states are dominant, but hybridization with Fe d states is also significant. The width of (ii) + (iii) to which the Fe d states make a large contribution is $\approx 3\,eV$, which is not particularly small. Therefore, the Fe d electrons in $Fe_{1/3}TiS_2$ should be treated as itinerant electrons. In part (iv) hybridization between Ti d and S p states is significant and the contribution from the Fe d states is comparatively small. By comparing Figs. 5.6, 8 it is seen that the nonbonding band of TiS_2 has been modified significantly by the intercalation and there are strong hybridization between the Fe d and Ti d states in parts (ii) and (iii) of $Fe_{1/3}TiS_2$. Further we may expect that parts (i) and (iv) of $Fe_{1/3}TiS_2$ have characters of bonding and antibonding states of the Ti dγ and S p orbitals. Detailed bonding characters of each part of $Fe_{1/3}TiS_2$ will be clarified in Sect. 5.3.

The gross features of the energy band structures of other intercalation compounds $M_{1/3}TiS_2$ (M = Ti, V, Cr, Mn, Co, Ni) are similar to that of $Fe_{1/3}TiS_2$. At the same time, however, there exist systematic differences. Figures 5.9, 10 show the density of states of $Cr_{1/3}TiS_2$ and $Ni_{1/3}TiS_2$, respectively. The energy separation between the center of the energy range (i) where S p is dominant and that of the energy range (ii) where M d is dominant becomes narrow as the atomic number of the intercalant M atoms increases. This causes an enhancement of the hybridization between the M d and S p states and a reduction of the hybridization between the M d and Ti d states as the atomic

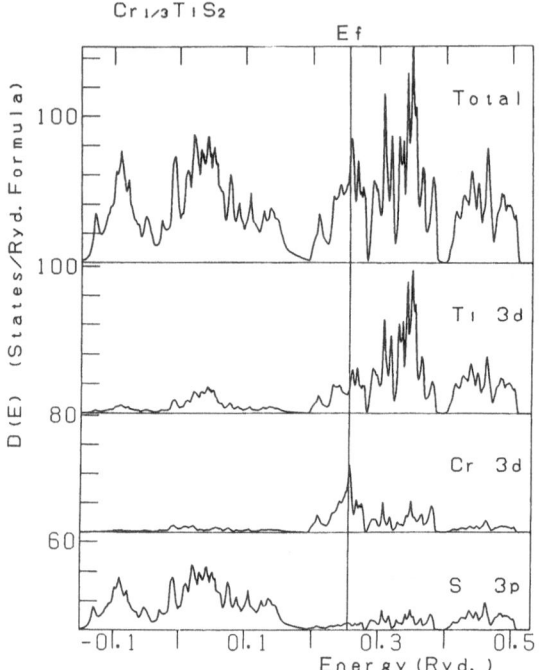

Fig. **5.9.** The density of states of $Cr_{1/3}TiS_2$

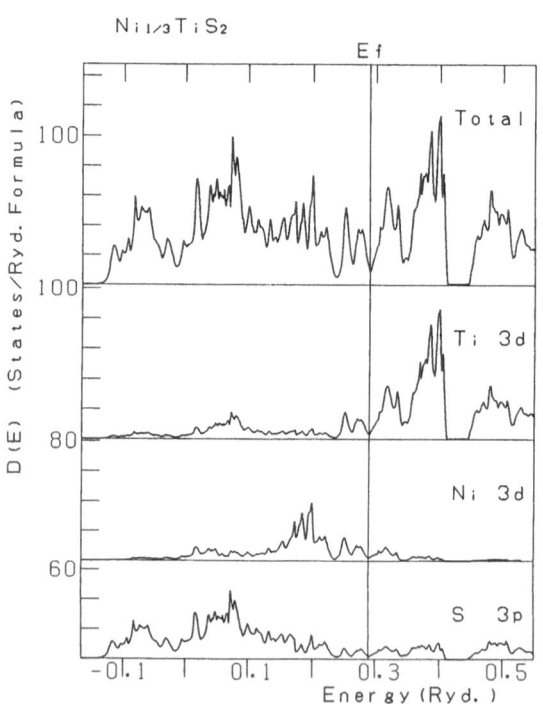

Fig. **5.10.** The density of states of $Ni_{1/3}TiS_2$

Table 5.2. The density of states $D(E_f)$ at the Fermi level of nonmagnetic $M_{1/3}TiS_2$ (M = Ti, V, Cr, Mn, Fe, Co, Ni). The total density of states, $D_\uparrow(E_f) + D_\downarrow(E_f)$, of ferromagnetic $M_{1/3}TiS_2$ (M = Cr, Fe, Co) is shown in parentheses. The theoretical value, γ_{theory}, of the electronic specific heat coefficient has been evaluataed by using the value of $D_\uparrow(E_f) + D_\downarrow(E_f)$ for $M_{1/3}TiS_2$ (M = Cr, Fe, Co). The experimental data, γ_{exp}, are taken from [5.36]. The units of $D(E_f)$ and γ are states/(Ryd. formula) and mJ/(mol K^2), respectively

M	$D(E_f)$	γ_{theory}	γ_{exp}
Ti	35.9	6.2	
V	41.1	7.1	50
Cr	70.9 (39.1)	6.8	70
Mn	63.8	11.0	90
Fe	54.6 (34.7)	6.0	30
Co	48.8 (25.8)	4.5	30
Ni	9.4	1.6	10

number of M atoms increases. These differences due to different M atoms can be ascribed to the difference of the atomic energy level of the M 3d states, i.e., with increasing the atomic number of M the atomic M 3d level is lowered and approaches to the atomic S 3p level.

The values of the density of states, $D(E_f)$, at the Fermi level of $M_{1/3}TiS_2$ (M = Ti, V, Cr, Mn, Fe, Co, Ni) are tabulated in Table 5.2. The value of $D(E_f)$ is quite large for Cr, Mn, Fe, and Co compounds, which indicates that the ferromagnetic state may be stabilized in these compounds. In fact, as shown in Sect. 5.2.3 $M_{1/3}TiS_2$ (M = Cr, Fe, Co) becomes ferromagnetic [5.35]. However, $Mn_{1/3}TiS_2$ does not show any magnetic ordering in spite of its large value of $D(E_f)$. The band structures of the ferromagnetic states of $M_{1/3}TiS_2$ (M = Cr, Fe, Co) will be given in the next section.

The gross feature of the density of states of fully intercalated compound, $FeTiS_2$, is similar to that of $Fe_{1/3}TiS_2$. However, there are a couple of differences between the two. The width of the Fe 3d bands of $FeTiS_2$ is wider than that of $Fe_{1/3}TiS_2$. Furthermore, the Fermi level is located at a position close to the sharp peak of the density of states and the value of $D(E_f)$ of $FeTiS_2$ is almost two times larger than that of $Fe_{1/3}TiS_2$. Such a large value of $D(E_f) = 120$ states/(Ryd. formula) is favorable for stabilization of the ferromagnetic state. In fact, $FeTiS_2$ is a ferromagnet with the Curie temperature $T_c = 160$ K while T_c of $Fe_{1/3}TiS_2$ is 55 K. It is noted finally that the density of states of the p–d bonding bands in $FeTiS_2$ has two peaks whereas the corresponding part in $Fe_{1/3}TiS_2$ or TiS_2 shows a three peak structure. The shape of the density of states of the p–d bonding bands in $FeTiS_2$ rather resembles the corresponding part of the density of states of transition-metal pnictides or chalcogenides, such as MnAs and VS, which have the NiAs-type crystal structure [5.36]. This similarity can be explained using the fact that the structure of $MTiS_2$ is basically the NiAs type;

that is, if M atoms are replaced by Ti atoms, the crystal structure of $MTiS_2$ is exactly of the NiAs type.

5.2.2 Ferromagnetic States

Electronic band structures for the ferromagnetic states of $M_{1/3}TiS_2$ (M = Cr, Fe, Co) and $FeTiS_2$ have been calculated also by Motizuki's group. Figure 5.11 shows the density of states of ferromagnetic $Fe_{1/3}TiS_2$ for each spin component. As seen from the figure the up- and down-spin bands of Fe 3d components show a large splitting. Moreover, this splitting cannot be described by a rigid splitting of the nonmagnetic band. This nonrigid splitting character is a common feature also for $Cr_{1/3}TiS_2$, $Co_{1/3}TiS_2$, and $FeTiS_2$. The up-spin Fe 3d states hybridize strongly with the S 3p states while the down-spin Fe 3d states hybridize mainly with the Ti 3d states. The up- and down-spin bands of Ti 3d or S 3p states do not show a large splitting, but nonnegligible moment is induced on both Ti and S ions.

The estimated magnitudes of spin moment inside each muffin-tin sphere and the total moment per formula are tabulated in Table 5.3. One of the characteristic features of the ferromagnetic state of $Fe_{1/3}TiS_2$ or $FeTiS_2$ is that the direction of moment of Ti ion is opposite to that of Fe ion. Furthermore,

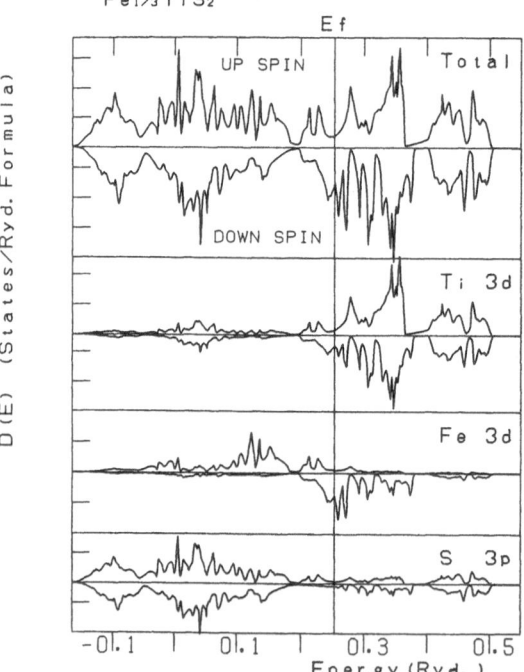

Fig. 5.11. The density of states of the ferromagnetic bands of $Fe_{1/3}TiS_2$

Table 5.3. The magnetic moment within each MT sphere and the total magnetic moment per formula of ferromagnetic $M_{1/3}TiS_2$ (M = Cr, Fe, Co) and $FeTiS_2$ in unit of μ_B. The experimental data for the total moment are taken from [5.38, 39]

	M	Ti	S	Total	Exp.
$Cr_{1/3}TiS_2$	2.60	0.21	-0.20	2.92	0.73
$Fe_{1/3}TiS_2$	2.36	-0.09	0.03	2.39	2.1
$Co_{1/3}TiS_2$	0.69	0.04	0.01	0.79	0.5
$FeTiS_2$	2.33	-0.36	0.03	2.03	

reflecting the large hybridization between the Ti 3d and Fe 3d states the induced moment on the Ti site is larger than that on the S site.

5.2.3 Comparison with Experimental Results

In this section the principal results of the experimental measurements performed on M_xTiS_2 (M = transition metal) are summarized and they are discussed in terms of the electronic band structures reviewed in Sects. 5.2.1, 2. For details of the experimental results, see [5.8].

(a) *Specific heat measurement [5.37]:*

Specific heat measurements for M_xTiS_2 (M = V, Cr, Mn, Fe, Co, Ni) have been made with a calorimetry technique. The specific heat for each intercalation compound shows temperature dependence like $\gamma T + \beta T^3$, except at low temperatures. The electronic specific heat coefficient γ depends on the species and concentration of the intercalant M atom and γ is found to be comparatively large (10–100 mJ/mol K^2). The value of γ is directly related to $D(E_f)$ by the relation $\gamma \equiv (1/3)\pi^2 k_B^2 D(E_f)$. We can evaluate the values of γ for $M_{1/3}TiS_2$ by making use of the calculated values of $D(E_f)$ in Table 5.2. The results are shown by open circles in Fig. 5.12 together with the experimental values (closed circles). For M = Cr, Fe, and Co we have used the values of $D(E_f)$ for the ferromagnetic state. In all compounds the experimental value of γ is larger by a factor five or more than the theoretical value. But it should be emphasized that the M-dependence of γ, i.e., the relative magnitude of γ among different $M_{1/3}TiS_2$ compounds is well explained by the band calculation. The discrepancy between the theory and experiment in the magnitude of γ may be partly remedied by taking into account the effect of mass enhancement due to the electron–electron or electron–phonon interaction. It should be noted here that the rigid-band model gives a value of $D(E_f)$ smaller even than the theoretical values given in Table 5.2. Therefore, we can say that the results of specific heat measurements indicate clearly the inapplicability of the rigid-band model and suggest modification of the electronic band structure of the mother crystal due to the intercalation.

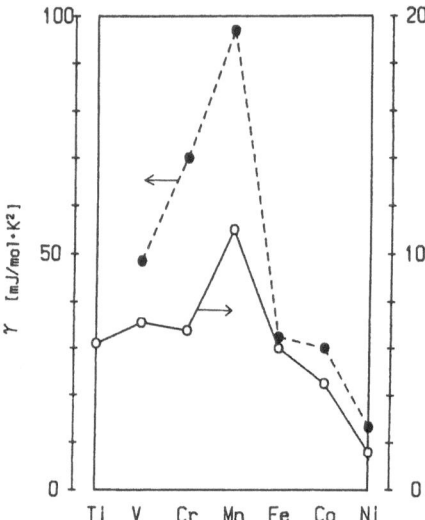

Fig. 5.12. The coefficient γ of electronic specific heat of $M_{1/3}TiS_2$ (M = Ti, V, Cr, Mn, Fe, Co, Ni). The open circles denote the theoretical values evaluated from the band calculations and the closed circles the experimental values obtained by Inoue et al. [5.37]

(b) *Magnetic measurements [5.38–46]:*

Depending on the kind of guest 3d metals and their concentration the intercalation compounds M_xTiS_2 show various types of magnetic phase such as spin-glass, ferromagnetic and antiferromagnetic ones. For example, in the case of Fe_xTiS_2 there appear a spin-glass phase in the low concentration region ($x \leqq 0.2$), a cluster-glass or micromagnetic phase in the intermediate region ($0.2 \leqq x \leqq 0.4$), and a ferromagnetic phase in the high concentration region ($0.4 < x$). In the case of Cr_xTiS_2 a ferromagnetic phase appears for $x \leqq 3/4$ and an antiferromagnetic behavior is found for $x = 1.0$. If we see magnetic properties of M_xTiS_2 with a particular x-value such as $x = 1/3$, Cr compound is a ferromagnet with $T_c = 15$ K [5.35], Fe compound shows a ferromagnetic cluster-glass behavior, Co compound is a ferromagnetic with small saturation moment, and other compounds are paramagnetic.

From magnetization measurements the saturation moment of $Cr_{1/3}TiS_2$, $Fe_{1/3}TiS_2$, and $Co_{1/3}TiS_2$ have been determined and the results are shown in Table 5.3. The values of saturation moment determined experimentally are much smaller than those expected from the free M^{2+} ions and this fact also indicates the itinerant character of the M 3d electrons. Compared with the total moment per formula obtained by the band calculations for the ferromagnetic state we see a good agreement for $Fe_{1/3}TiS_2$ and $Co_{1/3}TiS_2$, but for $Cr_{1/3}TiS_2$ there is a large discrepancy.

(c) *Transport measurements [5.47, 48]:*

The electrical resistivities of all kinds of M_xTiS_2 show metallic behaviors. The Hall coefficients measured for Mn_xTiS_2 ($x \leqq 0.25$) and Co_xTiS_2 ($x \leqq 0.175$) are both negative, which indicates that carriers are electron-like. Though the Hall coefficients of $Fe_x TiS_2$ ($x \leq 1/3$) are positive, the carriers are considered to be electron-like by taking into account the results of the thermoelectric power. The positiveness of the Hall coefficient of $Fe_x TiS_2$ has been ascribed to the anomalous Hall effect. In $Ni_{1/3} TiS_2$ both the thermoelectric power and the Hall coefficient change their sign from negative to positive at $x = 0.3$ with increasing x. This fact indicates that the carriers change from electron-like to hole-like at $x = 0.3$ or that both electron and hole pockets exist for $x > 0.3$. Since the rigid-band model predicts only electron-like carriers, the results of measurements of the Hall coefficient and the thermoelectric power in $Ni_{1/3} TiS_2$ also indicate the inapplicability of the rigid-band model.

Since the shape of the Fermi surface is an important factor in determining the sign of the Hall coefficient, *Suzuki* et al. [5.16] have constructed the Fermi surfaces of $Ni_{1/3} TiS_2$. Among 38 p–d mixing bands three bands intersect the Fermi level of $Ni_{1/3} TiS_2$. One of them forms a complicated electron Fermi surface and the remaining two bands form hole Fermi surfaces with rather simple shape. By assuming that the argument of the Hall coefficient in cubic systems [5.49] is applicable qualitatively to $Ni_{1/3} TiS_2$, *Suzuki* et al. [5.16] have shown that the Hall coefficient of $Ni_{1/3} TiS_2$ can be positive in total.

(d) *Optical measurements [5.50–52]:*

Both the ultraviolet photoemission spectroscopy (UPS) and the X-ray photoemission spectroscopy (XPS) have been done on the intercalation compounds $M_x TiS_2$. The results of UPS measurements indicate strong hybridization between the intercalant M 3d states and the host TiS_2. The results of XPS measurements show also hybridization between the M 3d and S 3p states and further they indicate the importance of the intraatomic coulomb and exchange energies for the M 3d electrons.

5.3 Bonding Nature in $M_x TiS_2$ (M: 3d Transition-Metal)

Analysis of the partial densities of states tells us the extent of mixing of the *l*-components within the muffin-tin (MT) spheres, but it cannot inform us of the nature of bonding between atoms. The bond orders defined by *Godby* et al. [5.53] are quantities useful for clarifying the bonding nature. In general, the bond order for a wave function ψ between two atomic or atomic-like orbitals ϕ_i and ϕ_j centered on different nuclei is defined by

$$\mu_{ij} \equiv \langle \phi_i | \psi \rangle \langle \psi | \phi_j \rangle . \qquad (5.7)$$

As explained in Sect. 5.2, the eigenfunction corresponding to the eigenvalue E_{nk} obtained in the APW band calculation (n and k being the band suffix and wave vector, respectively) is expressed by a linear combination of APW, $\chi(k_i, r)$, as follows

$$\psi_{nk}(r) = \sum_i c_i(nk)\chi(k_i, r) \,. \tag{5.8}$$

Within muffin-tin spheres each $\chi(k_i, r)$ is given in a linear combination of atomic-like orbitals as expressed by (5.3) and hence the eigenfunction ψ_{nk} can be expressed in the νth MT sphere as follows

$$\psi_{nk}(r) = \sum_{lm}\left[\sum_i c_i(nk) A_{lm}^{\nu}(k_i)\right]\phi_{lm}^{\nu}(r - R_{\nu}) \,. \tag{5.9}$$

Here $\phi_{lm}^{\nu}(r - R_{\nu})$ represents an atomic-like wave function centered on the νth MT sphere whose position vector is R_{ν} and it is expressed as a product of a spherical harmonic function $Y_{lm}(\hat{\rho})$ and a radial wave function $u_l(\rho)$ with $\rho = r - R_{\nu}$ and $\hat{\rho} = \rho/\rho$. Since the atomic-like orbital ϕ_{lm}^{ν} is defined within the νth MT sphere and there is no overlap between different MT spheres, two atomic-like orbitals on different MT spheres are orthogonal. Therefore, the bond order between ϕ_{lm}^{ν} and $\phi_{l'm'}^{\nu'}$ $(\nu \neq \nu')$ for a state ψ_{nk} is obtained from (5.7) and (5.9) as follows

$$\mu_{\nu lm,\,\nu'l'm'}(nk) = \langle \phi_{lm}^{\nu} | \psi_{nk}\rangle\langle\psi_{nk} | \phi_{l'm'}^{\nu'}\rangle$$

$$= \left[\sum_i c_i(nk) A_{lm}^{\nu}(k_i)\right]\left[\sum_i c_i(nk) A_{l'm'}^{\nu'}(k_i)\right]^* \,. \tag{5.10}$$

The bond order $\mu_{\nu lm,\,\nu'l'm'}$ represents correlation between ϕ_{lm}^{ν} and $\phi_{l'm'}^{\nu'}$ in the state $|\psi_{nk}\rangle$, and it is clear that the magnitude of $\mu_{\nu lm,\,\nu'l'm'}$ is large when both ϕ_{lm}^{ν} and $\phi_{l'm'}^{\nu'}$ are important in ψ_{nk}. Further, the phase of $\mu_{\nu lm,\,\nu'l'm'}$ carries information about the bonding and antibonding character of the wave function ψ_{nk}.

It is rather convenient to use real atomic-like orbitals ϕ_{la}^{ν} such as dγ- and dε-like wavefunctions, which are given in general by a linear combination of ϕ_{lm}^{ν}:

$$\phi_{la}^{\nu} = \sum_m d_{ma}\,\phi_{lm}^{\nu} \,. \tag{5.11}$$

Then the bond order between ϕ_{la}^{ν} and $\phi_{l'a'}^{\nu'}$ $(\nu \neq \nu')$ is expressed as

$$\mu_{\nu la,\,\nu'l'a'}(nk) = \sum_{mm'} d_{ma}^* d_{m'a'}\,\mu_{\nu lm,\,\nu'l'm'}(nk) \,. \tag{5.12}$$

Considering the degeneracy of two states (n, k) and $(n, -k)$ we can redefine the bond order between ϕ_{la}^{ν} and $\phi_{l'a'}^{\nu'}$ as follows:

$$\beta_{\nu la,\,\nu'l'a'}(nk) = \tfrac{1}{2}[\mu_{\nu la,\,\nu'l'a'}(nk) + \mu_{\nu la,\,\nu'l'a'}(n - k)] \,. \tag{5.13}$$

The bond order $\beta_{\nu la,\,\nu'l'a'}(nk)$ defined in this way is a real quantity because $\psi_{nk} = \psi_{n-k}^*$ and the bonding and antibonding nature is related to the sign of $\beta_{\nu la,\,\nu'l'a'}(nk)$. We can choose appropriately signs of atomic-like orbitals ϕ_a^{ν} so that $\beta > 0$ means bonding character, $\beta < 0$ antibonding character, and $\beta \simeq 0$ nonbonding character.

It should be noted here that the absolute value of μ or β depends on the normalization of the wave functions ϕ_{lm}^v and ψ_{nk}. Hence comparison of the absolute values of bond orders should be made carefully. In the APW band calculation, ϕ_{lm}^v has been normalized for all vlm to be unity in each MT sphere and ψ_{nk} has been normalized for all nk to be unity in a unit cell. Therefore, it makes sense definitely to compare the values of bond orders for different m or a by fixing vl and $v'l'$. However, it may lose its meaning to compare the magnitudes of bond orders for different sets of (vl; $v'l'$), e.g., to compare the magnitude of the bond orders for ($v =$ Ti, $l = 2$; $v' =$ S, $l' = 1$) with those for ($v =$ Ti, $l = 2$; $v' =$ Fe, $l' = 2$).

Since it is too intricate to illustrate the bond order for each band, it is rather convenient to introduce the averaged bond orders:

$$\beta_{va,\,v'a'}(k) = \frac{1}{N_j} \sum_{n\,\in(j)} \beta_{va,\,v'a'}(nk) , \qquad (5.14)$$

where n runs over energy bands belonging to an energy range (j) which is one of (B), (N), and (A) in Fig. 5.6 or one of (i), (ii), (iii), and (iv) in Fig. 5.8, and N_j denotes the number of the energy bands in the energy range (j). Six, three, and two energy bands belong respectively to (B), (N), and (A) of TiS$_2$, and eighteen, five, nine, and six energy bands belong respectively to (i), (ii), (iii), and (iv) of Fe$_{1/3}$TiS$_2$.

(1) Bond order in TiS$_2$

First, we show the averaged bond orders in TiS$_2$ calculated along the ΓM line in the first BZ. Figure 5.13(a) shows the bond order, $\beta_{\text{Ti}dx',\,\text{S}px'}(k)$, between $\phi_{3x'^2-r^2}^d$ orbital of the Ti(0) ion and the $\phi_{x'}^p$ orbital of the S(1) ion. Here, the locations of the Ti(0) and S(1) ions are shown in Fig. 5.1 and x' represents the direction towards S(1) from Ti(1). $\phi_{3x'^2-r^2}^d$ represents the Ti(0) dγ orbital directed to S(1) and $\phi_{x'}^p$ denotes the S(1) p orbital directed to Ti(0). The value of

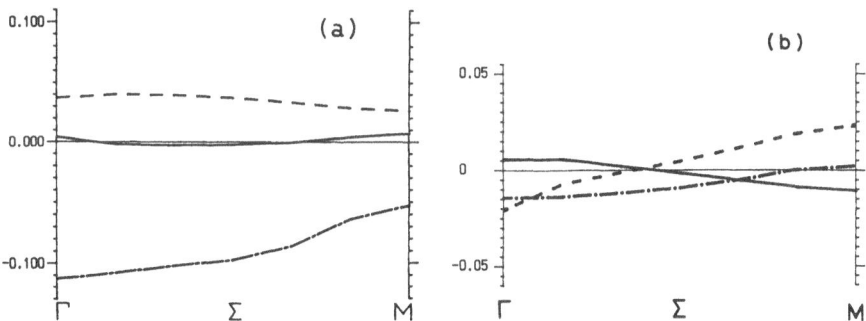

Fig. 5.13. The averaged bond order in TiS$_2$ as a function of k along the ΓM line: (a) $\beta_{\text{Ti}dx',\,\text{S}px'}$; (b) $\beta_{\text{Ti}dx'y',\,\text{S}px'}$. Solid (——), dotted (- - - - -), and dot-dashed (-·-·-) curves denote the mean value averaged over the energy range of (B), (N), and (A), respectively

$\beta_{\text{Ti}dx', \text{S}px'}(k)$ is positive and large for the lower-energy region (B), negative and large for the higher-energy region (A), and very small for the intermediate energy region (N).

Figure 5.13(b) shows the bond order, $\beta_{\text{Ti}dZ', \text{S}px'}(k)$, between the $\phi^d_{Z^2}$ orbital of Ti(0) and the $\phi^p_{x'}$ orbital of S(1) ion. Here, $\phi^d_{Z^2}$ denotes the orbital constructed by a superposition of three $d\varepsilon$ orbitals of the Ti(0) ion, $\phi^d_{Z^2} \equiv (\phi^d_{y'z'} + \phi^d_{z'x'} + \phi^d_{x'y'})/\sqrt{3}$. As can be seen from the figure, the magnitude of $\beta_{\text{Ti}dZ', \text{S}px'}(k)$ is small for all the energy regions, (B), (N), and (A), in contrast with that of $\beta_{\text{Ti}dx', \text{S}px'}(k)$. Hence, we can say that the chemical bond between the Ti $d\varepsilon$ and S p states is weak compared with that between the Ti $d\gamma$ and S p states.

From these results, we can conclude that, as *Inglesfield* [5.30] pointed out, parts (B) and (A) in Fig. 5.6 can be regarded as the bonding and antibonding states, respectively, between the Ti $d\gamma$ and S p states, and part (N) has a nonbonding states character consisting mainly of the Ti $d\varepsilon$ components.

(2) Bond orders in $Fe_{1/3}TiS_2$

For $Fe_{1/3}TiS_2$ the following four types of averaged bond orders have been calculated along the $\Gamma K'$ line which corresponds to the ΓM line in the BZ of TiS_2.

(a) $\beta_{\text{Ti}(1)dx', \text{S}(3)px'}(k)$ between $\phi^{\text{Ti}(1)}_{3dx'^2 - r^2}$ and $\phi^{\text{S}(3)}_{px'}$

(b) $\beta_{\text{Fe}(0)dx'', \text{S}(4)px''}(k)$ between $\phi^{\text{Fe}(0)}_{3dx''^2 - r^2}$ and $\phi^{\text{S}(4)}_{px''}$

(c) $\beta_{\text{Ti}(1)dx'y', \text{Ti}(2)dx'y'}(k)$ between $\phi^{\text{Ti}(1)}_{3dx'y'}$ and $\phi^{\text{Ti}(2)}_{3dx'y'}$

(d) $\beta_{\text{Fe}(0)dZ, \text{Ti}(1)dZ}(k)$ between $\phi^{\text{Fe}(0)}_{3dZ^2}$ and $\phi^{\text{Ti}(1)}_{3dZ^2}$

Here Fe(0) denotes the Fe atom at the 0 site in Fig. 5.3. The locations of Ti(1), Ti(2), S(3), and S(4) and the local coordinate systems (x', y', z') and (x'', y'', z'') are also shown in Fig. 5.3. $\phi^{\text{Ti}(1)}_{3x'^2 - r^2}$ and $\phi^{\text{Fe}(0)}_{3x''^2 - r^2}$ represent, respectively, the Ti(1) $d\gamma$ orbital directed to S(3) and the Fe(0) $d\gamma$ orbital directed to S(4). The ϕ_{Z^2} orbital of Ti(1) and that of Fe(0) represent the superposition of three $d\varepsilon$-orbitals, $\phi^d_{x'y'}$, $\phi^d_{y'z'}$, and $\phi^d_{z'x'}$ of the Ti(1) ion and $\phi^d_{y''z''}$, $\phi^d_{z''x''}$, and $\phi^d_{x''y''}$, of the Fe(0) ion, respectively. It should be noted that $\phi^d_{Z^2}$ of Ti(1) or Fe(0) is an orbital directed along the c-direction.

The calculated results of the above four types of bond orders (a)–(d) are shown in Fig. 5.14. The results of type (a) bond order, $\beta_{\text{Ti}(1)dx', \text{S}(3)px'}$, illustrated in Fig. 5.14(a) show that the nature of bonding and antibonding of Ti $d\gamma$ and S p is still kept clearly in the low and high energy ranges (i) and (iv), respectively. From Fig. 5.14(b) which shows $\beta_{\text{Fe}(0)dx'', \text{S}(4)px''}$ we find that the Fe $d\gamma$ orbital and the S p orbitals also form the bonding and antibonding states. The bonding states are included mainly in part (i) while the anitbonding states are spread over in parts (iii) and (iv). Figure 5.14(c) shows the bond order $\beta_{\text{Ti}(1)dx'y', \text{Ti}(2)dx'y'}$ between $d\varepsilon$ orbitals of the nearest neighboring Ti ions in the c-plane. This bond

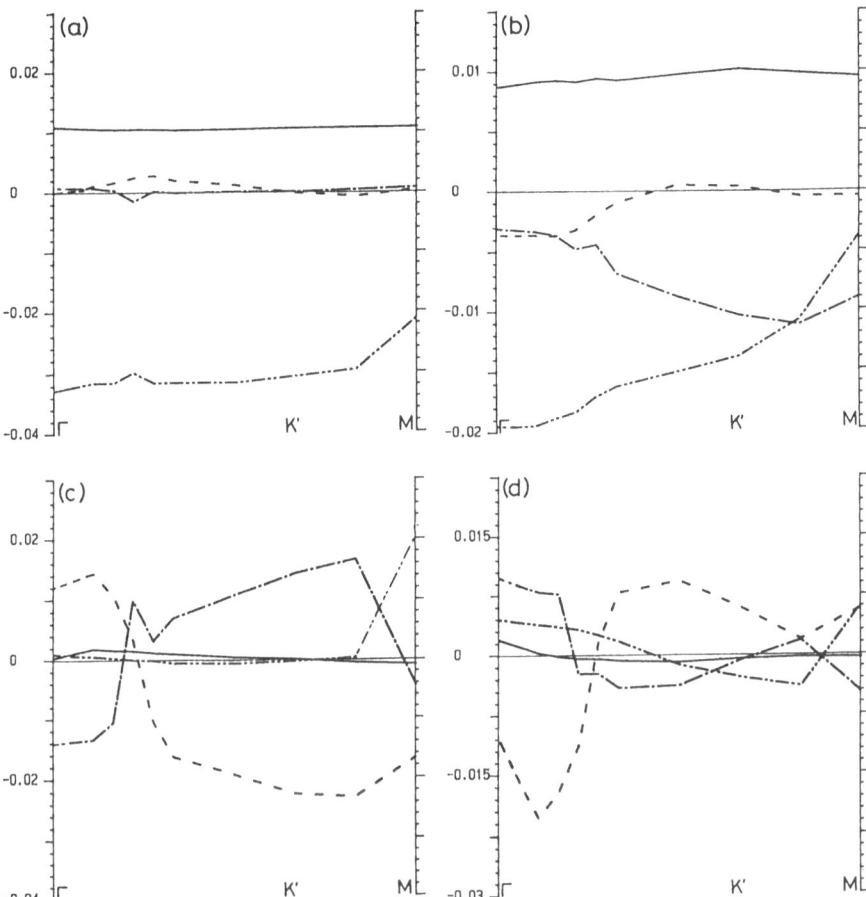

Fig. 5.14. The averaged bond order in Fe$_{1/3}$TiS$_2$: (a) $\beta_{Ti(1)dx',S(3)px'}$ (k); (b) $\beta_{Fe(0)dx'',S(4)px''}$ (k); (c) $\beta_{Ti(1)dx'y',Ti(2)dx'y'}$ (k); (d) $\beta_{Fe(0)dZ,Ti(1)dZ}$ (k). Solid (——), dashed (- - - -), dot-dashed (-·-·-), and double-dot-dashed (-··-) curves are used respectively for the parts denoted as (i), (ii), (iii), and (iv) in Fig. 5.8

order takes very small value in parts (i) and (iv) and it exhibits fairly large dispersion in parts (ii) and (iii). This results mean that the dε orbitals of the nearest neighboring Ti ions in the c plane hybridize each other and form energy bands in the energy range (ii) and (iii). Figure 5.14(d) which illustrates the calculated result of $\beta_{Fe(0)dZ, Ti(1)dZ}$ shows that the dε orbitals of the n.n. Ti and Fe ions along the c-direction hybridize each other in the energy ranges (ii) and (iii).

From these results the bonding nature in Fe$_{1/3}$TiS$_2$b is summarized as follows:

(1) Part (i) corresponds to the bonding bands of Fe dγ and S p mixing and of Ti dγ and S p mixing.

(2) Part (ii) consists mainly of Fe $d\varepsilon$ states, but hybridization of Ti $d\varepsilon$ states is fairly large.
(3) Part (iii) consists of the hybridization bands of Ti $d\varepsilon$ and Fe $d\varepsilon$ mixing and the antibonding bands of Fe $d\gamma$ and S p mixing.
(4) Part (iv) corresponds to the antibonding bands of Ti $d\gamma$ and S p mixing and of Fe $d\gamma$ and S p mixing.

Similar bonding nature have been found for $FeTiS_2$ as well as for other $M_{1/3}TiS_2$ compounds. Therefore, it is suggested that intercalation proceeds by M atoms occupying the octahedral sites in the van der Waals gap layer which seem to be suitable for formation of covalent bonds with S ions and favorable energetically.

5.4 Electronic Band Structures of Ag_xTiS_2

Silver intercalation compound, Ag_xTiS_2, is another interesting material [5.54–56]. Silver atoms also occupy the sites coordinated octahedrally with S atoms of neighboring TiS_2 sandwiches, that is, the octahedral sites in the van der Waals gap layers. There are three intercalated phases, a gas phase for small concentration, a second-stage phase for intermediate concentration ($x = 1/6$ being an ideal concentration) and a first-stage phase for large concentration ($x = 1/3$ being an ideal concentration). The upper limit of the silver concentration is $x = 0.42$. For samples with nominal composition $x > 0.42$, the powder patterns show lines of metallic silver.

In the first-stage intercalate, $Ag_{1/3}TiS_2$, all the van der Waals gap layers are partially filled by Ag atoms and a super-structure due to three-dimensional order of the Ag atoms, $\sqrt{3}a_0 \times \sqrt{3}a_0 \times 2c_0$, is present at low temperatures. The stacking of Ag atom layers along the c-axis is ABAB . . . in contrast with the ABC stacking of M atom layers in $M_{1/3}TiS_2$ (M = transition metal). The transition to the disordered structure at $T_c = 300\,K$ is of the second-order nature. In the second-stage phase the van der Waals gap layers are alternatingly partially occupied and unoccupied by silver atoms. In a partially-occupied layer the Ag atoms form also a $\sqrt{3}a_0 \times \sqrt{3}a_0$ super-structure at low temperatures, but along the c-axis there is no ordering at whole temperatures.

The staging structures are characteristic to the silver intercalate, Ag_xTiS_2, compared with the transition-metal intercalate, M_xTiS_2. Furthermore, the temperature-induced order–disorder transition of intercalant atoms is also characteristic to Ag_xTiS_2. The origin of these differences between Ag_xTiS_2 and M_xTiS_2 may be seeked for the differences in the electronic structures or the bonding nature between the intercalant atoms and the host material.

To get a deeper insight in the differences between Ag intercalates on one side and transition-metal intercalates on the other, the electronic band structures of $Ag_{1/3}TiS_2$ have been recently calculated also by Motizuki's group [5.57] with use of the linearized APW method of *Takeda* and *Kübler* [5.58]. The MT approximation and the local density approximation by *Gunnarson* and *Lundqvist* have been used also. In order to compare the obtained results directly with the electronic structures of $M_{1/3}TiS_2$ (M = transition metal) in which M atom layers stacks as ABCABC . . ., the band calculations have been carried out for the hypothetical ABC-stacking $Ag_{1/3}TiS_2$ as well as for the first-stage AB-stacking $Ag_{1/3}TiS_2$. The space group of ABC-stacking $Ag_{1/3}TiS_2$ is C_{3i}^1 and the primitive unit cell contains one Ag, three Ti and six S atoms while the space group of the first stage $Ag_{1/3}TiS_2$ is D_{3d}^2 and the unit cell contains twenty atoms in total, i.e., two Ag, six Ti, and twelve S ions. The hexagonal lattice constants used in the calculations are $a = 5.917$ Å and $c = 18.858$ Å in the case of ABC-stacking, and $a = 5.917$ Å and $c = 12.572$ Å in the case of AB-stacking. The value of $l_{max} = 7$ has been used for both the cases. The value of $|k + G|_{max}$ is taken to be $5.1 \times (2\pi/a)$ for ABC-stacking and $5.5 \times (2\pi/a)$ for AB-stacking, and thus the number of basis functions used in the calculation is about 790 in the former case and about 1950 in the latter case. The radii of the MT spheres are 1.243 Å for Ag, 1.065 Å for Ti, and 1.361 Å for S in the case of ABC-stacking $Ag_{1/3}TiS_2$, and 1.124 Å for Ag, 0.888 Å for Ti, and 1.479 Å for S in the case of AB-stacking $Ag_{1/3}TiS_2$. The potential and the charge density of the crystal have been determined self-consistently by using four special points.

The density of states obtained for ABC-stacking $Ag_{1/3}TiS_2$ is shown in Fig. 5.15. Compared with the band structures of $M_{1/3}TiS_2$ the most striking aspect is the Ag 4d band, i.e., it hybridizes strongly and overlaps energetically

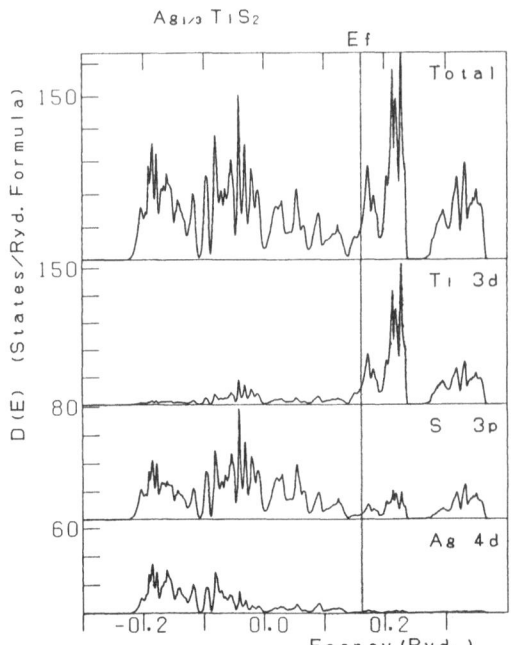

Fig. 5.15. The density of states of ABC-stacking $Ag_{1/3}TiS_2$

with S 3p bands. As the results, the interaction between Ti 3d and S 3p is diminished, or in other words, the covalent Ti–S bond is weakened by intercalation of Ag atoms. These characteristic features have been obtained also for first-stage AB-stacking $Ag_{1/3}TiS_2$. Here, it should be noted that the strong interaction or hybridization between Ag 4d and S 3p states is not favorable for energy gain because almost all the states of the hybridized bands of Ag 5d and S 3p states are occupied as seen from Fig. 5.15. This characteristic feature of the electronic band structure or the bonding nature of Ag_xTiS_2 may be strongly related to the fact that it is impossible to intercalate Ag atoms more than $x = 0.42$ in TiS_2 and also to the fact that the Ag atoms show an order–disorder transition in the van der Waals gap layer with increasing temperature.

For assumed intercalation compound $AgTiS_2$ the ASW (augmented spherical wave) band calculation has been done by *Dijkstra* et al. [5.59]. The obtained hybridization behaviors among the Ti 3d, S 3p, and Ag 4d states are almost similar to those of $Ag_{1/3}TiS_2$ obtained by our calculation.

5.5 2H–Type TX_2 (T = Nb, Ta; X = S, Se) Intercalated with Transition-Metals

Intercalation of 3d transition metals into van der Waals gaps of 2H–type TX_2 (T = Nb, Ta; X = S, Se) has been studied also extensively and various interesting physical properties have been observed [5.2, 3, 60–65]. In the mother crystal 2H–TX_2 (space group D_{6h}^4), the transition metal (T) ions form triangular lattices, each T ion being surrounded by six chalcogen (X) ions coordinated prismatically. The crystal structure of 2H–TX_2 is shown in Fig. 5.16. When 3d transition metal (M) ions are intercalated, the basic structure of the mother crystal is unchanged, and M ions occupy octahedral sites between the S–T–S sandwiches. The M ions form the ordered $2a_0 \times 2a_0$ and $\sqrt{3}a_0 \times \sqrt{3}a_0$ superstructures in the c-plane for concentration 1/4 and 1/3, respectively, where a_0 is the lattice constant of the hexagonal basal plane of the mother crystal. Figure 5.17 shows the crystal structure of $M_{1/4}TiS_2$ (space group D_{6h}^4). The unit cell contains twenty-six ions in total (two M, eight T, and sixteen X ions). There are two types of T and X ions (two T1 and six T2 atoms; four X1 and twelve X2 atoms): T2 atoms have no direct M neighbors, whereas each T1 is linearly coordinated by two M atoms along the c-axis; X1 atoms have no M neighbors while X2 has one M neighbor. The shape of the first BZ of 2H–TX_2 and M_xTX_2 is the same as that of 1T–TiS_2.

Among the intercalation compounds M_xTX_2, the magnetic properties of $Mn_{1/4}TaS_2$ have been investigated in detail. This intercalation compound shows a ferromagnetic transition at $T_c \sim 80$ K and the saturation moment is estimated experimentally to be 3.9–4.2 μ_B per Mn atom [5.60, 63], which is 20% smaller than that expected from a free Mn^{2+} ion. The measurement of magnetization

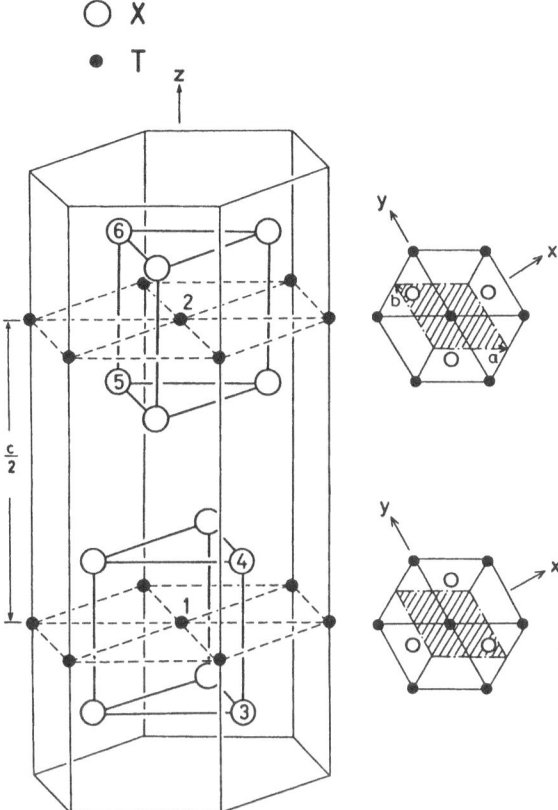

Fig. 5.16. The crystal structure of 2H–TX$_2$

density distribution by *Parkin* et al. [5.64] shows that the magnetization is principally localized on the Mn sites, but there is a significant spin polarization at the Ta sites. *Parkin* et al. [5.64] concluded that the 3d electrons of Mn atoms are localized and that the reduction of the Mn moment and the conduction band spin polarization are caused by the s–d mixing which gives rise to the RKKY interaction.

Recently, inelastic neutron scattering measurements have been performed on Mn$_{1/4}$TaS$_2$ [5.65]. According to the results, the spin wave dispersion along the c-axis is much larger than that in the c-plane. In order for the observed spin wave dispersion to be reproduced within the usual spin wave theory of the localized spin system, one must assume that the exchange interaction along the c-axis is much larger than that in the c-plane. On the other hand, according to the theoretical estimation of the RKKY interaction on the basis of the realistic electronic band structure of TaS$_2$, the inter-plane RKKY interaction is smaller by one order of magnitude than the in-plane RKKY interaction [5.66]. These facts suggest that the Mn moment cannot be regarded as fully localized and that the Mn 3d electrons are more itinerant than expected previously.

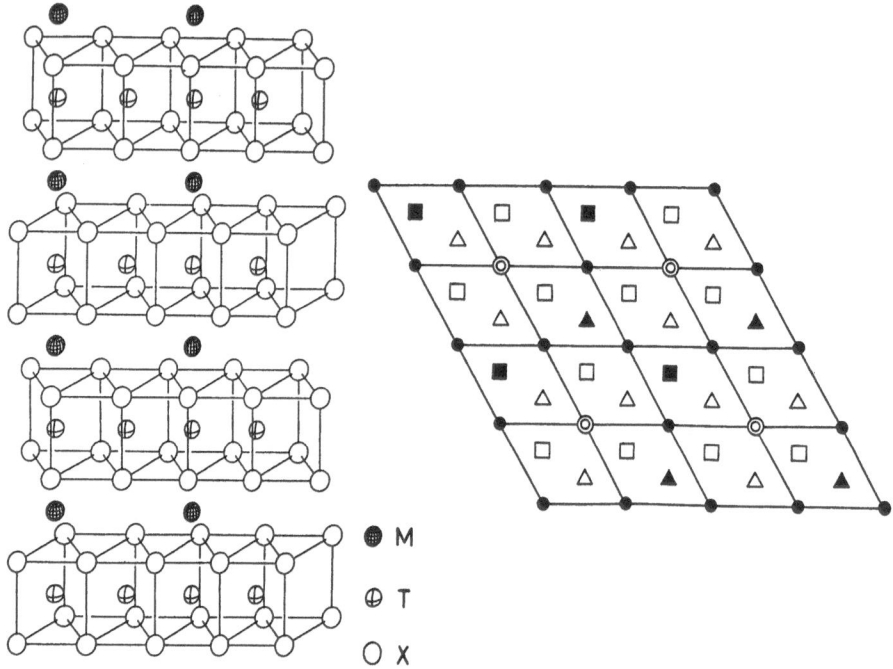

Fig. 5.17. The crystal structure of $M_{1/4}TX_2$. The left-hand figure shows the relationship of the M atoms to the TX_2 layers. The right-hand figure represents the (0 0 1) projection of the $M_{1/4}TX_2$ unit cell: $M = \copyright$ $(z = 0, 1/2)$; $T1 = \copyright$ $(z = 1/4, 3/4)$; $T2 = \bullet$ $(z = 1/4, 3/4)$; $X1 = \blacktriangle$ $(z = 1/8, 3/8)$ and $\blacksquare$ $(z = 5/8, 7/8)$; $X2 = \triangle$ $(z = 1/8, 3/8)$ and $\square$ $(z = 5/8, 7/8)$. T2 atoms have no direct M neighbors, whereas each T1 is linearly coordinated by two M atoms along the c-axis. X1 atoms have no M neighbors while X2 has one M neighbor

As for the electronic band structures of intercalation compounds of $2H–TX_2$, there has been only an ASW band calculation for the ferromagnetic state of $Fe_{1/3}TaS_2$ and $Mn_{1/3}TaS_2$ by *Dijkstra* et al. [5.67]. Their results show that the electronic structure of these intercalates can be understood only in first approximation in terms of the rigid band model.

Very recently, band calculations for $Mn_{1/4}TaS_2$ have been done for the ferromagnetic state as well as for the nonmagnetic state by *Tomishima* et al. [5.68] with use of a self-consistent LAPW method of *Takeda* and *Kübler* [5.58]. The lattice constants of $Mn_{1/4}TaS_2$ are $a = 6.64$ Å and $c = 12.54$ Å [64]. The distances of S layers within one sandwich and between neighboring sandwiches are 3.02 Å and 3.25 Å, respectively. The MT approximation has been used and the radii of MT spheres used for the calculation are 1.244 Å for Mn, 1.171 Å for Ta and 1.274 Å for S. Exchange and correlation are treated in the local density approximation by *Gunnarson* and *Lundqvist* [5.24]. The starting charge density for the nonmagnetic states has been constructed by superposing the self-consistent charge densities of the neutral atoms:

Mn: $(3d)^5 (4s)^2$,   Ta: $(5d)^3 (6s)^2$,   S: $(3s)^2 (4p)^4$.

For the ferromagnetic state, the starting spin polarization was assumed as follows:

Mn: $(3d \uparrow)^3 (3d \downarrow)^2$,   Ta: $(5d \uparrow)^2 (5d \downarrow)^1$.

Inside the MT spheres, the angular momentum expansion is truncated at $l = 7$ for the wave functions. The LAPW functions with the wave vector $|k + G| \leqq K_{max} = 4.7 \times (2\pi/a)$ are usd in the expansion of the eigenfunctions, leading to about 1130 basis functions. The potential and the charge density of the crystal have been determined self-consistently by using four k points, Γ, A, M and L, for the nonmagnetic state and by using two k points, Γ and A, for the ferromagnetic state.

Figures 5.18(a), (b) show the dispersion curves and the density of states of 2H–TaS$_2$ calculated by the LAPW method. These results are in agreement with the LMTO calculations by *Guo* and *Liang* [5.69] and with the ASW calculations by *Dijkstra* [5.70]. The Ta 5d states, except $5d_{z^2}$ state, hybridize with the S 3p states and form bonding and antibonding bands. The Ta $5d_{z^2}$ band is located between the bonding and antibonding bands and the Fermi level lies in this $5d_{z^2}$ band.

Figure 5.19 shows the total density of states and the partial density of states in each muffin-tin sphere for the nonmagnetic state of Mn$_{1/4}$TaS$_2$. It is clearly seen that the Mn 3d states hybridize strongly with the Ta $5d_{z^2}$ state. The electronic structure near the Fermi level of host 2H–TaS$_2$ is modified substantially and hence the rigid band model is not applicable. The hybridization between the Mn 3d and the S 3p states is very small. This situation is quite different from the transition-metal intercalation compounds of 1T–TiS$_2$, in which the 3d states of the intercalated transition-metal atoms hybridize well with the S 3p states as we have seen in Sects. 5.2, 3.

The density of states of the spin polarized bands in the ferromagnetic state is illustrated in Fig. 5.20 for both the up-spin and the down-spin bands. Large spin splitting occurs only for the Mn 3d bands. Through hybridization effects, however, magnetic moments parallel to that of Mn atoms are also induced on Ta and S atoms. The magnitude of the spin moment within each MT sphere is as follows (in unit of μ_B):

Mn: 3.61,   Ta1: 0.039,   Ta2: 0.054,   S1: 0.012,   S2: 0.016.

The total magnetic moment per Mn (MnTa$_4$S$_8$) is 4.35 μ_B, which is in good agreement with the observation 3.9–4.2 μ_B [5.60, 63]. The magnitude of the total moment is a little bit larger than that of Mn$_{1/3}$TaS$_2$ (4.18 μ_B) obtained by *Dijkstra* et al. [5.59].

In summary, the Mn 3d states in Mn$_{1/4}$TaS$_2$ hybridize strongly with the Ta $5d_{z^2}$ states which form the conduction bands of the host compound TaS$_2$ and hence the rigid band model is not applicable for Mn$_{1/4}$TaS$_2$. Therefore, the

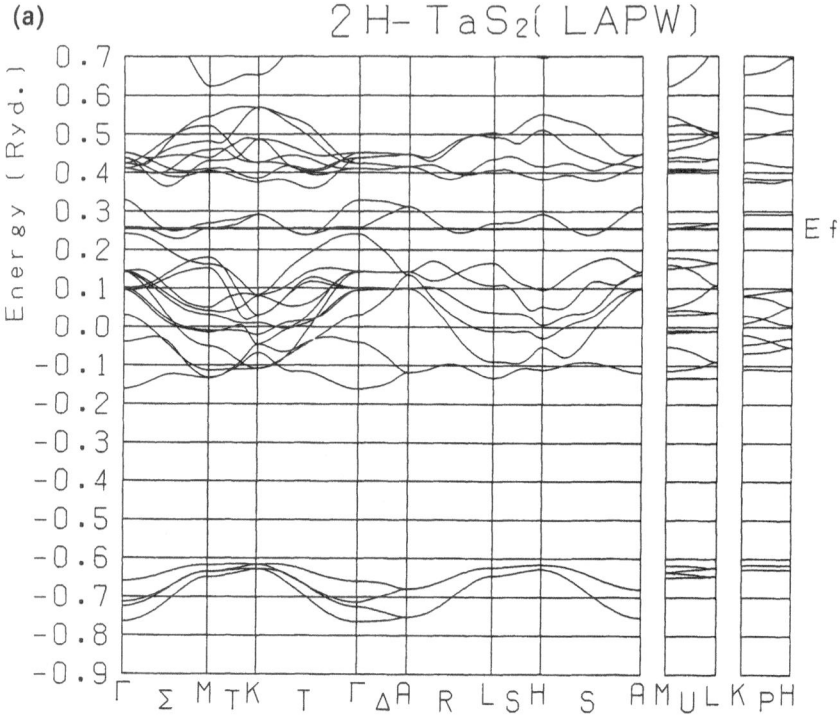

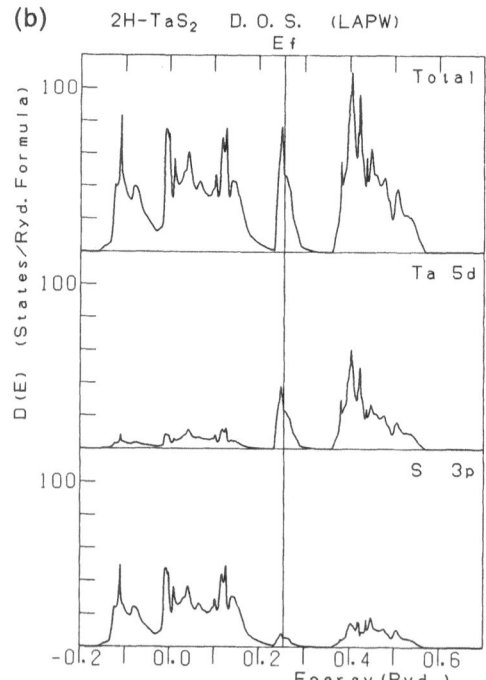

Fig. 5.18. (a) The energy dispersion curves and (b) the density of states of 2H-TaS₂

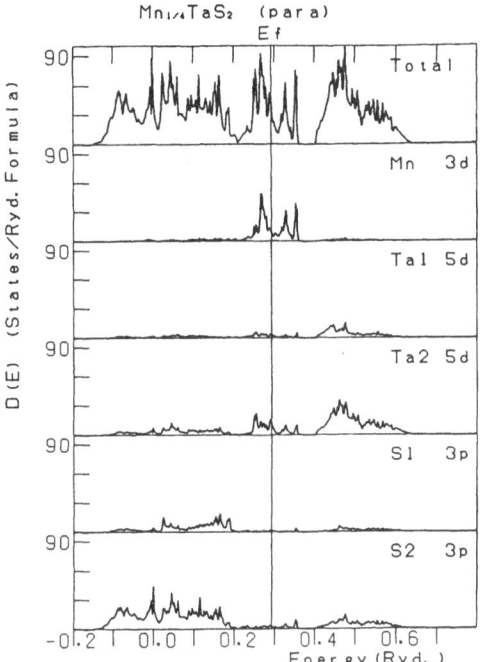

Fig. 5.19. The density of states of the nonmagnetic bands of Mn$_{1/4}$TaS$_2$. Partial density of states is calculated in each MT sphere. Note that the number of Ta2 atoms in the unit cell is three times that of Ta1 atoms

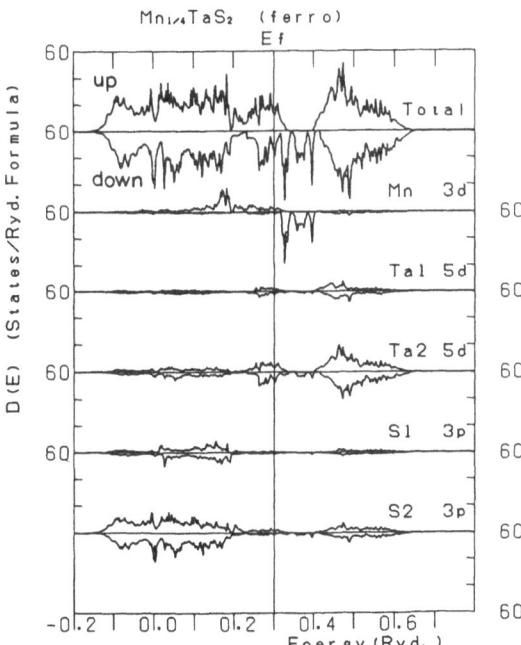

Fig. 5.20. The density of states of the ferromagnetic bands of Mn$_{1/4}$TaS$_2$

Mn 3d moments cannot be regarded as fully localized although the net magnetic moment per Mn atom is fairly large. Finally, it should be noted that the strong hybridization between the Mn 3d and the Ta $5d_{z^2}$ states may cause a strong Mn–Mn interaction via Ta atoms along the c-axis. Calculation of spin wave excitations on the basis of the obtained electronic band structures will be one of the most interesting problems left to be solved in the future.

The detailed results of the band calculations for $Mn_{1/4}TaS_2$ will be reported in [5.71].

5.6 Discussion

Recent systematic APW band calculations have yielded many important and interesting consequences about the electronic structure and bonding nature of the transition-metal intercalation compounds, M_xTiS_2 (M = Ti, V, Cr, Mn, Fe, Co, Ni). Principal results and consequences obtained by the band calculations are summarized as follows:

(a) The 3d states of the intercalant transition-metal atoms hybridize strongly with the Ti 3d and S 3p states and the original band structures of the mother crystal TiS_2 are modified considerably by the intercalation and hence the so-called rigid-band model cannot be applied. Reflecting the different atomic energy level of the M 3d states the hybridization between the M 3d and S 3p states is enhanced and the hybridization between the M 3d and Ti 3d states is reduced as the atomic number of M increases.

(b) The magnitude of the spin moments obtained from the band calculations for the ferromagnetic states of $M_{1/3}TiS_2$ (M = Cr, Fe, Co) is much smaller than that expected from a free M^{2+} ion.

(c) In TiS_2 the suggestion by Inglesfield on the basis of the tight-binding picture has been confirmed, i.e., the Ti $3d\gamma$ and S 3p states form bonding and antibonding bands while the Ti $3d\varepsilon$ states form a nonbonding band between the two bands.

(d) In M_xTiS_2 the covalent-like bond between the Ti $3d\gamma$ and S 3p states is weakened while the M $3d\gamma$ states also make fairly strong covalent-like bonds with the S 3p states. As for the $3d\varepsilon$ states of the Ti and M ions, they hybridize each other and form bonding and antibonding states. Thus, it is suggested that intercalation of M atoms into TiS_2 proceeds by M atoms occupying the octahedral sites which seem to be suitable for formation of covalent bonds with S ions and favorable energetically.

Inapplicability of the rigid-band model to M_xTiS_2 has been verified experimentally through measurements of specific heats and Hall coefficients. Strong hybridization of the M 3d states with the Ti 3d and S 3p states has been confirmed by analyzing valence states photoemission spectra. Further, the observed saturation moment of $M_{1/3}TiS_2$ (M = Cr, Fe, Co) is much smaller than that expected from a free M^{2+} ion, which is consistent with the results of band calculatioins for the ferromagnetic state of $M_{1/3}TiS_2$.

Thus many physical properties of the intercalation compounds $M_{1/3}TiS_2$ can be understood on the basis of the electronic band structures obtained by the APW method with use of the local density approximation. It should be noted, however, that a couple of experimental results indicate a necessity of taking account of the electron correlation effects which are not included properly in the band calculations:

(1) As pointed out already in Sect. 5.2.3 the values of electronic specific heat coefficient γ observed by *Inoue* et al. [5.37] are larger by a factor more than five compared with those expected from the band calculations. This fact may indicate importance of the effect of mass enhancement due to the electron correlation as well as the electron–phonon interaction.

(2) From analysis based on a cluster model *Fujimori* et al. [5.52] pointed out that the photoemission spectra of $Ni_{1/3}TiS_2$ can be properly understood by taking into account the intraatomic correlation energy of the intercalated transition-metal atoms and they proposed a picture of correlated bands of the intercalated M 3d states as a result of hybridization between the M 3d states and the host Ti 3d and S 3p states. If we need to take into account the intraatomic correlation effects, the Anderson model will be suitable. Investigations on the electron correlation effects in M_xTiS_2 are further problems left in the future.

The electronic structure and bonding nature of M_xTiS_2 are making a striking contrast with those of silver intercalation compound of TiS_2. In Ag_xTiS_2 the Ag 4d states hybridize strongly with the S 3p state and destroy the covalent-like bond between Ti 3d and S 3p states, but the strong hybridization between Ag 4d and S 3p states does not yield an energy gain because almost all the states of the hybridized bands of Ag 4d and S 3p states are occupied. These features seem to be strongly related to the fact that there is a maximum value for the Ag concentration and also to the fact that the Ag atoms show a temperature-induced order–disorder transition in the van der Waals gap layer.

The electronic and bonding properties of alkali intercalation compounds of TiS_2 are different from those of Ag_xTiS_2 as well as M_xTiS_2. The electronic band structures of Li intercalation compound $LiTiS_2$ have been calculated by *Umrigar* et al. [5.31] and *Dijkstra* et al. [5.59]. According to their results none of the occupied levels in $LiTiS_2$ has an Li-character, and the rigid-band model that the Li atom donates simply an electron to the S–Ti–S sandwiches is applicable fairly well. As a result, the bonding between the Li ions and the S–Ti–S sandwiches is weak, and in fact diffusive motion of the Li^+ ions in the van der Waals gap layers has been observed in $Li_{0.94}TiS_2$ by nuclear magnetic resonance measurements [5.72].

Acknowledgements. We are very grateful to the members of our group who have collaborated with us in the study of topics reviewed in this article: Dr. T. Yamasaki, Mr. Y. Yamazaki, Dr. T. Teshima, Mr. S. Tomishima, and Mr. M. Nogome. We are deeply indebted to Professor M. Inoue and Dr. H. Negishi of Hiroshima University for useful discussions about their experimental data. We also express our sincere thanks to Professor A. Yanase for providing us with the computer program for the APW band calculation.

This work is supported by a Grant-in-Aid from the Ministry of Education, Science, and Culture of Japan. Financial support from the Kurata Foundation is also acknowledged greatly. One of the authors (K.M.) would like to express her sincere thanks to the Yamada Science Foundation for financial support.

References

5.1 For example, *Intercalated Layered Materials*, ed. by F.A. Levy (Reidel, Dordrecht 1979) pp. 320–562
    H. Zabel, S.A. Solin (eds.): *Graphite Intercalation Compounds I, II*, Springer Ser. Mater. Sci., Vols. 14, 18 (Springer, Berlin, Heidelberg 1990, 1992)
5.2 G.V. Subba Rao, M.W. Shafer: Intercalation in layered transition metal dichalcogenides, in *Intercalated Layered Materials*, ed. by F.A. Levy (Reidel, Dordcrecht 1979) pp. 99–199
5.3 A.R. Beal: The first row transition metal intercalation complexes of some metallic group VA transition metal dichalcogenides, in *Intercalated Layered Materials*, ed. by F.A. Levy (Reidel, Dordrecht 1979) pp. 251–305
5.4 A.D. Yoffe: Solid State Ionics 9–10, 59 (1983)
5.5 R.H. Friend, A.D. Yoffe: Adv. Phys. 36, 1 (1987)
5.6 F.R. Gamble, F.J. DiSalvo, R.A. Klemm, T.H. Geballe: Science 168, 568 (1970)
5.7 M.S. Whittingham: Science 192, 1126 (1976)
5.8 M. Inoue, H.P. Hughes, A.D. Yoffe: Adv. Phys. 38, 565 (1989)
5.9 N.J. Doran: Physica 99B, 227 (1980)
5.10 C.Y. Fong, M. Schlüter: Electronic structure of some layer compounds, in *Electrons and Phonons in Layered Crystal Structures*, ed. by T.J. Wieting, M. Schlüter (Reidel, Dordrecht 1979) pp. 145–315
5.11 K. Motizuki, N. Suzuki: Microscopic theory of structural phase transitions in layered transition-metal compounds, in *Structural Phase Transitions in Layered Transition Metal Compounds*, ed. by K. Motizuki (Reidel, Dordrecht 1986) pp. 1–134
5.12 T. Yamasaki, N. Suzuki, K. Motizuki: J. Phys. C 20, 395 (1987)
5.13 N. Suzuki, T. Yamasaki, K. Motizuki: J. Mag. Mag. Mater. 70, 64 (1987)
5.14 N. Suzuki, T. Yamasaki, K. Motizuki: J. Phys. C 21, 6133 (1988)
5.15 N. Suzuki, T. Yamasaki, K. Motizuki: J. Phys. (Paris) 49 C8, 201 (1988)
5.16 N. Suzuki, T. Yamasaki, K. Motizuki: J. Phys. Soc. Jpn. 58, 3280 (1989)
5.17 N. Suzuki, T. Teshima, K. Motizuki: In *Proc. 4th Asia Pacific Physics Conference*, ed. by A.H. Ahn, S.H. Choh, Il-T. Cheon, C. Lee (World Scientific, Singapore 1991) pp. 448–451
5.18 T. Teshima, N. Suzuki, K. Motizuki: J. Phys. Soc. Jpn. 60, 1005 (1991)
5.19 M. Inoue, H. Neġishi: J. Phys. Chem. 90, 235 (1986)
5.20 T.L. Loucks: *Augmented Plane Wave Method* (Benjamin, New York 1967)
5.21 L.F. Mattheiss, J.H. Wood, A.C. Switendick: A procedure for calculating electronic energy bands using symmetrized augmented plane waves, in *Methods in Computational Physics*, ed. by B. Alder, S. Fernbach, M. Rotenberg (Academic, New York 1968) pp. 63–147
5.22 J.C. Slater: Phys. Rev. 51, 846 (1951)
5.23 W. Kohn, L.J. Sham: Phys. Rev. 140, A1133 (1965)
5.24 O. Gunnarsson, B.J. Lundqvist: Phys. Rev. B13, 4274 (1976)
5.25 D.D Koelling, B.N. Harmon: J. Phys. C 10, 3107 (1977)
5.26 D.J. Chadi, M.L. Cohen: Phys. Rev. B8, 5747 (1973)
5.27 H.J. Monkhorst, J.D. Pack: Phys. Rev. B13, 5188 (1976)
5.28 O. Jepson, O.K. Andersen: Solid State Commun. 9, 1763 (1971)
5.29 G. Lehmann, M. Taut: Phys. Status Solidi (b) 54, 469 (1972)
5.30 J.E. Inglesfield: J. Phys. C 13, 17 (1980)

5.31 C. Umrigar, D.E. Ellis: D. Wang, H.K. Krakauer, M. Posternak: Phys. Rev. B 26, 4935 (1982)
5.32 A. Zunger, A.J. Freeman: Phys. Rev. B15, 9064 (1977)
5.33 C.H. Chen, W. Fabian, F.C. Brown, K.C. Woo, B. Davies, B. Delong, A.H. Thompson: Phys. Rev. B21, 615 (1980)
5.34 J.J. Barry, H.P. Hughes, P.C. Klipstein, R.H. Friend: J. Phys. C 16, 393 (1983)
5.35 H. Negishi, A. Shoube, H. Takahashi, Y. Ueda, M. Sasaki, M. Inoue: J Mag. Mag. Mater. 67, 179 (1987)
5.36 K. Motizuki, K. Katoh, A. Yanase: J. Phys. C 19, 495 (1986)
5.37 M. Inoue, Y. Muneta, H. Negishi, M. Sasaki: J. Low. Temp. Phys. 63, 235 (1986)
5.38 M. Inoue, M. Matumoto, H. Negishi, H. Sasaki: J. Mag. Mag. Mater. 53, 131 (1985)
5.39 M. Inoue, H. Negishi: J. Mag. Mag. Mater. 70, 199 (1985)
5.40 H. Negishi, M. Koyano, M. Inoue, T. Sakakibara, T. Goto: J. Mag. Mag. Mater. 74, 27 (1988)
5.41 H. Negishi, S. Ohara, M. Koyano, M. Inoue, T. Sakakibara, T. Goto: J. Phys. Soc. Jpn. 57, 4083 (1988)
5.42 Y. Tazuke, T. Endo: J. Mag. Mag. Mater. 31–34, 1175 (1983)
5.43 T. Yoshioka, Y. Tazuke: J. Phys. Soc. Jpn. 54, 2088 (1985)
5.44 T. Satoh, Y. Tazuke, T. Miyadai, K. Hoshi: J. Phys. Soc. Jpn. 57, 1743 (1988)
5.45 F. Matsukura, Y. Tazuke, T. Miyadai: J. Phys. Soc. Jpn. 58, 3355 (1989)
5.46 S. Shibata, Y. Tazuke: J. Phys. Soc. Jpn. 58, 3746 (1989)
5.47 M. Koyano, H. Negishi, Y. Ueda, M. Sasaki, M. Inoue: Phys. Status Solidi (b) 138, 357 (1986)
5.48 M. Koyano, H. Negishi, Y. Ueda, M. Sasaki, M. Inoue: Solid State Commun. 62, 261 (1987)
5.49 C.M. Hurd: The Hall Effect in Metals and Alloys (Plenum, New York 1972) Chap. 2
5.50 Y. Ueda, H. Negishi, M. Koyano, M. Inoue: K. Soda, H. Sakamoto, S. Suga: Solid State Commun. 57, 839 (1986)
5.51 Y. Ueda, K. Fukushima, H. Negishi, M. Inoue, M. Taniguchi, S. Suga: J. Phys. Soc. Jpn. 56, 2471 (1987)
5.52 A. Fujimori, S. Suga, H. Negishi, M. Inoue: Phys. Rev. B38, 3676 (1988)
5.53 R.W. Godby, G.A. Benesh, R. Haydock, V. Heine: Phys. Rev. B32, 655 (1985)
5.54 G.A. Scholz, R.F. Frindt: Mat. Res. Bull. 15, 1703 (1980)
5.55 A.G. Gerards, H. Roede, R.J. Haange, B.A. Boukamp, G.A. Wiegers: Synth. Metals 10, 51 (1984/85)
5.56 Y. Kuroiwa, K. Ohshima, Y. Watanabe: Phys. Rev. B 42, 11591 (1990)
5.57 M. Nogome, T. Teshima, N. Suzuki, K. Motizuki: Mater. Sci. Forum 91–93, 381 (1992)
5.58 T. Takeda, J. Kübler: J. Phys. F 9, 661 (1979)
5.59 H. Dijkstra, C.F. van Bruggen, C. Haas: J. Phys. Condens. Matter 1, 4297 (1989)
5.60 S.S.P. Parkin, R.H. Friend: Phil. Mag. B 41, 65 (1980)
5.61 S.S.P. Parkin, R.H. Friend: Phil. Mag. B 41, 95 (1980)
5.62 S.S.P. Parkin, A.R. Beal: Phil. Mag. B 42, 627 (1980)
5.63 Y. Onuki, K. Ina, T. Hirai, T. Komatubara: J. Phys. Soc. Jpn. 55, 347 (1986)
5.64 S.S.P. Parkin, E.A. Marseglia, P.J. Brown: J. Phys. C 16, 2749 (1983)
5.65 L.D. Cussen, E.A. Marseglia, D. Mck. Paul, B.D. Rainford: Physica B 156 & 157, 712 (1989)
5.66 N. Suzuki, Y. Yamazaki, T. Teshima, K. Motizuki: Physica B 156 & 157, 286 (1989)
5.67 J. Dijkstra, P.J. Zijleman, C.F. van Bruggen, C. Haas, R.A. de Groot: J. Phys. Condens. Matter 1, 6363 (1989)
5.68 S. Tomishima, N. Suzuki, K. Motizuki, H. Harima: J. Phys. Soc. Jpn. 59, 1913 (1990)
5.69 G.Y. Guo, W.Y. Liang: J. Phys. C. 20, 4315 (1987)
5.70 H. Dijkstra: The electronic structure of some transition metal chalcogenides and intercalates, Dissertation, University of Groningen (1988)
5.71 N. Suzuki, S. Tomishima, K. Motizuki: Electronic band structures and magntic properties of Mn-intercalated compounds of 2H–type transition-metal dichalcogenides, in New Horizons in Low Dimensional Electron Systems – A Festschrift in Honour of Professor H. Kamimura, ed. by H. Aoki, M. Tsukada, M. Schulter, F. Levy (Kluwer, Dordrecht 1991) pp. 55–69
      K. Motizuki, N. Suzuki, S. Tomishima: J. Mag. Mag. Mater. 104–107, 681 (1992)
5.72 T. Eguchi, C. Marinos, J. Jonas, B.G. Silbernagel, A.H. Thompson: Solid State Commun. 38, 919 (1981)

5.73 Y. Nishio, M. Shirai, N. Suzuki, K. Motizuki: J. Phys. Soc. Jpn. **63**, 156 (1994)

5.74 K. Motizuki, Y. Nishio, M. Shirai, N. Suzuki: Physica B **219, 220**, 83 (1996) (*Proc. of 4th Int'l. Conf. on Phonon Phys. and 8th Int'l. Conf. on Phonon Scat. in Condens. Matter*)

5.75 K. Motizuki, Y. Nishio, M. Shirai, N. Suzuki: J. Phys. Chem. Solids **57**, 1091 (1996)

5.76 A. Kimura, S. Suga, T. Matsushita, S. Imada, S. Shino, Y. Saitoh, H. Shigeoka, H. Daimon, T. Kinoshita, A. Kakizaki, S.-J. Oh, H. Negishi, M. Inoue: Jpn. J. Appl. Phys. **32**, Suppl. 32-3, 255 (1993)

5.77 T. Matsushita, S. Suga, Y. Tanaka, H. Shigeoka, T. Nakatani, T. Okuda, T. Terauchi, T. Shishidou, A. Kimura, H. Daimon, S.-J. Oh, A. Kakizaki, T. Kinoshita, H. Negishi, M. Inoue: J. Elect. Spec. Rel. Phenom. **78**, 477 (1996) (*Proc. 11th Int'l. Conf. Vacuum Ultraviolet Radiation Physics*)

5.78 A. Kimura: Electronic Structures of Mn-Based Alloys and Transition Metal Intercalated 1T–TiS$_2$, Dissertation, Osaka University (1995)

6 Structural Phase Transformation

F.E. Fujita

Examples of phase transformations (transitions) in various materials are first introduced and the difference between the continuous and discontinuous transformation is outlined. The difference is theoretically analyzed by combining Ehrenfest's criterion and Landau's formulation and graphically shown by the G–T–η diagrams. Second, the precursor phenomena associated with the weak first-order transformation, which appear in the martensitic transformation of β-phase shape memory alloys, perovskite type ferroelectric materials and A-15 type superconductors, are explained by using a statistical thermodynamic theory. Transformations of other useful new materials such as ceramics and diamond are also described from the physical view point.

6.1 General View

6.1.1 Discoveries of Phase Transformations

Since old times when people learned the skill of producing ceramics, it has been known that the temperature of the kiln must be raised especially slowly when passing through the range between 550° and 600°C. Otherwise, the green ware would be self-cracked in that range, even if they have been dried enough before they are put into the kiln. Mineralogy and crystallography, which had already been developed before the discovery of X-ray diffraction in 1912, could explain why cracking takes place by rapid heating in that temperature range: It is due to a large volume expansion associated with the crystal structure change of silica (SiO_2) particles, usually contained in an appreciable amount in clay, from α-quartz to β-quartz at 573°C. This was probably the first knowledge on the phase transition of materials obtained by the people in industry. Cooling down the kiln must also be done carefully, but in this case, the reverse transition to α-quartz doesn't always take place; for instance, a high temperature phase, β-cristobalite, doesn't go back to β-quartz but goes to α-cristobalite, and from the molten silica silica glass is obtained as is well known. Further studies have clarified other structural changes of silica with the temperature and pressure variations, as the P–T diagram for silica in Fig. 6.1 shows. The high pressure phases, coesite, stishovite, and their transitions, are closely related to earth

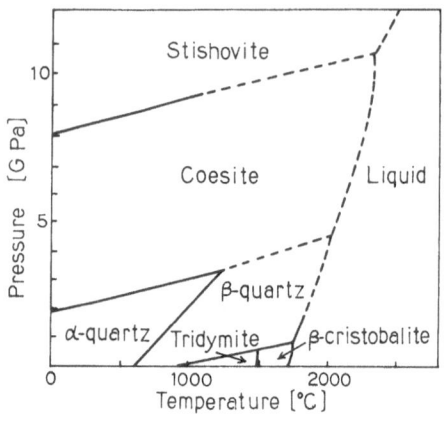

Fig. 6.1. The P–T phase diagram for silica (SiO$_2$). Metastable-cristobalite mentioned in the text is not shown

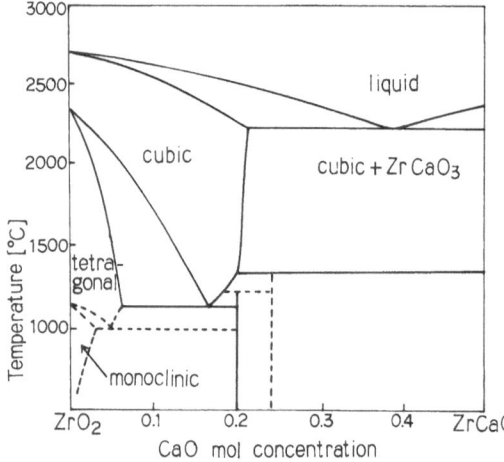

Fig. 6.2. A part of the phase diagram for ZrO$_2$–CaO system

science, because SiO$_2$ is one of the main constituent minerals exposed to high pressures and temperatures deep within the earth.

In the ceramic work and research, many other phase transitions have been known. For instance, calcium phosphate, Ca$_3$(PO$_4$)$_2$, which is one of the main components of the famous ceramic material, bone china, has complicated phase transitions, and the newer ceramic material, zirconia (ZrO$_2$), has an undesirable phase transition with a large volume change from the tetragonal to the mono-clinic structure at about 1000°C when cooled, as the extreme left side of Fig. 6.2 shows. In order to prevent cracking due to the cooling, addition of about 0.2 mol of calcia to make the stable cubic phase is effective at the sacrifice of lowering the material's strength. However, recently the better solution for it was found by adding a smaller amount of calcia and/or yttria to partially stabilize the structure by producing the constrained tetragonal precipitate particles. This is a new ceramic material, the partially stabilized zirconia. When a large stress is

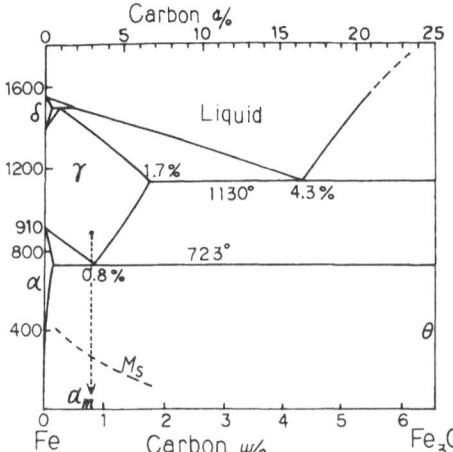

applied, the $t \rightarrow m$ transformation, of the precipitate particles, which is the stress induced phase transformation, takes place, suppressing the crack propagation and, at the same time, yields plastic deformation instead of the slip mechanism by dislocations. The stress induced transformation of the new deformable ceramic will be mentioned later in relation to the martensitic transformation.

The martensite structure in steel was first identified and defined by Osmond in 1895.[1] This was probably the first phase transition of solid material to be understood and effectively utilized in industry. As the arrow in Fig. 6.3 shows, a carbon steel containing, say, 3 atom.% carbon, transforms from the austenite (γ) phase to the quasi-equilibrium martensite (α_m) phase by water-quenching from, say, 900°C, which is a phase transition from fcc to bct structure, the latter containing supersaturated interstitial carbon atoms. Other typical martensitic transformations in alloys are of Fe–Ni ($\sim 30\%$) and Au–Cd ($\sim 50\%$) alloys, for which transformation temperatures, M_s (martensite starts from austenite under cooling), M_f (martensite finishes), A_s (austenite conversely starts under heating), and A_f (austenite finishes) are shown by the electrical resistivity changes in Fig. 6.4 [6.1]. The large hysteresis in the phase transformation of the Fe–Ni system arises from a strong resistance against the nucleation and/or propagation of the martensite phase in the parent phase, austenite, in cooling, and vice versa. On the other hand, the Au–Cd alloy exhibits much smaller hysteresis, suggesting the easy formation of martensite in the parent matrix. It is worthy of note that most of the β-phase alloys, which have the bcc or the CsCl type ordered structure, show small hystereses in martensitic transformation and the shape memory effect. The crystallographic difference in the mechanism, and thereby the products of transformation between the cases of the large and the small hystereses, will be discussed later to explain the basic mechanism of the shape memory.

---

[1] Martensite, was named after Martens, because Osmond thought that the structure he found had already been observed by Martens.

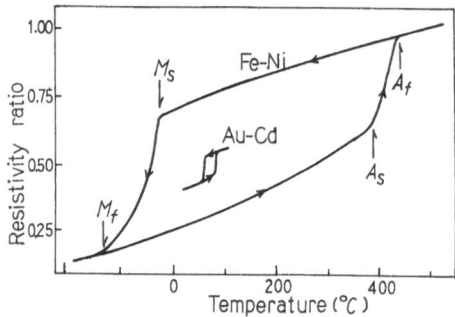

Fig. 6.4. Comparison of martensite transformation temperatures of Fe–Ni and Au–Cd alloy systems. Note that there is a large difference in the size of hysteresis loops

6.1.2 Continuous and Discontinuous Transformation

Similar hysteresis phenomena are found in ferromagnetic materials when the direction of magnetization is reversed by converting the applied magnetic field, and in ferroelectric materials when the transition takes place between the ferro- and paraelectric phase or between the different ferro phases. The former's hysteresis is not associated with the phase transition but arises from the difficulty or easiness of nucleation of the magnetic domains of opposite direction or propagation of the domain walls. By enhancing the difficulty, new strong permanent magnets, like the Fe–Nd–B magnet mentioned in Sect. 1.2, have been made, and by pursuing the easiness, new soft magnets, like the amorphous magnetic ribon, have been developed. As a physics of new materials, the magnetic hysteresis loop is an interesting and important subject, but it is not that of the phase transition to be discussed in this chapter. On the other hand, the latter's behaviour is quite the same as that of β-phase alloys, exhibiting a small hysteresis loop, a small or sometimes no appreciable latent heat and volume change, gradual changes of lattice parameters below the transition temperature and the pretransition or the precursor phenomena. A typical example of this type is seen in a ferroelectric material, barium titanate ($BaTiO_3$), which has the perovskite ($CaTiO_3$) structure as shown in Fig. 6.5. Its successive

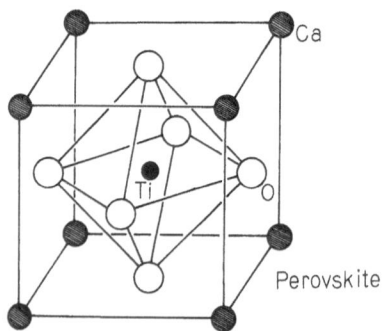

Fig. 6.5. Perovskite ($CaTiO_3$) structure

phase transitions,

$$\text{trigonal} \rightleftharpoons \text{orthorhombic} \rightleftharpoons \text{tetragonal} \rightleftharpoons \text{cubic} \rightleftharpoons \text{hexagonal} \rightleftharpoons \text{liquid},$$

173 K   278 K   393 K   1733 K   1885 K

are partially shown by the changes of the dielectric constant [6.2] in Fig. 6.6. The ferro- to paraelectric transition takes place at 117°C, below which the spontaneous polarization by the relative displacement of anions and cations in the lattice occurs, reducing the lattice symmetry and changing the lattice parameters with the temperature, as Fig. 6.7 shows. The extremely high values of the dielectric constant near the transitions arise from the easily induced polarization by the applied electric field. The atomic displacement, and thereby the degree of asymmetry of the lattice, can be considered as the degree of order, or more specifically the order parameter, which was already discussed in Chap. 5 in order to express to what extent the atomic distribution on the lattice points in an alloy is ordered. In this chapter, including the case of the order–disorder alloys, the phase transition will generally be treated by a phenomenological theory utilizing the free energy and the order parameter. The degree of order gradually changes below the transition point, and at that point it shows a sudden jump in some cases accompanying a discontinuous volume change, an entropy change, an enthalpy change, i.e., a heat evolution, etc. This kind of transition is called the first order or the discontinuous transition, since the first derivative of the Gibb's

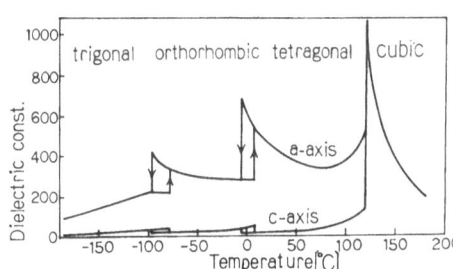

Fig. 6.6. Changes of dielectric constants of $BaTiO_3$ associated with structural phase transitions

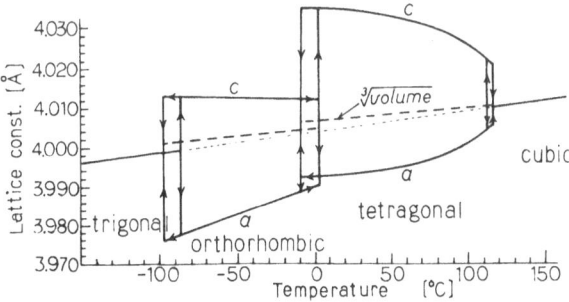

Fig. 6.7. Changes of lattice parameters of $BaTiO_3$ associated with structural phase transitions

energy by, for instance, temperature is discontinuous, as the thermodynamical expression of entropy, $S = -\partial G/\partial T$ shows. On the other hand, in some other cases, the above physical quantities show no discontinuity nor the order parameter, but instead the quantities corresponding to the second derivatives of Gibb's energy, such as the specific heat, show sudden jumps. This kind of transition is called the second order or the second kind transition. Furthermore, according to the mathematical or thermodynamical definition, transitions of higher orders could exist, and in addition there actually exist transitions with very small, sometimes almost unappreciable, discontinuities like those of β-phase alloys. These problems will also be discussed in the next section.

6.1.3 Various Types of Phase Transitions

Phase transitions occur in various materials in the universe; in intersteller materials, planetary materials, globe materials including all kinds of minerals and ice, biological substances, and industrial materials. In addition, there are different types of phase transitions in condensed matters; one is the crystal or atomic structure transition due to the systematic displacement of atoms including melting (solidification), glass transition, liquid crystal transition, order–disorder in substitutional and interstitial[2] alloys, martensitic transformation, stacking layer polytype transition, and other lattice structure transitions[3]. Others are the spin (magnetic) configuration transition, such as ferro–para–antiferro magnetic transitions and spin glass transition, the ionic valence transition, the superconductive transition of metals, ceramics, and molecular materials, and the superfluid transition of quantum liquid (helium). Except for the very last one, the last three kinds of transitions arise from the changes of the electronic states, which will not be mentioned in this chapter but will be discussed in other chapters; for instance, the magnetic and superconductive transitions are discussed in Chap. 2. The order–disorder transitions of alloys are also precisely mentioned in Chap. 7. The phase transitions to be discussed in this chapter are those of minerals, metals, and ceramics related to industrial products and developed rather recently as new materials. For instance, diamond is a well-known natural product, which has various phase transitions, but today artificial diamond including that of the CVD film is a new industrial material and therefore will be introduced in the last part of this chapter.

The term, structural phase transition, is usually used for the transition with simple and small displacements of atoms such as from the cubic to the tetragonal structure, as shown in Fig. 6.6. The title employed in this chapter,

---

[2] A typical example is hydrogen in metals, which makes transitions between the disordered, the ordered, and the hydride state with temperature or concentration change.

[3] Even some quasi-crystals exhibit the structural changes, for instance, from the icosahedral to the decagonal phase.

"Structural Phase Transformation", however, has a slight different meaning: Solidification, the order–disorder transition, the martensitic transformation of first order, etc., which are associated with large atomic displacements and discontinuous change of lattice symmetry, are included in it. It is worth noting that the word, transformation, is used in metallurgy instead of transition, like martensitic transformation.

In physical metallurgy, precipitation, spinodal decomposition and other phase separation phenomena are also meant by the term, phase transformation. In this chapter, however, these phenomena associated with the long range atomic diffusion will not be discussed.

6.2 A Phenomenological Theory and a Statistical View of Phase Transition

6.2.1 Degree of Order and Landau's Formulation of Phase Transition

As mentioned in the preceding paragraph, in any case of phase transition, the degree of order of the structure, η, can be defined and used to describe the state of the phase during the process of transition. In this respect, Landau's classification of phase transitions to the first and the second kind, or the discontinuous and continuous type, using the expansion of the free energy, G, by η, is a simple and successful one. Making a simple power series expansion of G by

$$G = G_0 + A\eta + B\eta^2 + C\eta^3 + D\eta^4 + \ldots \tag{6.1}$$

near the transition point and taking account of the change of lattice symmetry, Landau took the term, $B\eta^2$ and $D\eta^4$, especially with $B = \beta(T - T_c)$ where $\beta > 0$, and qualitatively expressed the second kind transition[4], in which η starts from zero in cooling, rather steeply rises up when passing through the transition temperature T_c, and gradually grows toward $\eta = 1$ at lower temperatures. It is easily understood from the description in the last section that $\eta = 0$ corresponds to the parent phase and $\eta = 1$ the perfectly ordered new phase. By choosing the appropriate values for β, T_c and D, respectively, the above treatment well explains the nature of the second order transition, as the temperature variation of G in Fig. 6.8a shows: The minimum point of G stays at $\eta = 0$ in cooling above T_c, spreads wide to allow a large fluctuation of η near T_c, and continuously slides to larger values of η toward $\eta = 1$ below T_c, as expected in the above. The temperature dependence of η given by the above calculation is not quite realistic, however, since the relation $\eta = \{\beta(T_c - T)/2D\}^{1/2}$ obtained from the above is

---

[4] A simple change of lattice symmetry, for instance, from cubic to tetragonal by a uniaxial elongation, is mathematically expressed by a transition between two crystal groups one of which is a partial group of the other and in such a case the odd terms, $A\eta$, $C\eta^3$, etc. drop from (6.1).

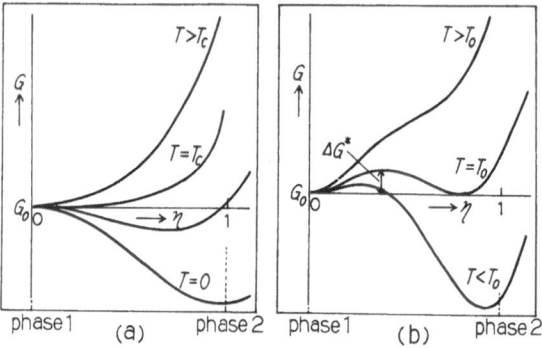

Fig. 6.8. The free energy vs order parameter curves for (a) the second order transition and (b) the first order transition

fairly different, for instance, from that of the well-known Brillouin function to describe the temperature dependence of the ferro–para magnetic transition, and from that of the Bragg–Williams or the Bethe approximation for the order–disorder phenomena. This is due to the fact that the above series expansion is applicable only near the transition point. On the other hand, when the third power, $C\eta^3$, is added, the second minimum of the free energy, G, is built up near $\eta = 1$, as Fig. 6.8b shows, and at T_0 (corresponding to T_c), η will jump from the original state $\eta = 0$, to a high η value of the new state through the activation process. In this way, the first order transition can be expressed by the free energy expansion, as well as the second order transition. Nevertheless, some unrealistic features exist in the above formulation as in the case of the second order transition. For instance, in the real first order transition, the degree of order often jumps from almost zero to almost unity, which is quite difficult to reproduce in the above calculation, as the Fig. 6.8b shows that η is still far less than unity below the transition temperature, T_0. Another unavoidable difficulty in the above treatment is that above T_0 the minimum of free energy at $\eta = 0$ stays still when the temperature is either lowered or raised. This doesn't agree with the fact that in both types of transition a certain degree of short range order is commonly observed above T_c (or T_0), corresponding to the energy minimum associated with a small value of η. In order to improve it, one could add the linear term, $A\eta$, to the free expression, which is usually not used because of the constraint from the lattice symmetry relation in the phase transition. In the case of the first order transition, like melting, in which the lattice symmetry completely disappears in the liquid structure, the above theory must depart from the lattice symmetry problem, and in addition the transient state between the two minima, at $\eta = 0$ and $\eta > 0$, shown in Fig. 6.8b can be regarded as a fractional order in a heterogeneous phase, which is none other than the two phase mixture and can be described by the linear term, $A\eta$, as an average of the two states in the phenomenological treatment. As will be mentioned later, the statistical thermodynamical treatment of the tweed structure in the pre-transformation structure in alloys and oxide ceramics is very close to the two phase mixture, so that there exists a reason apart from the symmetry consideration to add a linear term in a macroscopic average free energy expression in the phenomenological theory. Real short range order in the microscopic scale often appears in the homogoneous parent phase above T_0. A typical example is the order–disorder alloys, in which local orders take place in the very smallest scale, that is the near neighbour relations. Such a case of short range order is substantially different from the above described, and, therefore, difficult to be exactly taken into account by the phenomenological theory. The statistical thermodynamical theory of short range order will be mentioned in Chap. 7.

6.2.2 Ehrenfest's Criterion and Landau's Picture in the G–T–η Diagram

For general theoretical understanding of the phase transition, it is quite useful to consider Ehrenfest's criterion for the order of phase transition, which has been regarded as containing a sort of contradiction in the case of the second order transition: According to Ehrenfest's definition, the first derivatives of free energy curves for two phases in problem must be continuous, or tangentially touching and crossing, at the transition point, as Fig. 6.9a shows, and, in order to fulfill this condition, at least one of the two G curves has to locally reduce the inclination, which is a local reduction of entropy near T_c with increasing temperature and certainly a contradiction. This apparent contradiction, however, can easily be removed by introducing the before mentioned order parameter, η. In order to explain it, let us first make a schematical G–T–η diagram for second order transition, as in Fig. 6.10a, which is a combination of Ehrenfest's picture and Landau's picture [6.3]; projection of the free energy curves for the parent ($\eta = 0$) and the new ($\eta = 1$) phase, respectively, to the G–T plane represents the former's picture, and the G–η plane projection of G curves in the cross sections perpendicular to T axis represents the latter's picture. The dot-and-

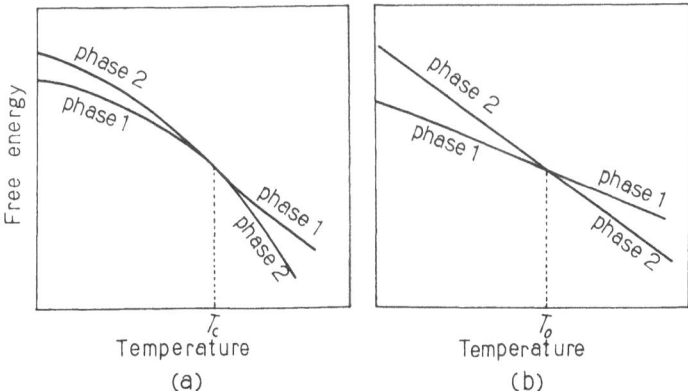

Fig. 6.9. A sketch of Ehrenfest's criterion for (a) the second order transition and (b) the first order transition

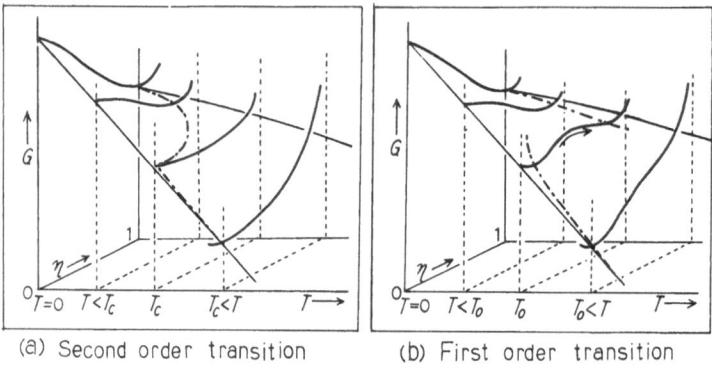

(a) Second order transition   (b) First order transition

Fig. 6.10. Free energy(G)-temperature(T)-order parameter (?) diagrams for (a) the second order transition and (b) the first order transition

dashed line on the curved G plane is the trajectory of the free energy minimum state at every temperature, which exhibits no jump of the state at T_c. Above T_c, η is not zero but has a small value corresponding to a short range order, which can be phenomenologically or mathematically expressed by adding a linear term, $A\eta$, to Landau's formula, as mentioned already. In the figure can be also given the projection of the trajectory to the T–η plane to visualize the before mentioned relation between the degree of order and the temperature. It is worth noting that the above mentioned contradiction of the local entropy reduction with increasing temperature near T_c actually doesn't exist, because in the G–T–η diagram of Fig. 6.10a one can draw the energy minimum trajectory across T_c without any local decrease of inclination or that of entropy with temperature increase.

Ehrenfest's criterion for the first order transition is seen in Fig. 6.10b, in which the two phases' free energy curves intersect at T_0 with different slopes, i.e., with a discontinuity of entropy. This can be combined with the Landau's phenomenological expression, as well as the second order transition, as in Fig. 6.10b. Different from Fig. 6.10a for the second order transition, there are two troughs on the free energy surface, between which the transition of first order takes place through an activation process. No contradiction of local reduction of entropy exists in this case. In the following section, the weak first order transition, which has a very small bump of the activation process, will be mentioned in relation to the martensitic transformation of the shape memory alloys.

6.2.3 Fine Heterogeneous Structure in the First Order Transition

It is quite instructive and useful to discuss the heterogeneous or mixed phase, which is associated with structural transition, in a finer scale than those described in the above phenomenological theory by employing a statistical theory, which enables us to count each atomic site in a fine hetero-phase structure. The result of this kind of calculation will be applied in the following paragraphs to explain various phenomena associated with the first order transition. It must be noticed that the short range order in Chap. 7 is considered in the otherwise homogeneous phase, while the local or partial order treated in the present chapter spreads to the range wider than the distance of nearest neighbour, giving rise to the heterogeneous phase so that the order structure treated here should be called the medium range or even the local long range order structure.

X-ray diffraction and electron microscope investigations have revealed that the heterogeneous pretransition structure of various materials is a mixed structure containing a number of embryo particles which could be regarded as already transmuted to the new phase. This kind of mixed structure often appears in various first order structural transformations; for instance, in the weak first order martensitic transformation of alloys, zirconia, ferroelectric materials, A-15 type superconductive materials, etc. Even in the strong first order transition, such an inhomogeneous structure could appear. For instance, the crystalline embryos may already appear in the molten phase near the solidification temperature or in the supercooled state, and by rapid cooling from the melt they could be retained as the medium range order in the amorphous structure. An example of the size of embryos determined or estimated by experiments will be shown later by a high resolution electron micrograph of an Fe–Pd martensite alloy. This gives us a rough idea of the number of atoms, say a few hundred, contained in one embryo particle in this case. Nevertheless, generally speaking, the size of embryos or the ordered regions ranges from the order of ten to one thousand atoms, depending upon the material, especially upon the difference in cohesive energy and other characters between the parent and the new phase, their boundary structure, the temperature, the pressure, the accompanying strain energy, etc. This will also be discussed later with some examples.

6.2.4 Statistical Calculation of Embryonic Structure

In this section, it is most important to treat the configuration of finely dispersed, tiny ordered regions by a statistical calculation as a basis of statistical thermodynamics for the quasi-mixed phase associated with the first order transition. The basic principle in the statistical calculation in the present problem is that there are a number of ways to distribute the embryos of the new phase of such a size as mentioned above in the parent phase, which would give rise to a noticeable contribution of configurational entropy to the free energy of the system.

As schematically shown in Fig. 6.11, consider the number of ways, W, to make n embryos with an average size of i atoms in the parent phase, in which $(N - ni)$ atoms will remain as the untransformed part, N being the total number of atoms in the system. After making $(p - 1)$ embryos, $N - i(p - 1)$ ways are

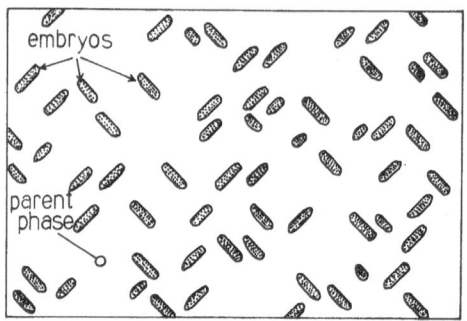

embryos

parent
phase

Fig. 6.11. Schematic presentation of distribution of a new phase particles

possible to assign the position of the first atom of the pth embryo among the untransformed atoms. The possibility of associating the neighbouring second atom in the embryo is $[1 - i(p - 1)/N]$. By repeating the associations $(i - 1)$ times, the pth embryo will be completed. The order of assigning the atoms in each embryo does not need to be taken into account. Therefore, the number of ways to form the pth embryo is

$$w_p = [N - i(p - 1)][1 - i(p - 1)/N]^{i-1} = i^i \xi / N^{i-1} \cdot [N/i - (p - 1)]^i, \quad (6.2)$$

where ξ is the number of shape diversity of embryos to take into account their different shapes. In the case of polymer solution, instead of solid materials, ξ will be a large number and will play an important role, because of the highly flexible shape of each molecule, as is seen in Flory's calculation for polymer solution. But, the embryos in the present case are all alike, usually having the shape of a small sphere or thin platelet, so that ξ is not far from unity. When thin platelet embryos can lie on any of six $[1\,1\,0]$ planes in a cubic parent lattice, ξ will be at least six. Finally, using the above calculated, w_p, the configurational entropy arising from the number of ways to distribute n embryos in the parent lattice is given as

$$k \ln(w) = k \ln(n!^{-1} \prod_{p=1}^{n} w_p) = k \ln \left[\frac{i^{ni} \xi^n}{n! \, N^{n(i-1)}} \left(\frac{(N/i)!}{(N/i - n)!} \right)^i \right]$$

$$= k \left[-n \ln \left(\frac{n}{N} \right) - (N - ni) \ln \frac{N - ni}{N} + n \ln \xi - n(i - 1) \right]$$

$$= -R \{ (C/i) \ln(C/i\xi) + (1 - C) \ln(1 - C) + C[1 - (1/i)] \}, \quad (6.3)$$

where k is the Boltzmann constant, R the gas constant, and, in the last line, the total concentration of new phase atoms contained in all embryos is written as $C = ni/N$, see [6.3]. Note that, when the above calculation is applied to the problem of mixing of i atoms molecules and mono-atoms, the well-known entropy of mixing, $S_{mix} = -R[(C/i) \ln C + (1 - C) \ln(1 - C)]$, is obtained, and when i is taken as 1 in the last line of (6.3), the ordinary entropy of mixing of two species of single atoms, $S_{ord} = -R[C \ln C + (1 - C) \ln(1 - C)]$, is obtained. On the other hand, when the shape diversity, ξ, is calculated for winding chain

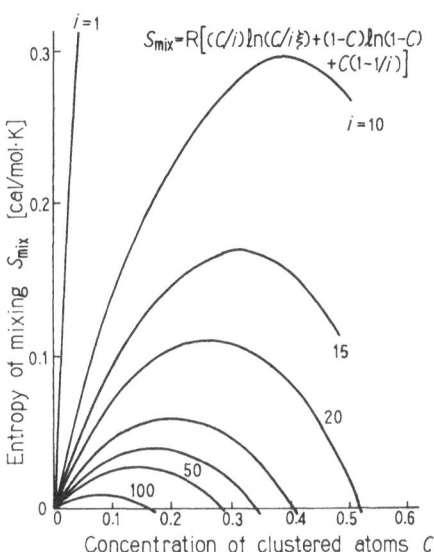

$$S_{mix} = R\left[(C/i)\ln(C/i\,\xi) + (1-C)\ln(1-C) + C(1-1/i)\right]$$

Fig. 6.12. Entropy of mixing of i-atoms embryos (clusters) of the new phase in the parent phase. n is the number density of embryos, i the average number of atoms contained in one embryo, and $C = ni/N$ the total concentration of the clustered atoms

molecules, the above formulation leads to the entropy of a mixing of a polymer solution, which was already deduced by Flory as mentioned before.

It is worthy to examine the size dependence of the above obtained entropy of mixing of embryos and compare it with that of the ordinary atomic mixing, S_{ord}. With the size i increasing, the entropy of embryo mixing given by (6.3) decreases; for instance, as shown in Fig. 6.12, when $i = 20$ and $C = 0.1$, it becomes a tenth of the ordinary entropy of mixing, and when $i = 100$ and $C = 0.1$, it is only a fiftieth. In addition, it always reaches a maximum with the increasing concentration and then decreases down to zero, beyond which the negative value of entropy loses its physical meaning. This is quite natural, because by increasing the number of embryos the whole volume will become so tightly occupied by them that there will no longer be free choice in placing the embryos. This may be called the configurational catastrophy, which is not important when the concentration C is low enough. Whether or not the above calculated entropy of embryo mixing plays an important role depends on the amount of enthalpy change and the temperature of phase transformation, as will be shown later.

6.3 Martensitic Transformation of Metals and Alloys

6.3.1 Martensitic Transformation of Steel

To find examples of useful applications of phase transformations in metals and alloys is not easy. On the contrary, phase transformation often hinders the application so that large efforts are necessary to suppress or distinguish the phase transformation in metals and alloys in industrial uses. For instance, metallic uranium to be used for nuclear fuel in atomic reactors has phase transformations,

$$\alpha(\text{orthor.}) \rightleftharpoons \beta(\text{tetrag.}) \rightleftharpoons \gamma(\text{bcc}) \rightleftharpoons \text{Liquid.}$$

940 K   1048 K   1406 K

Among them, the α–β transformation is associated with a large volume and anisotropic structural change, which deforms the solid material during heat treatment or heat cycles. In addition, during the nuclear fission reactions of ^{235}U in α-uranium, the material gets a self irradiation effect by the produced high energy fission fragments and fast neutrons and makes a large dimensional change or cracking. Therefore, to use α-uranium as a fuel element, as it is, is not practical as the structure must be transformed to another crystal structure by alloying or making an oxide. Today, the uranium fuel elements are usually bars or plates with fcc structure, being alloyed with aluminium or niobium, or oxide pellets of UO_2 form. A similar trouble with phase transitions occurs in plutonium, in which five structural changes take place before melting, as follows:

$$\alpha(\text{monoc}) \rightleftharpoons \beta(\text{bc. monoc}) \rightleftharpoons \gamma(\text{orthr}) \rightleftharpoons \delta(\text{fcc}) \rightleftharpoons \delta'(\text{tetrag}) \rightleftharpoons \varepsilon(\text{bcc}) \rightleftharpoons \text{Liquid.}$$

396 K   475 K   590 K   729 K   755 K   913 K

Historically, the largest disaster due to the transformation may be the sudden disintegration and dusting of tin wares in the Queen Ekaterina II's royal court by tin pests in a cold Russian winter in the eighteenth century, which was no other than the α–β transition of tin. To mention further examples of undesirable phase transitions is probably unnecessary.

As above exemplified, most of the structural phase transformations of metals were not desirable in daily and industrial use in the past. A well-known exception is the before mentioned martensitic transformation of steel which was an excellent method of strengthening steel, especially for swords and spears, used more than one thousand years ago[5]. This old technique in steel work was suddenly found to be useful again in a new material, that is the shape memory alloys. Before mentioning about the shape memory alloys and the physical aspects of the mechanism of shape memory, the martensitic transformation itself must be introduced rather precisely. The martensitic transformation is a diffusionless structural transformation of first order mainly by shear deformation of successive atomic planes[6], which starts at M_s temperature and propagates in the parent phase, producing a peculiar surface relief pattern known as the crystallographic martensite structure, as shown in Fig. 6.13. The transformation doesn't

---

[5] The Japanese swords of the highest quality were already made by fold-forging and water quenching in the middle of the Heian era, which was about 1000 years ago.

[6] Even the lattice deformation from cubic to tetragonal with very small dimensional changes, which often occurs in martensitic transformation, can be decomposed into two uniform shear deformations crossing each other, although there is no evidence for separative occurrence of the two components.

25µm

Fig. 6.13. Optical micrograph of martensite structure of Fe-29.8%Pd alloy (by courtesy of R. Oshima)

finish until a lower temperature M_f is reached, as already mentioned before by using Fig. 6.3, since the lattice deformation for transformation, for instance from fcc to bcc, shown in Fig. 6.3, unavoidably induces a strong constraint from the parent lattice, which hinders or arrests the nucleation or further propagation of the martensite phase. This means also that, at the expected transition temperature, T_0, in Fig. 6.9b, the phase transformation will not take place but needs an excess Gibbs' energy or motive force to surmount the extra strain energy arising from the above lattice constraint in the first nucleation of martensite. Accordingly, the M_s temperature must be lower than T_0, as is shown in Fig. 6.14, in which the free energy of martensite, G_m, plus extra strain and boundary energy for nucleation, g_m, balances the free energy of the parent (austenite) phase, G_a, at M_s. M_f is naturally even lower as mentioned before. Likewise, in the reverse transformation from martensite to austenite, a similar extra energy g_a, which may be nearly the same as g_m, is required for starting the austenite phase, as is also shown in the same figure. A study on the alloy concentration dependence of M_s and A_s temperature of Fe–Ni alloys by *Kaufman* and *Cohen* [6.1] seems to perfectly coincide with the above idea. They carried out a thermodynamical calculation, without taking into account the nucleation energy, to compare the free energies of the martensite and austenite phases of the alloys, and determined the equal free energy line, i.e., T_0, as shown by a full line in the middle of Fig 6.15. Agreement between the T_0 line and the experimentally determined midpoint

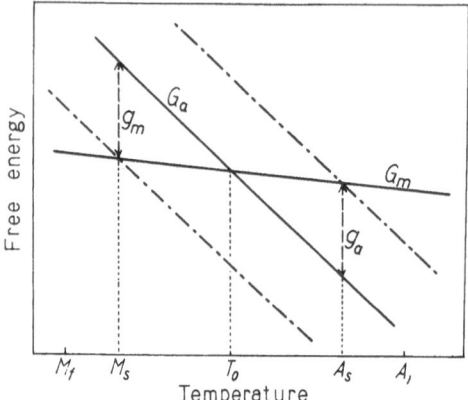

Fig. 6.14. Schematic presentation of the relations between the free energy curves, G_a and G_m, and the transformation temperatures

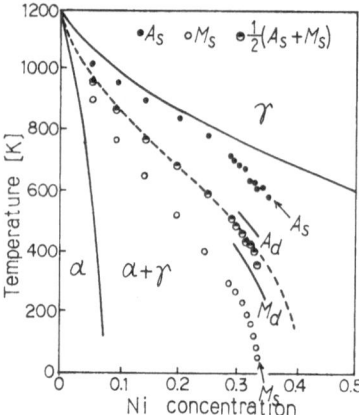

Fig. 6.15. Agreement between the thermodynamically calculated T_0 and the experimentally determined $(M_s + A_s)/2$ for Fe–Ni alloys

values between M_s and A_s, i.e., $(M_s + A_s)/2$ represented by full circles, is excellent, verifying the validity of the above explanation of the hysteresis of martensite transformation. In this case, the necessary driving force g_m in Fig. 6.14 was obtained to be $6.09(T_0 - M_s)$ J/mol, depending on the nickel concentration, and in the case of Fe–C alloys it was about 1300 J/mol independent from the carbon concentration. The theory of nucleation of martensite is an interesting problem of physics in relation to phase transition, but it will not be described in this article.

6.3.2 Lattice Deformation in Martensitic Transformation

As for the lattice deformation in phase transformation, another interesting subject exists. Let us consider, as an example, a phase transformation from an orthorhombic to a monoclinic structure, as schematically shown by the outer shapes of the material before and after the transformation in Fig. 6.16a and b,

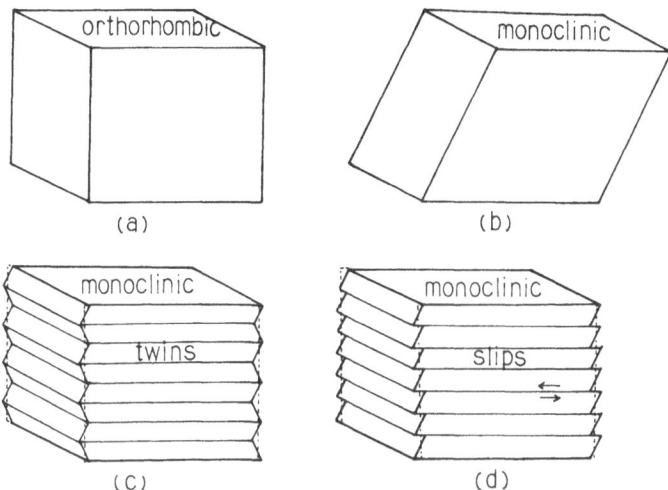

Fig. 6.16. Structural transformation from (a) orthorhombic to (b) monoclinic without any lattice invariant deformation, (c) with twinning deformation and (d) with slip deformation

respectively. This deformation from a square to an oblique one will certainly induce the before mentioned distortion of the surrounding parent lattice, which constrains conversely the transformed part. If the transformation is associated with multiple twinning as the Fig. 6.16c shows, the outer shape of the transformed part will roughly keep the original square one except for the end faces of the body, as the dotted lines show. Another way to eliminate the outer shape deformation is to produce laminar slips by normal dislocations along the shear plane to restore the shear, as shown in the Fig. 6.16d. Both types of plastic deformation accompanying the martensitic transformation are actually often observed: Fig. 6.17 is a transmission electron micrograph showing the microtwinning structure in a martensite plate of Kovar (Fe–27Ni–17Co) alloy.

Crystallography of martensitic transformation is another important problem. The oldest typical example is found again in the fcc–bcc martensitic transformation of steel. An fcc austenite structure containing an interstitial carbon atom in an octahedral interstice is shown in Fig. 6.18. In 1924, A. C. Bain suggested that by stretching the x and y axes of the fcc lattice by about 11% and contracting the z axis by about 21% in the transformation, a bcc or bct lattice would be obtained, as broken lines show, and the carbon would be automatically transferred to the bcc octahedral interstice, stretching the z axis of the new lattice, as expected from some experimental facts such that the tetragonality or the axial ratio, c/a, of the martensite lattice is proportional to the carbon concentration. This Bain's distortion process, however, couldn't be employed to explain the mechanism of the lattice transformation, because the lattice orientation relationship between the fcc parent and the bcc martensite,

$$(0\,0\,1)_f /\!/ (0\,0\,1)_b \quad \text{and} \quad [1\,0\,0]_f /\!/ [1\,1\,0]_b,$$

expected from the above simple elongation and contraction of axes was different

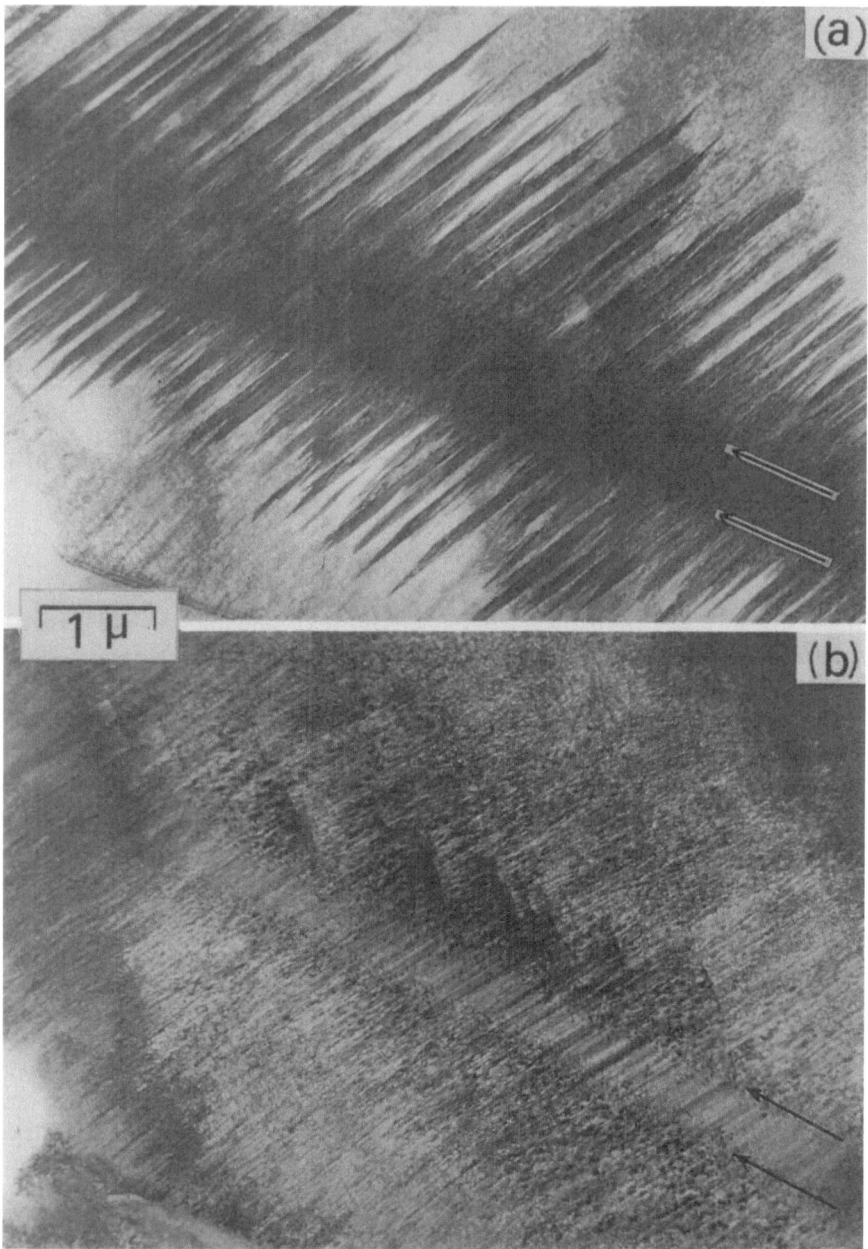

Fig. 6.17. Transmission electron micrograph of microtwinning structure in a martensite plate of Kovar (Fe–27Ni–17Co) alloy: (a) and (b) are taken in different diffraction conditions. (By courtesy of K. Shimizu)

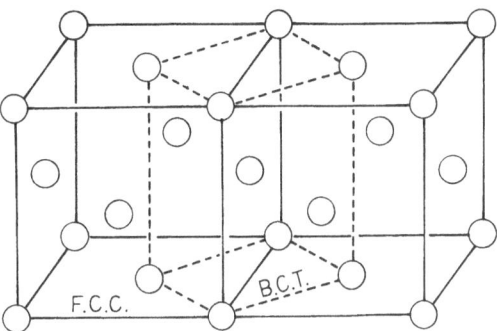

Fig. 6.18. Bain's suggestion for the martensite transformation of steel from fcc (austenite) to bct (martensite)

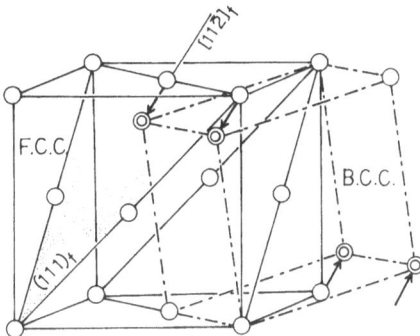

Fig. 6.19. A sketch of the deformation process of fcc–bcc(bct) transformation by the $a/12$ $[1\,1\,2]_f/(1\,1\,1)_f$ uniform shear associated with other small deformations

from what were actually observed. In 1930, *Kurdjumov* and *Sachs* [6.4] found by using the X-ray diffraction technique that the orientation relationship in the transformation of carbon steel was

$$(1\,1\,1)_f /\!/ (0\,1\,1)_b \quad \text{and} \quad [1\,0\,\overline{1}]_f /\!/ [1\,1\,\overline{1}]_b,$$

and a few years later *Nishiyama* [6.5] found another relationship in Fe–Ni alloys,

$$(1\,1\,1)_f /\!/ (0\,1\,1)_b \quad \text{and} \quad [1\,1\,\overline{2}]_f /\!/ [0\,1\,\overline{1}]_b.$$

Later, *Greninger* and *Troiano* [6.6] found that the above orientation relationships were actually a few degrees off from the exact parallelisms. Based upon their respective experimental results, *Kurdjumov* and *Sachs*, and *Nishiyama* proposed similar mechanisms of transformation, both of which consisted of a common main uniform shear of $19°28'$ in $[1\,1\,\overline{2}]_f$ direction in $(1\,1\,1)_f$ plane, the second uniform shear and axial expansions or contractions being somewhat different between them.

How to reach the bcc (or bct) martensite lattice from the fcc austenite lattice by operating the above $[1\,1\,\overline{2}]_f/(1\,1\,1)_f$ main shear and other necessary deformations is illustrated in Fig. 6.19. The displacement vector of the main shear, $a/12 \cdot [1\,1\,\overline{2}]_f/(1\,1\,1)_f$ where a is the fcc lattice constant, shown by a thick arrow, corresponds to half of the Burgers vector of half (or partial) dislocation, $a/6 \cdot [1\,1\,\bar{2}]_f/(1\,1\,1)_f$, which would produce a stacking disorder, such as AB-CACABC ..., in the regular ABCABCAB ..., stacking of the close packing $(1\,1\,1)_f$ planes, if it worked instead of the uniform shear. New atomic arrangement produced by the uniform shear is shown by small open circles in the figure, from which is readily seen a prototype of the bcc martensite structure as the broken lines show. The orientation relationship in the figure is $(1\,1\,1)_f//(0\,1\,1)_b$ and $[1\,1\,\bar{2}]_f // [0\,1\,\bar{1}]_b$, in accordance with Nishiyama's relation, although the above mentioned additional deformation components are necessary to complete the transformation.

The uniform shear deformation cannot always be treated by the theory of dislocation, but there is a close relationship between the two deformation modes and they can be linked with the shuffling mode to generally discuss the mechanism of structural transformations of various types. For instance, cobalt and some Co–Ni and Fe–Mn alloys transform from fcc to hcp on cooling, which is regarded as the result of operation of the $a/6 \cdot [1\,1\,\bar{2}]_f/(1\,1\,1)_f$ half dislocation on every second $(1\,1\,1)_f$ plane, changing the stacking mode from ABCABC ... to ABABAB .... On the other hand, the bcc–hcp transformation of Titanium and Zirconium can be carried out by the alternate shear motion of $(0\,1\,1)_b$ (or $(0\,0\,0\,1)_h$) successive planes in the opposite directions, $+ a'/6 \cdot [0\,1\,\bar{1}]_b$ and $- a'/6 \cdot [0\,1\,\bar{1}]_b$, where a' is the bcc lattice constant, which corresponds to the vector $\pm a/12 \cdot [1\,1\,\bar{2}]_f/(1\,1\,1)_f$, mentioned before. This is called the shuffling mode and can also be described by the $a/6 \cdot [1\,1\,\bar{2}]_f/(1\,1\,1)_f$ sliding of every second $(1\,1\,1)_f$ plane in the ABABAB ... stacking, which is no other than the hcp structure. The mechanism to reach the bcc structure from the hcp structure is shown in Fig. 6.20, in which a prototype bcc cell is drawn by broken lines.

6.3.3 Martensitic Transformation of β-Phase Alloys

Various types of martensitic transformation starting from the bcc structure are found in a large number of alloys generally called the β-phase alloys. They

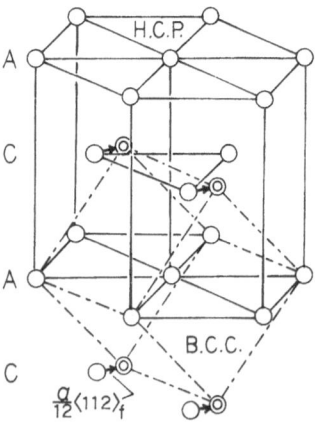

Fig. 6.20. A sketch of the deformation process of hcp–bcc transformation by shuffling of basal planes

are, for instance, Au–Cd, Au–Ag–Cd, Cu–Al–Zn, Cu–Al–Ni, Ni–Al, Ni–Zn–Cu, and Ti–Ni. The last one, with Ni content of about 50%, is a typical shape memory alloy and the others more or less have the shape memory effect too. Their martensite structures are various, but can be regarded in general as the stacking modifications, such as 18R, 9R and 2H (hexagonal), produced by modified shuffling modes of $(0\,1\,1)_b$ planes accompanied by other additional small deformation components. An example shown in Fig. 6.21 is of Au–Cd alloy which transforms from bcc (ordered DO_3) to 18R with a shuffled stacking sequence of $(0\,1\,1)_b$ planes corresponding to a $(1\,1\,1)_{fcc}$ or $(0\,0\,0\,1)_{hcp}$ stacking modification.

Martensitic transformations of β-phase alloys are second order-like, or more exactly, weak first order, as will be mentioned later. Another group of the weak first order transformation can be seen in the fcc–fct transformation of In–Tl, Fe–Pd and Fe–Pt alloys, which is similar to the cubic–tetragonal transition of many intermetallic and non-metallic compounds, such as Nb_3Sn, $BaTiO_3$, etc. The lattice softening phenomenon associated with the second order or second order-like transitions are closely related with the above mentioned shuffling mode and will be discussed in the next section. In the present paragraph, however, let us complete a general crystallographic map of all above introduced martensitic transformations in terms of the three deformation modes, uniform shear, dislocation and shuffling [6.7]. In Fig. 6.22 are shown all the above described crystal structures connected to each other by the main deformation modes necessary for martensitic transformation between them. For instance, from fcc to bcc (bct), the process of deformation can take either the route of

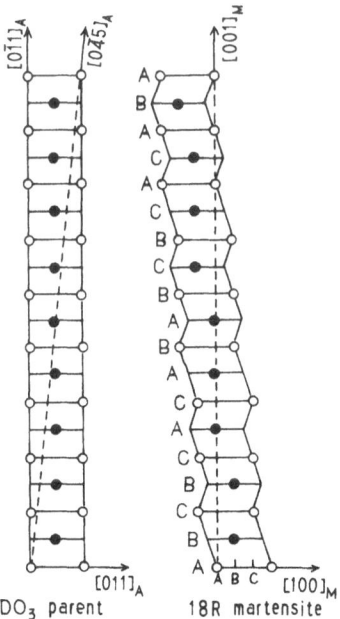

Fig. 6.21. A schematic representation of the lattice deformation of bcc-18R transformation by $(0\,1\,1)_b$ plane shuffling

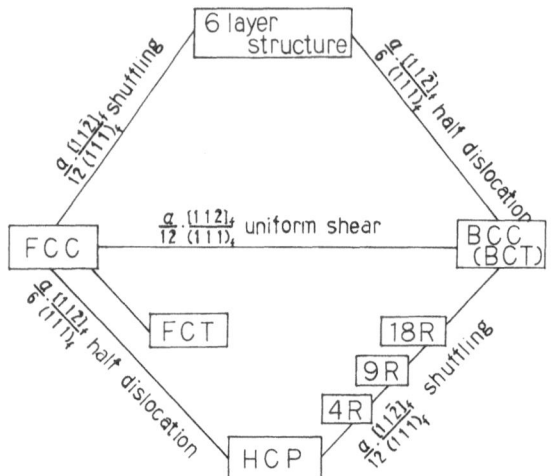

Fig. 6.22. A generalized diagram of lattice deformation paths connecting the fcc, bcc(bct), hcp and their modificatioin structures by the half dislocation, uniform shear and shuffling mechanism

$a/12 \cdot [1\,1\,\bar{2}]_f/(1\,1\,1)_f$ uniform shear or of $a/6 \cdot [1\,1\,\bar{2}]_f/(1\,1\,1)_f$ half dislocation on every second plane plus $a/12 \cdot [1\,1\,\bar{2}]_f/(1\,1\,1)_f$ regular shuffling through the hcp structure. Other transformations can be followed by similar ways. The bcc–bct transformation is not observed, but the fcc–fct one is quite common and generally appears as second order or second order-like. When the above half dislocation and shuffling mechanism are applied in the reverse order, that is, shuffling first and then half dislocation, one will get a special six layer structure of $(1\,1\,1)_f$ or $(0\,1\,1)_b$ stacking as an intermediate phase. Actually when an Fe–Mn–C alloy was quenched to below 170 K for martensitic transformation, a special distorted structure exhibiting the 1/6 satellite diffraction spots was obtained and it further changed to a normal bcc martensite when warmed up to room temperature [6.8]. This uppermost route in the map proposed by the author [6.7] has not been further studied and is still controversial. Nevertheless, in this way the linkage map of martensite transformation is completed. The paths in the middle and upper part are first order and athermal, while those in the lower part are second order or second order-like and thermoelastic. The shape memory effect appears along this route.

6.4 Shape Memory Effect and Premartensitic Phenomena

6.4.1 Mechanism of Shape Memory

The mechanism of the shape memory effect of alloys is closely related with the thermoelastic martensite transformation which has a very small hysteresis loop as in Fig. 6.4, the mode of mechanical deformation like twinning and slips shown

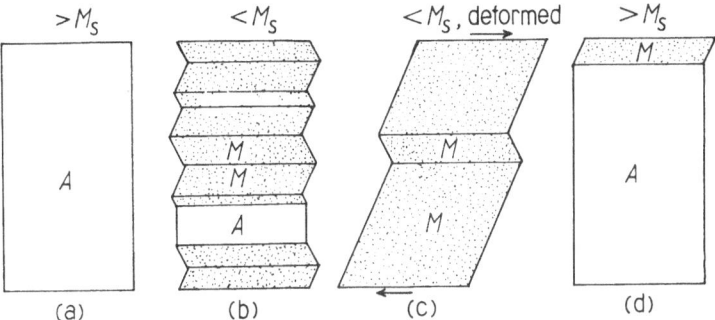

Fig. 6.23. A simple sketch to show how the martensitic transformation works in the shape memory effect

in Fig. 6.16, and the deformation mode of transformation like shuffling in Fig. 6.21.

Macroscopically, it is easy to understand how the shape memory effect takes place in the working alloys, by using a simple sketch in Fig. 6.23. Let us first take a bar or any other shape of a shape memory alloy made of, for instance, Ti–Ni or Cu–Al–Ni, as shown in Fig. 6.23a. By cooling it down to below M_s, the transformation will occur and the martensite phase will easily take over the whole volume as in the Fig. 6.23b, producing at the same time twins with many variants which can keep the original outer shape by cancelling the deformations of crystals as already explained by Fig. 6.16a–c. When it is mechanically deformed by an applied stress as Fig. 6.23c shows, this material can relax the produced internal stresses by the movement of twin boundaries or the impingement among the twin variants instead of the normal slip motion by dislocations shown in Fig. 6.16d. The twin variants impingement is clearly seen by three successive photographs of a Cu–Zn–Ga alloy specimen under the increasing uniaxial applied stress, as in Fig. 6.24a–c [6.9]. By warming the transformed and mechanically worked material to above M_s (A_f), the reverse transformation will take place, by which all existing martensite variants will go back exactly to the original austenite through their respective unique transformation paths so that the original shape of the material will be restored as Fig. 6.23d shows. This behaviour to reproduce the original shape is called the shape memory effect.

By enlarging the pictures of the above process, as in Fig. 6.25, it may be easy to analyze the mechanism microscopically, since the four figures in Fig. 6.25 correspond exactly to the four in Fig. 6.23, respectively. It must be noted that in one of the twin variants in Fig. 6.25b or c, the crystal structure is a shuffled one, like in Fig. 6.21, and the twin boundary motion can easily be done by reshuffling the martensite structure. It is now clear that, in order to realize the shape memory effect, the temperature range of transformation, i.e., between A_f, A_s, M_s and M_f must be small, twinning must be easier than dislocation motion in mechanical deformation, the twin boundary motion to switch from one variant to another must be easy, and the reverse transformation must follow the unique

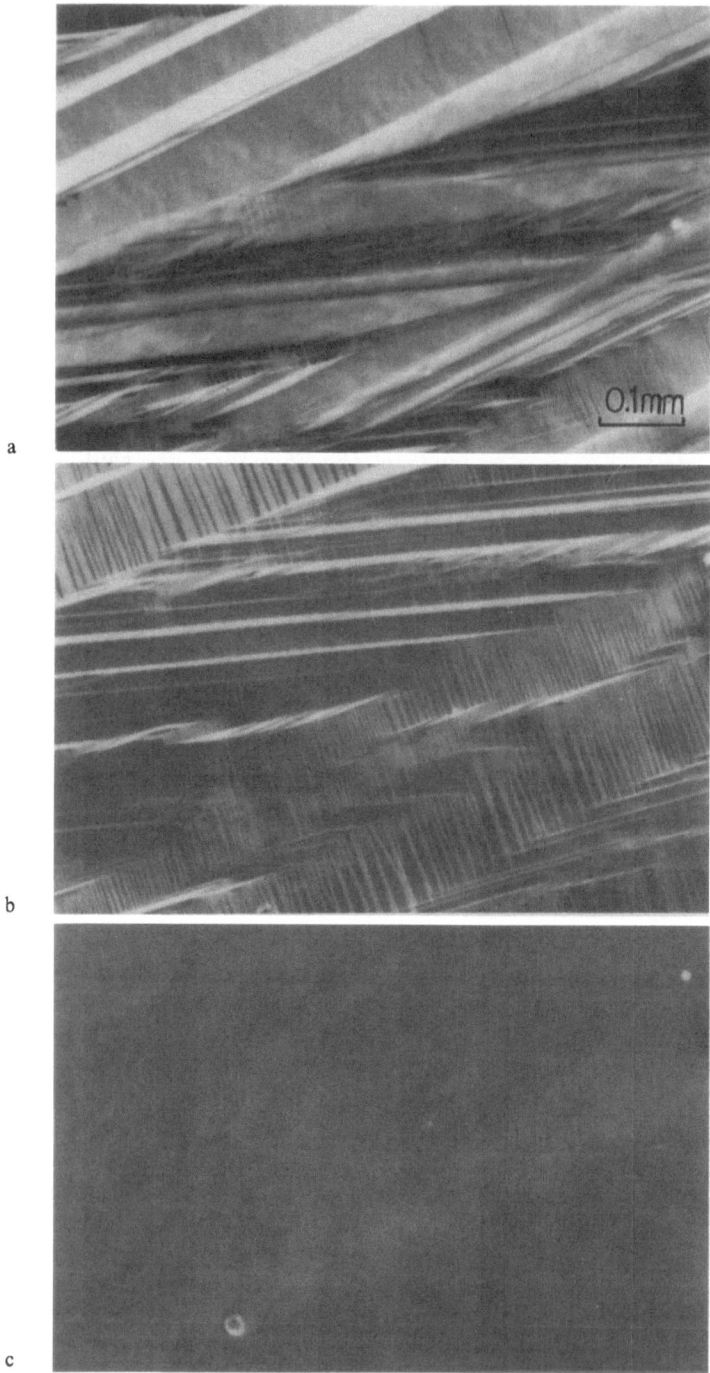

Fig. 6.24a–c. Three successive optical micrographs showing the twin variants impingement in a Cu–Zn–Ca alloy specimen under the increasing uniaxial applied stress (by courtesy of T. Saburi)

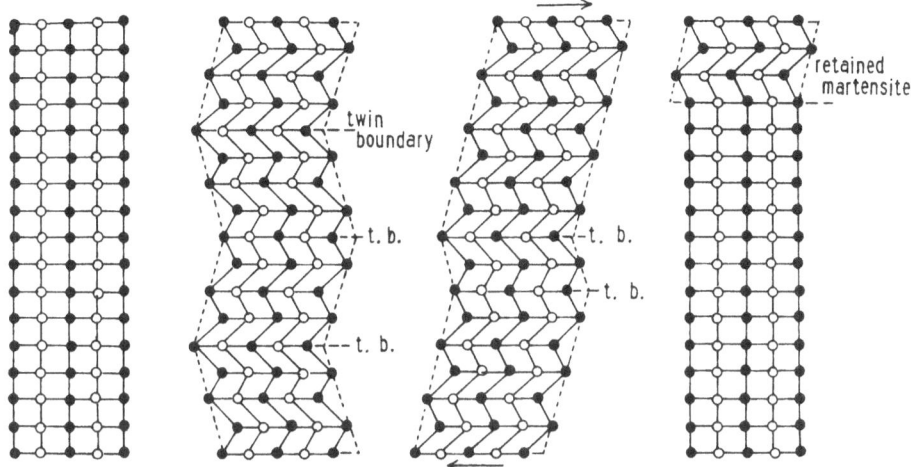

(a) austenite (b) martensite (c) martens. deformed (d) austenite

Fig. 6.25a–d. A microscopic picture of shape memory effect corresponding to Fig. 2.23

reversible path. The thermoelastic martensite transformations, which belong to the group in the lower part of the map of Fig. 6.22, where shuffling or similar soft type lattice deformations take place, may fulfill the above requirements. Technologically, there are some more requirements to develop the useful shape memory alloys: For instance, they must be perfectly shape restoring, fatigue resistant, capable of large mechanical deformation, desirable transformation and working temperatures, etc., and sometimes it is further required to have the two way memory, which means that under the cyclic temperature change the shape must change in cycle between the high temperature one and the low temperature one. Nevertheless, the mechanism underlying various functions of shape memory is not much more than that described in this paragraph. For further study of shape memory alloys, it is recommended to read more specified articles or books introduced in the reference [6.10].

6.4.2 Superplasticity and Ferroelasticity

Another interesting and useful behaviour closely related with the shape memory effect is superelasticity or rubber-like deformation as large as several percent in strain. Figure 6.26 shows how largely a thermoelastic $Au_{26}Cu_{30}Zn$ alloy deforms in a strange manner by a stretching load, and fully restores the length when unloaded [6.11]. This behaviour appears above M_s, where the martensitic transformation usually doesn't take place. When the stress is applied, in order to release the elastic strain energy, martensitic transformation will occur if the critical stress for plastic deformation by dislocations, σ_{cd}, is higher than that for

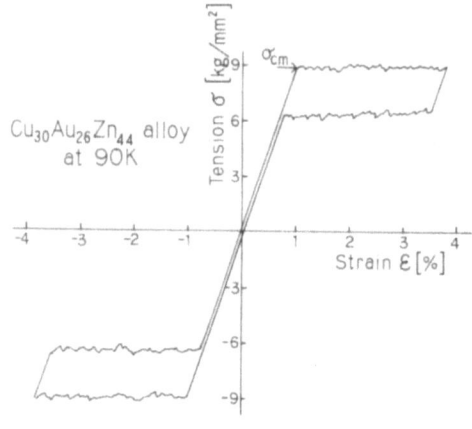

Fig. 6.26. A stress–strain diagram of the thermoelastic Cu–Au–Zn alloy showing superelasticity (by courtesy of N. Naka-nishi)

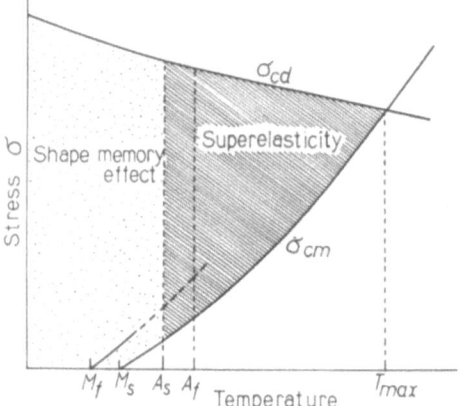

Fig. 6.27. An applied stress-temperature diagram to show the region of stress induced martensite which gives rise to super-elasticity

martensitic transformation, σ_{cm}, as shown in Figs. 6.26, 27. This is the stress induced martensitic transformation. It doesn't appear at temperatures below M_s, where normal martensite appears instead as a stable phase, and once it appears at a temperature between M_s and A_f, it doesn't fully disappear when unloaded. On the other hand, σ_{cm} increases with the temperature increase, finally crossing σ_{cd} at T_{max}, as is shown in Fig. 6.27, above which the stress induced martensite doesn't appear. Accordingly, in the hatched area in the figure, the stress induced martensite appears under a stress between σ_{cd} and σ_{cm} and disappears when the stress is removed, giving rise to the restoration of deformation. This is the mechanism of superelasticity, arising from easy generation and retrogression of the stress induced martensite. As long as the plastic deformation by dislocation motion is fully suppressed, the superelasticity is repeatable as well as the shape memory effect. The easy generation of martensite is by the lattice softening which occurs near T_0 or M_s as will be mentioned in the following.

A similar behaviour called the ferroelasticity appears below the transformation temperature, where the whole volume of the material is occupied by the martensite phase with various variants. Just like the case of shape memory effect, when a tensile stress is applied, deformation takes place by the twin boundary motion as already mentioned. With the stress released and then reversed to compression the deformation compliance doesn't trace the original forward process back but makes a delay, ultimately producing the stress–strain hysteresis loop in the stress cycle just like the $B–H$ loop of ferromagnetic material. Actually, the ferromagnetic and ferroelastic hysteresis loops equally arise not from the phase transition itself but from the domain boundary motion in the ferro-phase. It is worth noting that there is also an analogy between the stress induced martensite and the electric field induced ferroelectricity: When an electric field is applied to a ferroelectric crystal slightly above the ferro–para transition point, for instance above 120°C for $BaTiO_3$ as mentioned earlier, the ferro phase is induced and a small loop quite similar to that in Fig. 6.26 appears in the applied field vs. the spontaneous polarization curve.

6.4.3 Lattice Softening and Soft Phonon Mode

From the discussions in the preceding paragraphs, it is clear that the crystal lattice softens, especially against a certain shear mode, before the phase transformation of second order or weak first order takes place. This may mean that the lattice becomes unstable as the transition temperature is approached and is already inclined to change the crystal structure. Actually, the stress induced martensitic transformation, which occurs at an extremely low stress level near T_0, strongly suggests that the lattice is soft enough to be deformed by martensitic transformation instead of dislocation motion. In the case of Nb_3Sn, $BaTiO_3$ and β-phase alloys, the main component of lattice deformation is $\langle 1\,1\,0 \rangle/\{1\,\bar{1}\,0\}$ shear, which therefore must be the shear mode to be softened and the reduction of the corresponding elastic constant, $C' = (C_{11} - C_{12})/2$, must be observed near the transition temperature. Figure 6.28 shows that the two elastic constants, C_{11} and C_{12}, of Nb_3Sn measured by ultrasonic method come close to each other as the temperature is lowered and finally meet at T_c, which means the complete softening, $C' = 0$ [6.12]. Another example of lattice softening is shown in Fig. 6.29, where C' of an In–Tl alloy has a very small value at T_0 [6.13]. Generally speaking, the phase transitions of second order, for instance, of Nb_3Sn, exhibit full softening at T_0, those of weak first order also exhibit softening but not fully, as in the case of In–Tl alloys, and there is no evidence of lattice softening in the normal sharp first order transitions except in the case of melting. This is in good agreement with the Landau's phenomenological theory mentioned in Sect. 6.2. Corresponding to the flat $G–\eta$ free energy curve near T_c in Figs. 6.8a, 10a, the potential well for the atomic displacement for lattice deformation in the second order transition must also be flat for certain amplitude range near T_c, as Fig. 6.30a shows, allowing a perfectly soft atomic or

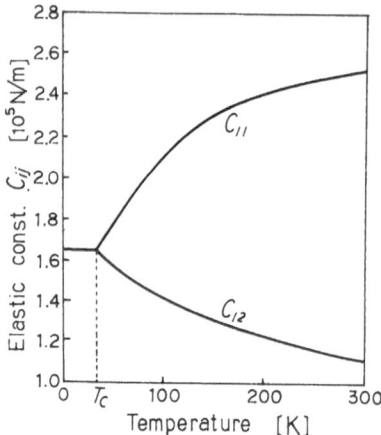

Fig. 6.28. Temperature dependence of two elastic constants, C_{11} and C_{12}, of Nb$_3$Sn measured by ultrasonic method. Coincidence of the two constants at T_c means the complete softening of the lattice deformation mode, 110/110, since $C' = (C_{11} - C_{12})/2$. 6.12]

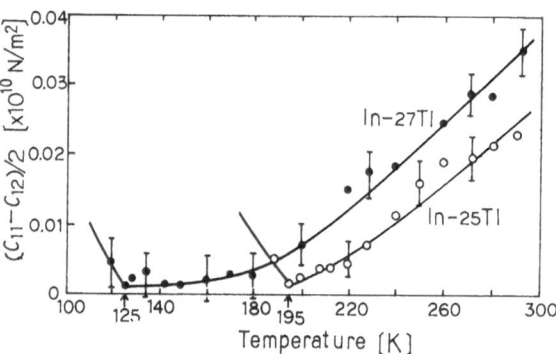

Fig. 6.29. Temperature dependence of the elastic constant, $C' = (C_{11} - C_{12})/2$, of In–Tl alloys, showing lattice softening [6.13]

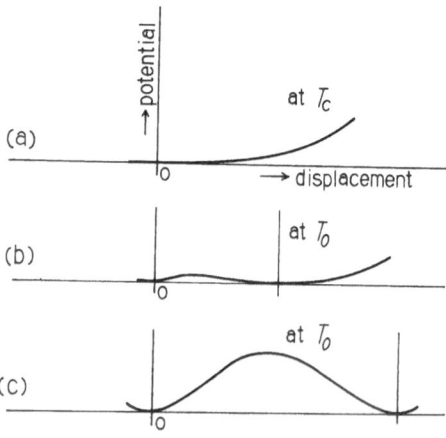

Fig. 6.30. Sketches of atomic potentials for displacement in phase transitions of (a) second order, (b) weak first order and (c) strong first order

lattice vibration and a large fluctuation of the state. This soft mode of lattice vibration is called the soft phonon mode, and, irrespective of its quantum or classical aspect emphasized, it is an essentially important subject for the study of the second order transitions. Likewise, the weak first order transition must have a fairly flat potential near T_0 to induce softening. Only a difference from the case of second order transition is that at T_0 a small shallow double well potential appears, as in Fig. 6.30b, giving rise to a tiny jump for the weak first order transition, or even the state of two phase mixture around T_0, as will be mentioned later. Since the pure $D\eta^4$ state in the free energy expression of (6.1) is not expected in this case, lattice softening is naturally imperfect. On the other hand, the strong first order transition, which has no flat potential against the atomic displacement at any temperature, as Fig. 6.30c shows, will not exhibit lattice softening or a soft phonon mode at all. The above condition generally well agrees with the experimental results on the soft phonon mode introduced below.

The soft phonon mode can be observed by various techniques; for instance, by neutron inelastic scattering, infrared spectroscopy, electron spin resonance, Brillouin scattering and Raman scattering, which must be chosen for different purposes and different materials. Consider, for instance, the transverse acoustic (TA) wave propagating in [1 1 0] direction in a crystal with the density ρ and the above mentioned elastic constant C'. Since the sound velocity v is given by $(C'/\rho)^{1/2}$, the frequency ω is related with the wave vector q by

$$\omega = v|q| = [(C_{11} - C_{12})/2\rho]^{1/2}|q|, \tag{6.4}$$

which is the dispersion relation of the TA phonon necessary to clarify its soft mode. When an incident neutron or laser light with a wave vector k_i and a frequency ω_i interacts with the phonon in the problem and scattered, k_i and ω_i will be changed to k_s and ω_s, respectively. The momentum conservation and the energy conservation are written, respectively, as

$$\hbar k_i - \hbar k_s = \hbar q \quad \text{and} \quad \hbar\omega_i - \hbar\omega_s = \pm\hbar\omega. \tag{6.5}$$

The former relation is simply drawn in Fig. 6.31, and by using the scattering angle θ, we obtain

$$|q| = 2|k_i|\sin(\theta/2), \tag{6.6}$$

by which the wave vector q can be measured, and with the measurement of the energy change, $\hbar\omega$, the q–ω dispersion relation can be found. Optical scattering

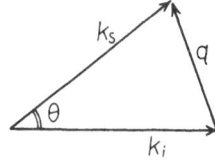

Fig. 6.31. Geometrical relationship between the incident probe wave vector k_i, the scattered wave vector k_s, the phonon wave vector q and the scattered angle θ

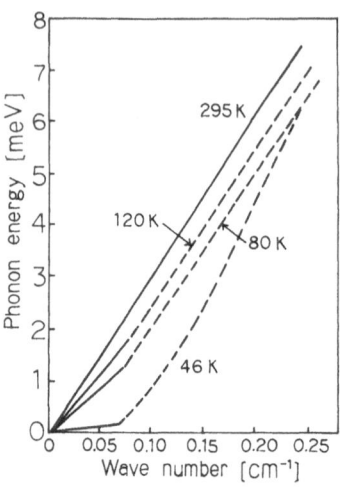

Fig. 6.32. TA$_1$ phonon dispersion relation for Nb$_3$Sn measured by neutron inelastic scattering. Note that near the transition temperature, $T_c = 45$ K, the curve goes down and C' comes close to zero [6.14]

methods are suitable to measure the dispersion in the region of $q \simeq 0$ and neutron scattering is useful to find the dispersion for q in the whole reciprocal space except around $q \simeq 0$. In Fig. 6.32 is shown a typical example of the neutron inelastic scattering measurement for Nb$_3$Sn carried by *Shirane* and *Axe* [6.14]. The dispersion relation goes down, exhibiting softening of the TA$_1$ phonon mode, as the transition temperature, 45 K, is approached, and finally the perfect soft phonon mode is realized in accordance with the $C' = 0$ line at $q = 0$. Soft modes also appear in different branches, around different q values, and in different materials, and the lattice instability has been discussed theoretically and experimentally in relation to various aspects of the soft phonon mode.

6.4.4 Premartensitic Structure and its Statistical Thermodynamic Theory

The above mentioned lattice softening is a typical pretransition or precursor phenomenon since it occurs before the transition temperature T_c (or T_0) is reached. Another remarkable precursor is the submicroscopic tweed or mottled structure, which is observed under electron microscope far above T_0 (or M_s) and always associated with diffuse streaks in the electron or X-ray diffraction patterns. In Fig. 6.33 is shown a transmission electron micrograph of a tweed structure of Fe–30%Pd alloy taken at 325 K, which is 55° above M_s where fcc–fct martensitic transformation of weak first order starts, as is shown in Fig. 6.34 [6.15]. Similar tweed structures have been observed for Fe–Pt, In–Tl, Nb$_3$Sn and many other systems. Analyses of the micrographs and diffraction patterns reveal that the tweed contrast arises from the small platelets of prototype martensite phase on {1 1 0} planes associated with shear strains of surrounding lattice mainly along $\langle 1\bar{1}0 \rangle / \{1 1 0\}$ and that their sizes are some atomic layers thick and some tens of atomic distances wide. At high temperatures, they look like nondirectionally dispersed tiny speckles, which are called

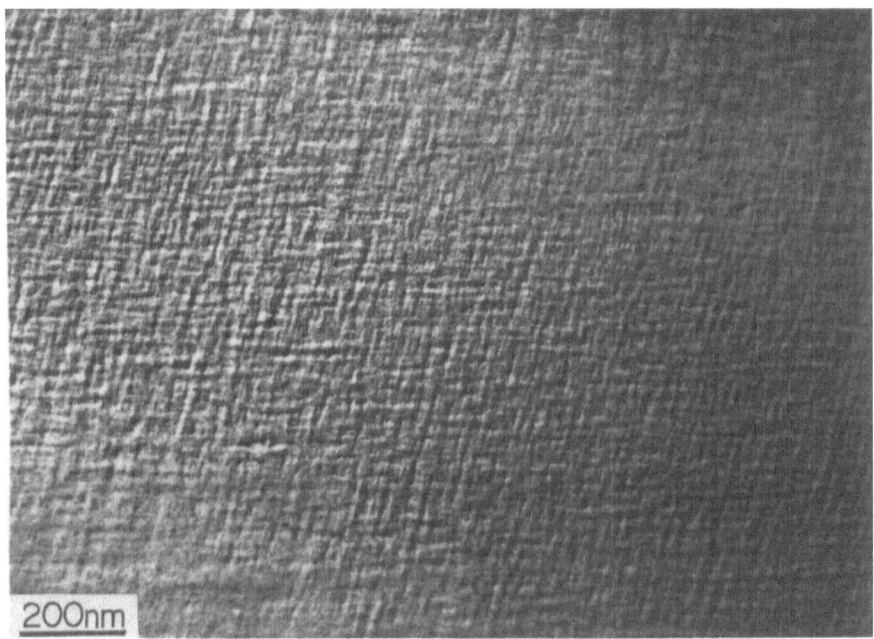

Fig. 6.33. A transmission electron micrograph of tweed structure of Fe-30%Pd alloy taken at 325 K, which is 55° above M_s [6.15]

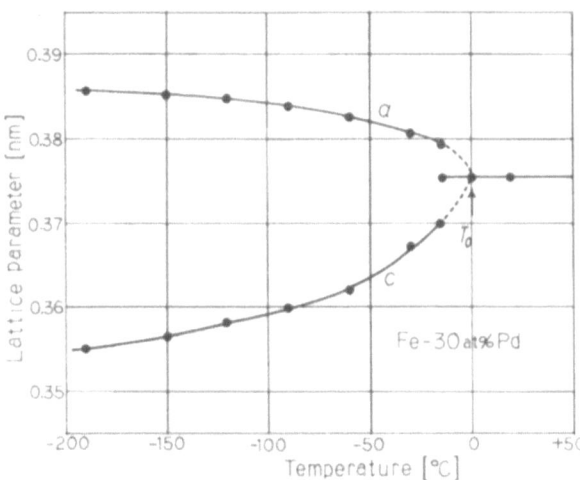

Fig. 6.34. Temperature dependence of lattice parameters of Fe–30%Pd alloy showing fcc–fct weak first order transition at 270 K [6.15]

the mottled structure. They become gradually directional and increase in the density and intensity as the temperature decreases, finally joining together to line up to form the martensite phase with a fine twin structure below M_s. The small platelet martensite prototype is called the embryo, which means in the theory of nucleation a subcritical size nucleus of new phase appearing and disappearing timewise and spacewise owing to the energy and configuration fluctuations of the system near the critical condition. In the present case, however, martensite embryos are not fluctuating but static and therefore their size is not always subcritical. The high resolution electron micrograph in Fig. 6.35 clearly shows the size and accompanying small static lattice distortion of martensite embryos [6.16].

The above observation gives us a rough idea of the number of atoms, say, a few hundred, contained in one embryo particle, which is already transmuted to the martensite phase. It looks like a contradiction that far above the transition point the new phase particles already appear. Nevertheless, if these very small regions with a certain value of order parameter are regarded as the short range or medium range order structure like those in disordered alloys, cooled or supercooled liquid, etc., the nature of the embryos which gives rise to the tweed structure will be understandable and its theoretical interpretation will also be possible as follows.

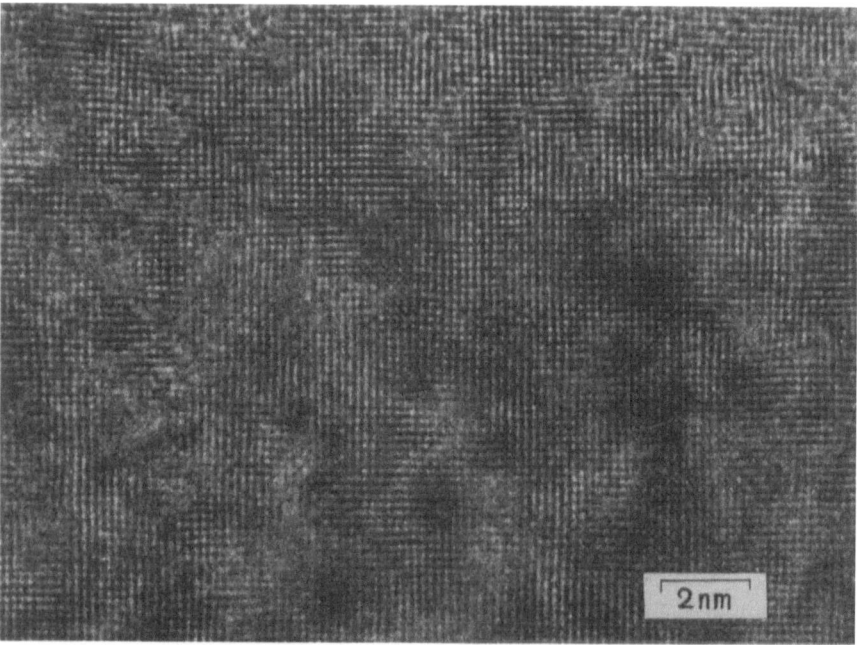

Fig. 6.35. A high resolution electron micrograph of tweed structure of an Fe–Pd alloy showing local static lattice distortions at martensite embryos [6.16]

The configurational entropy arising from the number of ways to distribute n embryos with the size i (atoms) is given by (6.3) in Sect. 6.2;

$$k \ln(w) = -R\{(C/i)\ln(C/i\xi) + (1 - C)\ln(1 - C) + C[1 - (1/i)]\}, \quad (6.7)$$

where $C = ni/N$ is the total concentration of the embryo atoms. The martensite embryos are expected to appear above M_s or T_0, if the total free energy of the system can be lowered by their existence. For the Fe–Pd alloy, the enthalpy change from fcc to fct, ε, is as small as 4×10^2 J/mol, and, likewise, the non-configurational entropy change, s, the strain energy around each embryo, χ, the phase boundary energy, and the embryo–embryo strain interaction energy are all small, especially the last three energies being considerably lowered by lattice softening. By taking account of the main contributing energy parts in the above, the total free energy change of the system arising from the embryo formation can be simply expressed as

$$\begin{aligned} G = C(\varepsilon - TS) + (C/i)\chi - RT[(C/i)\ln(C/i\xi) \\ + (1 - C)\ln(1 - C) + C(i - 1)/i] , \end{aligned} \quad (6.8)$$

and with the plausible parameter estimation from the experimental data, the equilibrium concentration C of the martensite embryos vs. temperature is calculated as in Fig. 6.36, where C can be as small as less than 2% at and above T_0 when $i = 400$, which wouldn't produce any detectable lattice parameter change in diffraction measurement but a tweed pattern with enough strain contrast on an electron micrograph [6.17]. Since the interaction energy between embryos and the size dependence of energies are not included in the calculation, the behaviour of embryos near M_s and their equilibrium size are not described. However, the origin and the nature of the tweed structure as a precursor of the phase transformation are well explained by the above theory.

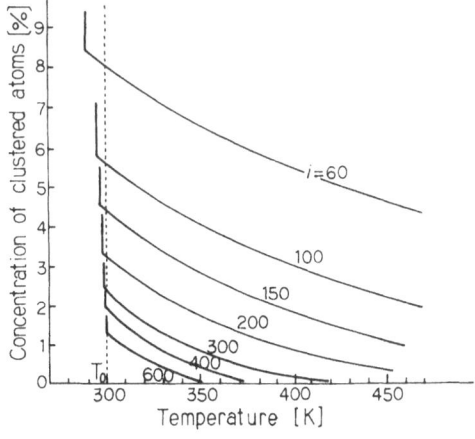

Fig. 6.36. Calculated temperature dependence of the concentration of small local ordered regions

6.5 Martensite and Other Problems in Ceramics

6.5.1 Martensitic Transformation of Zirconia

Among other problems which recently arose in relation to the development of new ceramics, utilization of martensitic transformation of zirconia is a noticeable one. As was mentioned in the first part of this chapter, when zirconia sintered in bulk is cooled from high temperature, following the cubic to tetragonal transformation at 2370°C, the martensitic transformation from tetragonal to monoclinic takes place at 950°C, shattering the bulk into pieces due to a volume expansion as large as 3.0% at the transformation and ultimately 4.9% when reaching the room temperature. As Fig. 6.2 shows, by adding calcia of, say, 0.1 mol, the tetragonal phase of ZrO_2–CaO solid solution partially appears, and if it is finely dispersed as precipitate particles and quenched down to room temperature, the tetragonal-monoclinic transformation will be suppressed. The stability of the tetragonal particles is determined by the matrix constraining, the chemical free energy affected by the composition, the nucleation barrier for transformation, etc., all of which depend upon the particle size and shape. Comparing with the full cubic phase material, which is realized by adding 0.2 mol of calcia or yettria, the mechanical properties such as toughness, thermal shock resistance and deformability are largely improved: When a critical stress for crack initiation is given by loading or thermal shock, the stress concentration at each crack tip will nucleate the tetragonal–monoclinic transformation of the stabilized particles. This is the stress induced martensitic transformation mentioned before, which retards the crack propagation by the strong compression field produced by the above mentioned volume change. It may be worthy of note that the precipitated tetragonal particles produce a tension field along their c-axis to the cubic matrix, while the same particles after the transformation to monoclinic associated with fine $(1\ 1\ 0)_m$ twinning produce a strong compression field perpendicular to the former tensile field. Needless to say, the initiation of the compression field is effective to retard or arrest the crack propagation, but the careful quantitative analysis of the combination of thermal expansions, crystallographic unit cell change, particle shape and overall shape change including twinning must be necessary.

The idea of the partially stabilized zirconia (PSZ) is not only applicable to the zirconia based materials but also to the alumina–zirconia composite material (ZTA). By dispersing the intergranular, irregularly shaped and constrained tetragonal particles of zirconia in the fine grained alumina matrix, the toughening mechanism the same as mentioned above can be exhibited [6.18]. Further extension of the idea seems to be possible for the development of other advanced tough ceramics. *Kriven* [6.19] has systematically and precisely studied the characters and properties of various ceramic materials, which have martensitic transformation associated with large positive volume changes and, therefore, have possibilities to be used as the transformation tougheners. Possible ones are, for instance, Dy_2O_3, $2CaO \cdot SiO_2$, NiS and $LuBO_3$. She also found that the polymorphic forms (crystallographic structures) and transformation temperatures of the lanthanide sesquioxides, including Dy_2O_3, systematically change with the atomic number and that the molar volume differences between the three structures, cubic, tetragonal, and monoclinic, also systematically change. She concluded that the possible alternative transformation tougheners to zirconia offer a wide range of chemical compatibilities with various matrices, but compared with metals, the compounds have received little study and far less is known about their transformation mechanisms. More systematic studies seem to be required for better understanding of the effects and applications of martensitic transformation to the mechanical properties of composite ceramics.

6.5.2 P–T Phase Diagram and Artificial Diamond

There are many other interesting subjects associated with the physical aspects of phase transition, which are applicable to the development of new advanced materials. However, because of space limitation and the wide spread of a variety of materials, introduction of all of them is not possible here. If to add only one more subject indispensable for this chapter is required, it may be that of artificial diamond, which is known, in the case of the bulk, to be closely related with the phase transition under high pressure and high temperature, but, in the case of the thin film made by the chemical vapour deposition (CVD) method, not always related with high pressure and high temperature. In its exact definition, diamond is not a ceramic, but development and fabrication of artificial diamond are already a new trend in ceramic industries so that to describe diamond in this section is not unappropriate.

Let us first examine the P–T phase diagram of carbon, including the phases of graphite, cubic diamond, hexagonal diamond and the melt, as shown in Fig. 6.37. It is widely known that at ambient pressure only graphite is stable and at ambient temperature diamond structure is not available from graphite by applying only a very high pressure of the order of, for instance, 100 GPa [6.20]. The barrier for this transition must be very high. Sometimes diamond is found at the bottom of a meteorite crater probably due to the extremely high pressure and temperature produced by the collision of the meteorite with the earth. Similar extreme conditions can be realized by the shock wave technique using detonation to artificially produce diamond from graphite. A typical case of formation of diamond under high pressure is that of natural diamond. Deep underground, the pressure and temperature rise up as a dotted line in the diagram shows, and the underground below 120 km is already in the region where the diamond phase is stable. Natural diamond is therefore considered to grow in the hatched area in the diagram and brought up to the ground surface by a local upward crustal motion with a relatively high speed before the transition to graphite takes place. Physics and chemistry of natural diamond, such as colour centres, lattice defects and thermal, electronic and mechanical

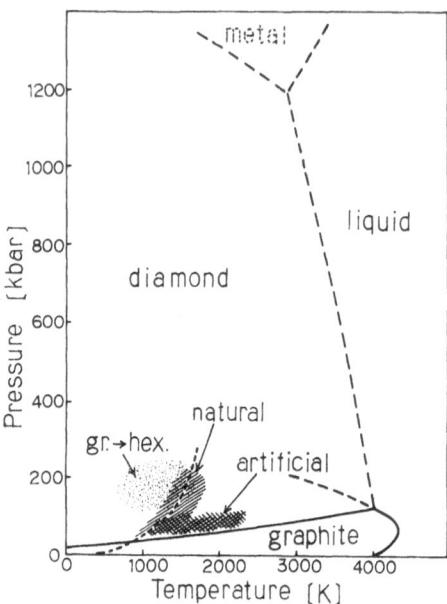

Fig. 6.37. P–T phase diagram for carbon. Broken lines are by supposition and the dotted line near the bottom which shows the pressure vs temperature relationship underground is also by guess

properties, have been extensively studied, especially by British researchers, and so many papers and books were published elsewhere. Reversion of natural diamond to graphite below 1500°C is extremely difficult as well as the above mentioned reaction from graphite to diamond under high pressure at room temperature. On the other hand, fabrication of artificial diamond is not difficult today since the first success by F.P. Bundy and others of General Electric Company in 1954 [6.21], who applied the flux method (usually called the solvent catalysis method) to graphite under high pressure and high temperature in the crosshatched region above the phase in Fig. 6.37. In the case of nickel flux, 1500°C/6 GPa is a suitable condition for fabrication of diamond.

Direct transformation from graphite to diamond is also industrially possible by applying a pressure of 13–16 GPa and a temperature of 3500–4000°C, sometimes beyond the triple point in Fig. 6.37, using the shock wave method or the direct electric current heating. It is noteworthy that a new material produced by the direct transformation technique is hexagonal diamond, which is called lonsdaleite as a minerological name. To understand the hexagonal diamond, Figs. 6.38, 39 will be helpful; the former shows the well-known diamond structure looked into nearly along the [1 1 0] direction, where hexagonal holes made of staggered network are clearly seen. This hexagonal network, consisting of tetrahedral sp^3 covalent bondings, could be directly formed from the s^2p^2 type flat hexagonal network of graphite structure as the latter is squashed along the c-axis in the above mentioned pressurizing process for direct transformation. Figure 6.39a shows the same cubic diamond structure with the [1 1 1] direction held vertical. Note that the trigonal atomic arrangements above and below

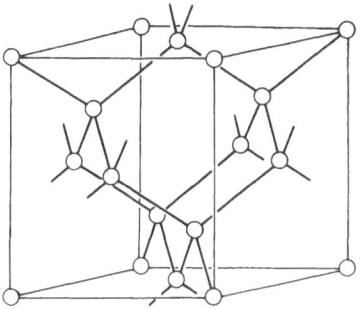

Fig. 6.38. Diamond structure viewed along nearly [1 1 0] direction

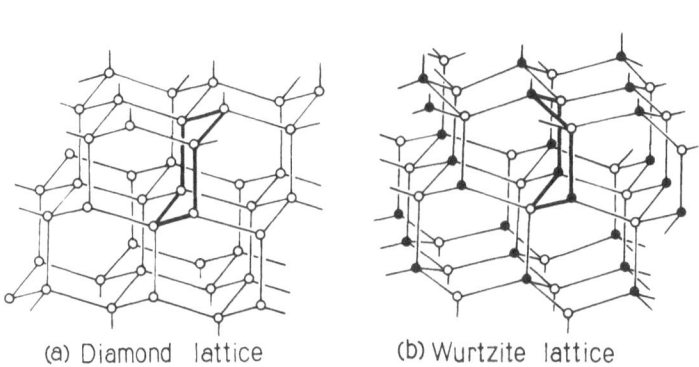

(a) Diamond lattice   (b) Wurtzite lattice

Fig. 6.39. (a) Diamond lattice and (b) wurtzite lattice to show the difference in their network structures. Compare the hexagonal network units drawn by thick lines in the two structures

every vertical bond axis are 60° (or 180°) twisted against each other. On the other hand, if the two trigonal atomic arrangements have no twist angle with each other, another lattice structure appears as in the Fig. 6.39b. This is the hexagonal diamond structure. Since the cubic and the hexagonal diamond structures are quite the same except for the above mentioned lattice symmetry, their densities, cohesive energies, and optical, thermal and mechanical properties are also very similar. A noticeable difference between the two network configurations is that, in the former all the staggered or bent hexagons are of the "tilde" shape, while in the latter the hexagons along the c-axis are buckled (or concaved) as are shown by thick lines in the Fig. 6.39a, b, respectively. The phase of hexagonal diamond cannot be exactly shown in the $P-T$ diagram probably because it is not a stable phase, but it is produced by keeping graphite in a high pressure vessel above 10 GPa and 1000°C, for instance, as the shaded area in the diagram shows. It is also produced by the shock pressure method. Hexagonal diamond is already commercially available as a new abrasive material, but its physical properties are little known. The bond configuration of the hexagonal diamond is quite the same as that of a wurtzite structure with tetrahedral bond configuration found in SiC, ZnS, CdS, CdTe, GaP and other covalent compounds. Note that most of them exhibit polymorphism, including not only the wurtzite structure but also the zinc blend structure, which has the same bonding network as the cubic diamond, and various $(1\,1\,1)_c$ layer stacking modifications as the mixture of the above two lattice structures, which is called the polytypism.

6.5.3 CVD Diamond

As mentioned in the above, the high pressure technique first established by P.W. Bridgman in the 1930's as a new field of physics, has been further developed and become a useful technique to produce industrial artificial diamond. On the other hand, it has been a long search to produce diamond without the high pressure process. It is said that the first indication was given by *Derjaguin* et al. of USSR in 1968 [6.22], who introduced hydrocarbon gas onto a diamond substrate intermittently heated by a xenon light source with mirrors, made the gas dissociate and condense on the substrate, and found fine particles of diamond. It is also said that W.G. Eversole of the U.S.A. was the first to grow diamond at low pressures; he achieved the growth of diamond from vapour on diamond seed crystals and gave proof of the growth in 1953, although the result was not found in ordinary scientific journals. It is a surprise that he did it one year earlier than the before mentioned success of the flux method using high pressure by Bundy's group [6.21]. Similar experiments were carried out by *Angus* et al. [6.23], and many other groups during those years too.

The modern era of low pressure diamond growth was begun in the mid 1970's by employing a supplementary but essentially important technique in the chemical vapour deposition method: It has been gradually clarified that active hydrogen and hydrocarbon radicals are necessary to induce the condensation of diamond or diamond-like layers on a substrate under low pressure, so that a hot filament, microwaves plasma, high frequency induction plasma or direct current plasma are used as a supplementary technique for that purpose.

Since 1982, M. Kamo and others at the National Research Laboratory for Inorganic Materials, have played a leading role in the development of this technique [6.24]. In Fig. 6.40a is shown a sketch of a CVD apparatus, in which the introduced mixed gas of, for instance, H_2 and a few % of CH_4 of 10^4 Pa, is heated and excited or dissociated by the filament at around 2000°C to 2800°C, and in Fig. 6.40b is the microwave apparatus, in which the activated gas is produced as well. The active or atomic hydrogen is considered to take certain definite sites of the substrate surface to cover it, react with the incident hydrocarbon radicals such as CH_3, CH_2, C_2H, CH, etc., and incorporate their carbons with the diamond structure. At the same time, it probably attacks and sweeps out the graphite layers deposited on the substrate. Undoubtedly graphite is more stable than diamond in the low pressure atmosphere in the process of CVD, but with the help of atomic hydrogen associated with the epitaxial growth

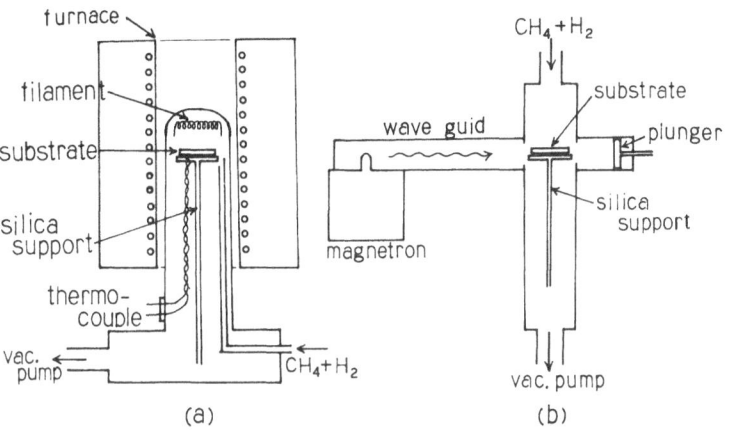

Fig. 6.40. Sketches of (a) the hot filament CVD method and (b) the microwave CVD method to make diamond coating films

mechanism, graphite deposition is suppressed or taken out and the growth of diamond is secured. Nevertheless, the molecular processes for nucleation and growth of diamond in the CVD method are scarcely known at present.

It is also possible to make the diamondlike hydrocarbon structure, in which carbon atoms form random networks, consisting of the sp^3 tetrahedral bondings most probably partially with sp^2 bondings, and hydrogen atoms of 0.2–0.6 mol concentration also take part in the structure, probably filling the dangling bonds of carbons and making local hydrocarbon molecular short range orders too. Diamond and diamondlike films made by CVD techniques are expected to be used for surface coating to protect tools, magnetic and optical disks, precision machine parts, etc. from chemical attacks, abrasion, wearing and other mechanical damages. They could be used as antireflective coatings too. It must be pointed out that the new hard material, diamondlike hydrocarbon, has a much higher atomic density and a stronger bonding than those of amorphous carbon, which has long been used as a useful soft material, although both of them have similar extremely disordered structures.

6.6 Conclusions

In conclusion of this chapter, by controlling the structural phase transformations in alloys, ceramics and other materials, some of which had already been useful from the beginning, and some which were undesirable for practical use of the materials, one can create new advanced materials, such as extremely hard or tough materials, new magnetic, optical, ferroelectric, conducting and superconducting materials, and new functioning materials. For that purpose, it is essentially advantageous and even necessary to understand the basic mechanisms of phase transitions, especially in terms of solid state physics. Fortunately, study of phase transition is one of the principal branches of physics, and physical theory of phase transition seems to be well established. This Chapter was aimed to introduce basic ideas of the physics of phase transitions related to the development of new materials. For real engineering applications, the researchers and engineers are recommended to study more specialized articles and papers.

References

6.1   L. Kaufman, M. Cohen: Prog. Met. Phys. **7**, 165 (1958)

6.2   J.W. Merz: Phys. Rev. **76**, 1221 (1949)

6.3   F.E. Fujita: In Proc. 5th Conf. R.Q.M. (Wurzburg), ed. by S. Steeb, H. Warlimont (Elsevier, 1985); pp. 585–588; J. Non-Cryst. Solids **106**, 286 (1988); Mater. Sci. Eng. A **127**, 243–248 (1990)

6.4   G.V. Kurdjumov, G. Sacks: Z. Phys. **64**, 325–343 (1930)

6.5   Z. Nishiyama: Sci. Rep. Tohoku Imp. Univ. I, **23**, 637–664 (1934)

6.6   A.B. Greninger, A.R. Troiano: Trans. AIME, **185**, 590–598 (1949)

6.7   F.E. Fujita: Metall. Trans. **8A**, 1727–1736 (1979)

6.8   R. Oshima, C.M. Wayman: Scr. Metall. **8**, 223–230 (1974)
      R. Oshima, H. Azuma, F.E. Fujita: Trans. J.I.M. **17**, 293–298 (1976)

6.9   T. Saburi, C.M. Wayman, K. Tanaka, S. Nenno: Acta Met. **28**, 15–32 (1980)

6.10 K. Shimizu, K. Otsuka: Int'l Metals Rev. **31**, 93–114 (1986)
      N. Nakanishi, Y. Murakami, S. Kachi, T. Mori, S. Miura: Phys. Lett. **37A**, 61 (1971)

6.11 H. Sakamoto, K. Otsuka, K. Shimizu: Scr. Metall. **11**, 607–611 (1977)

6.12 K.R. Keller, J.J. Hanak: Phys. Rev. **154**, 628 (1967)

6.13 D.J. Gunton, G.A. Saunders: Solid State Commun. **12**, 569 (1973); ibid. **14**, 865 (1974)

6.14 G. Shirane, J.D. Axe: Phys. Rev. Lett. **27**, 1803 (1967)

6.15 R. Oshima, M. Sugiyama, F.E. Fujita: Metall. Trans A, **19A**, 803–810 (1988)

6.16 S. Muto, S. Takeda, R. Oshima, F.E. Fujita: Jpn. J. Appl. Phys. **27**, L1387 (1988)

6.17 F.E. Fujita: Mater. Sci. Eng. A **27**, 2453–2458 (1990)

6.18 W.M. Kriven: The transformation mechanism of spherical zirconia particles in alumina, in *Advances in Ceramics*, **12**, Sci. & Tech. Zirconias II, ed. by N. Clausen et al. (American Ceramic Soc., Columbus 1984) pp. 64–77

6.19 W.M. Kriven: J. Am. Ceramic Soc. **71**, 1021–1030 (1988)

6.20 A.B. Aust, H.G. Drickamer: Science **140**, 817 (1963)

6.21 F.B. Bundy, H.T. Hall, H.M. Strong, R.J. Wentorf, Jr.: Nature (London) **176**, 51–54 (1955)

6.22 B.V. Deryaguin, D.B. Fedoseev: Zh. Fiz. Khim A**2**, 2360 (1968); Sci. Am. **233**(5), 102–109 (1975); Dkl. Akad. Nauk USSR **231**, 333 (1976); Ch. 4 Izd. Nauka, Moscow, USSR (1977)

6.23 J.C. Angus, H.A. Will, W.S. Stanko: J. Appl. Phys. **39**, 2915–2922 (1968)

6.24 S. Matsumoto, Y. Sato, M. Kamo, N. Setaka: Jpn. J. Appl. Phys. **21**, 183 (1982)

6.25 H.W. Kroto, J.R. Heath, S.C. O'Brien, R.F. Curl, R.E. Smalley: Nature **318**, 162 (1985)

6.26 S. Iijima: Nature **347**, 56 (1991); S. Iijima, T. Ichihashi, Y. Ando: Nature **356**, 776 (1992); R. Saito, G. Dresselhaus, M.S. Dresselhaus: J. Appl. Phys. **73**, 494 (1993)

6.27 S. Muto, G. Van Tendeloo, S. Amelinckx: Phil. Mag. B **67**, 443 (1993); G. Van Tendeloo, S. Amelinckx, S. Muto, M.A. Verheijen, P.H.M. Van Loosdrecht, G. Meijer: Ultramicrosc. **51**, 168 (1993)

6.28 H. Kuzmany, J. Fink, M. Mehring, S. Roth (eds): *Electrical Properties of Fullerenes*, Springer Ser. Solid-State Sci., Vol. 117 (Springer, Berlin, Heidelberg 1993)

6.29 S. Hasegawa, T. Nishiwaki, H. Habuchi, S. Nitta, S. Nonomura: Fullerene Science and Technology **3**, 163 (1995)

6.30 R. A. Jishi, D. Inomata, K. Nakao, M. S. Dresselhaus, G. Dresselhaus: J. Phys. Soc. Jpn. **63**, 2252 (1994)
6.31 See, for instance, U. Gonser: Mössbauer Spectroscopy in Materials Science, Chap. 9
6.32 Y. Hirose, Y. Terasawa: Jpn. J. Appl. Phys. **25**, L519 (1986)
6.33 Y. Hirose, S. Amanuma, K. Komaki: J. Appl. Phys. **68**, 6401 (1990)
6.34 F. Banhart, P. M. Ajayan: Nature **382**, 433 (1996); F. Banhart: Diamantbildung in "Kohlenstoffzwiebeln". Phys. Bl. **53**, 33–35 (1997)

7 The Place of Atomic Order in the Physics of Solids and in Metallurgy

R.W. Cahn

Order is ubiquitous in the world of materials, as is disorder. Both concepts are thoroughly ambiguous, and highly important, like "democracy" or "explanation". This chapter is devoted to just one of the manifestations of order: it is variously named *atomic order*, *chemical order*, *spatial order* or *compositional order*, but here only the first of these alternative terms will be used. To put the matter anthropomorphically, as it often is, this kind of order stems from a preference by atoms to have unlike atoms as nearest neighbours.

Mostly – though not exclusively – we shall be concerned here with atomic *long-range order*. The term denotes the regular distribution of atoms in a crystalline solid solution among the atomic sites, so that chemically distinct sublattices come into existence. The absence of atomic order is one of the many forms and degrees of *disorder* in the description of solids: it is a more substantial form of disorder than that represented by a point defect, and less substantial than that implied by the absence of crystallinity. A solid solution without any order at all is described as being *random*. Randomness is the exception rather than the rule, though there is a very widespread half-way house between long-range order and randomness; this is *short-range order*. In short-range order, the preference for unlike neighbours is merely statistical and does not entail the creation of distinguishable sublattices.

7.1 Historical Development

Until X-ray diffraction was discovered in 1912, it was only possible to guess at the structure of even the simplest crystals, and no-one suspected the existence of atomic order. The first indications that something peculiar could happen in metallic solid solutions emerged in a pioneering study in 1916 by *Kurnakow* et al. [7.1], a group of Petrograd chemists. They were working at a time when the nature of an intermetallic compound was still very unclear, and they sought elucidation by studying the entire range of alloys of copper and gold (which form a continuous range of mutual solid solutions). They discovered thermal arrests during continuous cooling at certain compositions (Fig. 7.1a), implying some kind of phase transformation, and then decided (a novel notion at the time) to compare the same alloys in the slowly cooled form and after water-quenching from a high temperature in the solid state. That way, they expected to be able to

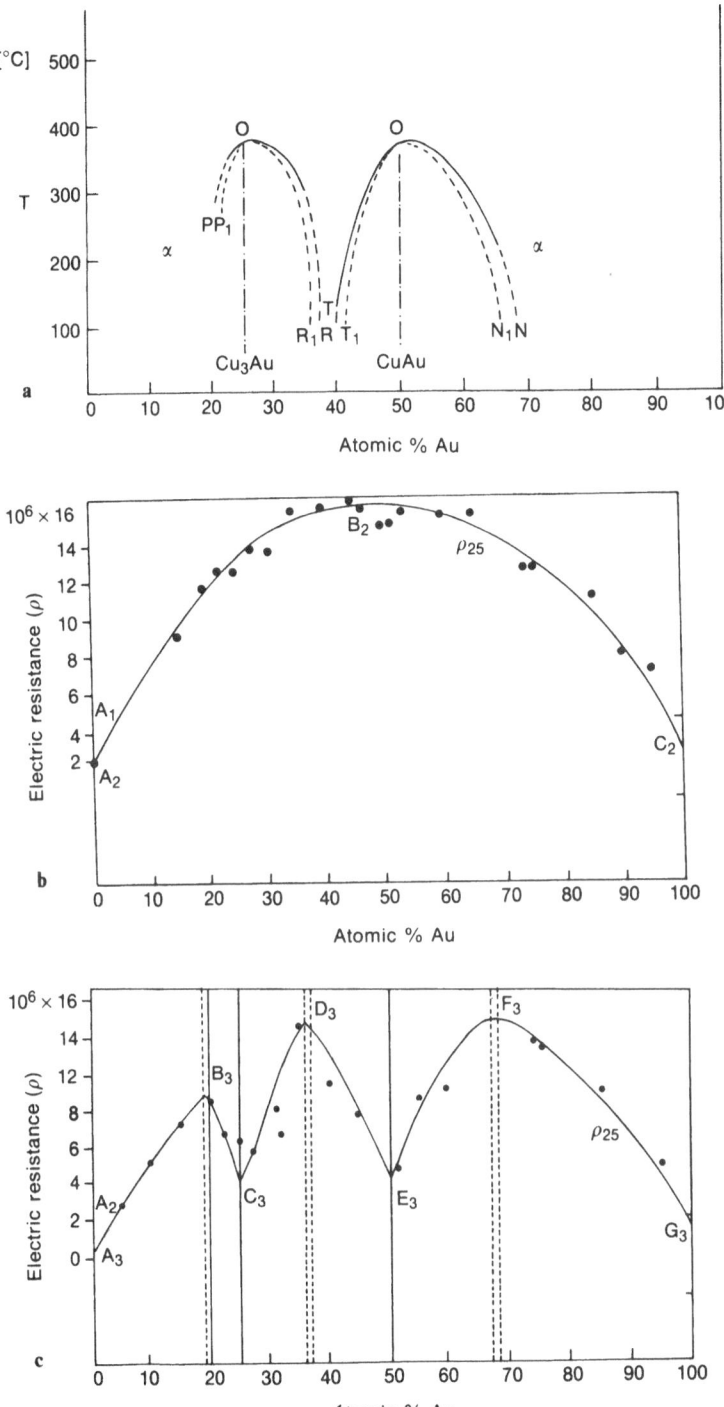

Fig. 7.1. (a) The "thermic diagram" of the Cu–Au system in the solid state; (b) The electrical resistance of Cu–Au alloys quenched from 800°C; (c) The electrical resistance of Cu–Au alloys annealed at 670°C and slowly cooled [7.1]

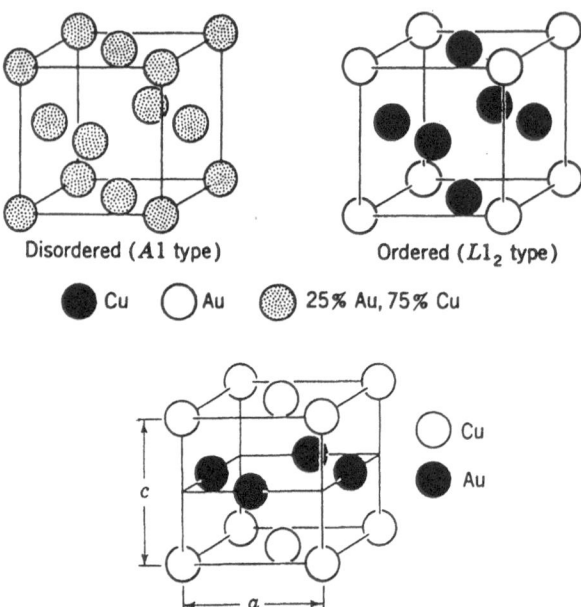

Disordered (*A*1 type)   Ordered (*L*1₂ type)

● Cu   ○ Au   ◉ 25% Au, 75% Cu

○ Cu

● Au

c

a

Fig. 7.2. The crystal structures of disordered and ordered Cu_3Au ($L1_2$) and ordered CuAu-I ($L1_0$)

compare the alloys in the transformed and the untransformed states. The quenched alloys showed no anomalies (Fig. 7.1b) but the slowly cooled alloys showed evidence of strongly pronounced anomalies at the Cu_3Au and CuAu compositions (Fig. 7.1c). Micrographs of the alloys in the two states showed differences which the authors were unable to interpret, but they were quite clear that the above-mentioned compositions were "compounds", which they called 'copper aurides'. And there the matter rested; the events of 1917 no doubt reduced the availability of gold in Petrograd, temporarily converted into Leningrad.

Although Kurnakow and his colleagues were working 4 years after von Laue had discovered X-ray diffraction, the new technique had not yet diffused widely among students of the solid state. It was not until 1925 that *Johansson* and *Linde* [7.2] applied X-ray diffraction to establish in detail the nature of the copper aurides; they observed extra diffraction lines, and thereby, the existence of *long-range order* (LRO). In the quenched – i.e., disordered – alloys, the copper and gold atoms were distributed at random on the face-centred cubic lattice sites; in the slowly cooled alloys, the two atom species segregated, to form two different structures (Fig. 7.2), each now based on a primitive lattice. The anomalies seen in Fig. 7.1a could now be perceived to represent the temperatures at which the disordered alloys began to order, while the other anomalies in Fig. 7.1c represented the lower resistivity of the ordered alloys; this was readily interpreted even on the simple models still being applied to the electrical resistivity of metals at that time. Figure 7.1a at once implied something important: ordering was not restricted to the precise stoichiometric compositions, Cu_3Au and CuAu, but could happen over a range of compositions in each case. The "copper aurides" had broad composition ranges; in other words, the ordered phases could dissolve extra copper or gold without losing all their order. This was the beginning of a modern way of looking at intermetallic phases: ordinary valency notions were of little use, and indeed at about this time *Hume-Rothery* [7.3] pointed out that the presence of metallic binding was likely to invalidate the ordinary valency concept. The British metallurgist Desch, in 1914, was unable to make sense of the formation of intermetallic compounds [7.4], but 20 years later he accepted that the new solid-state physics of that time cast an entirely new, clarifying light on the problem [7.5].

7.1.1 Superlattices

Soon, the term *superlattice* came to be applied to the crystal structure of an ordered intermetallic compound. It was firmly in use in the first edition of Barrett's classic text, *The Structure of Metals*, in 1943 [7.6]; this book contains probably the clearest early overview of superlattices. According to Barrett, the great American metallurgist Bain was in fact the first to see "superlattice diffraction lines", in 1923 [7.7], even before Johansson and Linde, and the early solid-state chemist, Gustav Tammann, had even speculated about the possibility of LRO as early as 1919. "Superlattice" is a curious term: it *seems* to imply the existence of an ordered unit cell *larger* than in the corresponding disordered lattice, which is most often inapposite: thus, the Cu_3Au unit cell is virtually the same size as the disordered lattice, though ordering does reduce the lattice constant by a fraction of one per cent, because the ordered atoms, being of two different sizes, pack more tidily into the available space. Nevertheless, the translation vectors for the ordered unit cell are larger than those for the disordered cell. (Other alloys, such as Fe_3Al and Ni_2AlTi, really do form superlattices with lattice parameters which are an approximate multiple, here twofold, of the lattice parameters in the disordered versions.)

In spite of this apparent objection, the term *superlattice* is actually appropriate, because the *translation vector* for the ordered structure is necessarily larger than that for the disordered structure, even if the unit cells are almost the same size in the two cases. Appropriate or not, the term is firmly established now. When a superlattice is present, the distinct atomic species occupy distinct, identifiable *sublattices*. As long as sublattices can be identified, we have to do with a state of long-range order; otherwise, the order is short-range.

It does not help clarity in nomenclature that the term *superlattice* has more recently been hijacked by applied physicists to denote artificially deposited multilayers of semiconductors or metals, periodically modulated in composition and with repeat distances large compared to unit cell sizes. Nevertheless, it must be admitted that this recent use of the term is more logical than the older, metallurgical, one.

Barrett pointed out that the term "superlattice" was at that time applied to the large number of metallic phases which were ordered at all temperatures up to the melting-point, as well as to those alloys, like the copper aurides, which became disordered at some temperature (termed the *critical temperature*) well below the melting-point. Nowadays, such *permanently ordered* phases are generally referred to simply as *intermetallic compounds*, or more briefly and commonly, just as *intermetallics*. It is implied thereby that 'real' intermetallic compounds have an ordered arrangement of their constituent atoms.

Superlattices come in three distinct families: The first is exemplified by Cu_3Au (Fig. 7.2) and CuZn (Fig. 7.3); starting from the disordered state, the constituent atoms are simply rearranged in a unit cell the shape and size of which remains unaltered except for the very slight contraction of the unit cell parameter(s) which accompanies ordering. The point group symmetry is unchanged, but the lattice type changes from centred (e.g., face-centred or body-centred) to primitive. The second family is exemplified by CuAuI (Fig. 7.2); the size and shape of the unit cell changes, and so does the symmetry (here, from cubic to tetragonal), and also the point group. This family is quite common, and the symmetry change may be considerable, as in the formation of orthorhombic Ni_4Mo from a cubic precursor. These first two types are given distinct structure symbols, originating from the prewar German compilation, *Strukturbericht*. Examples are $L1_2$ (Cu_3Au type), $L1_0$ (CuAu type), DO_3 (Fe_3Al type).

The third family is the so-called *long-period superlattice*. Perhaps the best-known example of this is CuAuII, stable between 385°C and the critical temperature at which order disappears, 410°C. Figure 7.4 shows what happens; after a number of unit cells measured parallel to the c axis, there is a displacement of half a lattice parameter transverse to that axis (this generates what is called an *antiphase domain boundary*) and then the new pattern continues. We shall have more to say about antiphase domain boundaries, below. The repeat distance, M, which is simply a multiple of the unit lattice parameter, is according to some investigators exactly 5 for the stoichiometric CuAuII, but others reported it to be fractional... which means that the value of

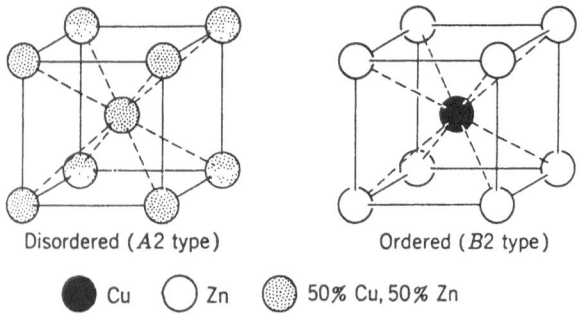

Disordered (*A2* type)   Ordered (*B2* type)

⬤ Cu   ◯ Zn   ◉ 50% Cu, 50% Zn

Fig. 7.3. The crystal structure of disordered and ordered CuZn (B2)

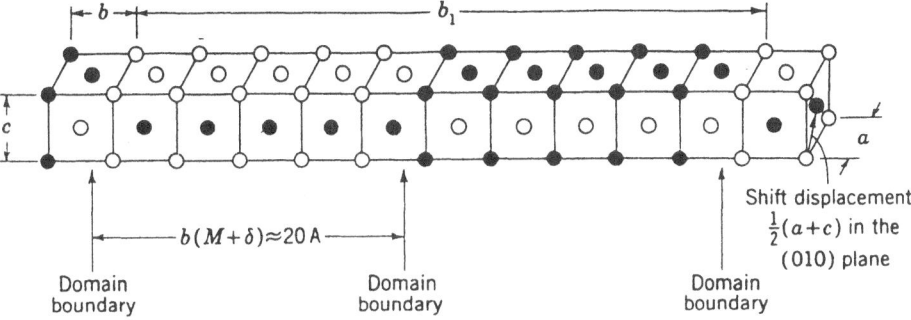

Fig. 7.4. The CuAu-II superlattice

M fluctuates somewhat from one repeat to another. There is a very well authenticated example of such a statistically variable long-period superlattice in $Ti_{28}Al_{72}$ [7.8, 9]; over long distances, it appears that this remarkable kind of variable repeat distance is statistically highly reproducible! Another variant of a long-period superlattice, observed in Cu–Pd, Au–Mn and other systems, has two-dimensional periodicity, with M_1 not necessarily equal to M_2.

7.1.2 Imperfect Long-Range Order

Long-range order is not necessarily perfect. Its perfection diminishes both with deviation from stoichiometry and with increase of temperature (even for a stoichiometric alloy). To measure this, an agreed *order parameter* is needed. The conventional order parameter was introduced by Bragg and Williams in what was probably the most influential paper in the history of research on atomic order [7.10]. Let there be N atoms in total in N positions (i.e., vacancies ignored), and let n of the sites be occupiable by either kind of atom. (In certain superlattices, a proportion of the sites are always occupied by one kind of atom, whether or not there is LRO). Suppose a fraction r of the n sites is occupied by A atoms in a state of perfect order; these rn sites are *right* positions for A atoms. If the order is imperfect, then some of these positions will be occupied by B atoms. If now p is the probability that a right position for an A atom is in fact occupied by an A atom, then the degree of LRO can be defined by the parameter S, defined by

$$S = (p - r)/(1 - r),\qquad(7.1)$$

which varies from 0 for a random alloy to 1 for a perfectly ordered one. S is known as the *Bragg order parameter*. It is quite straightforward to determine S by obtaining the ratio of the integrated intensities of a fundamental diffraction line (one which is unaffected by changes in degree of order) and a superlattice line, which is sensitive to order, with corrections for the temperature-dependent

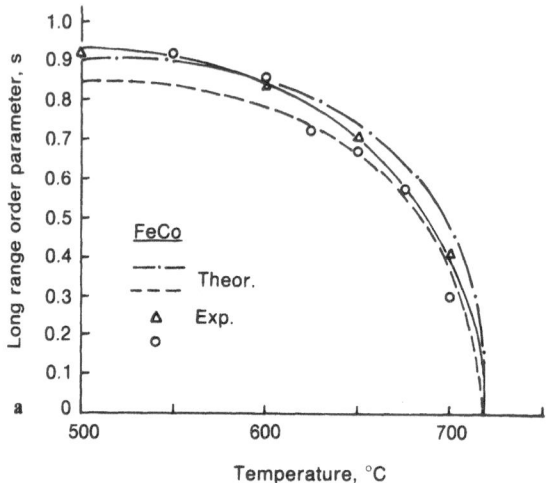

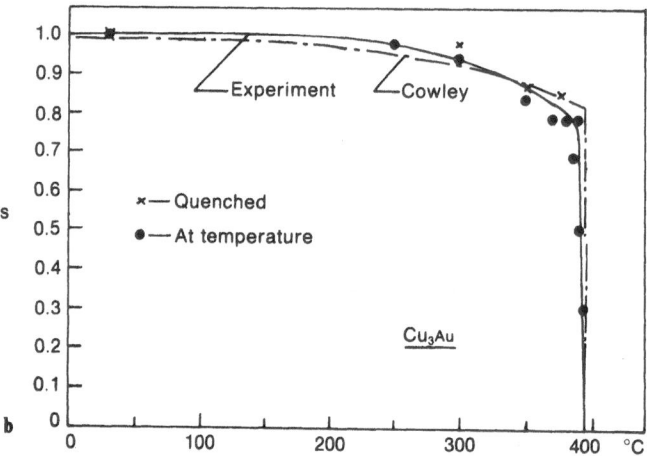

Fig. 7.5. The Bragg order parameter S, as a function of temperature, as measured for (a) FeCo [7.12]; (b) Cu₃Au [7.11]

Debye–Waller factor. Figure 7.5 shows S as a function of temperature for two superlattices, as determined by high-temperature X-ray diffractometry [7.11–7.13]. It can be seen that S either declines continuously to zero as the critical temperature, T_c, is approached (type 1), or else suffers a discontinuous drop at T_c (type 2). At one time it was thought that type 1 is necessarily characteristic of alloys with composition AB and type 2 of alloys with composition A_3B, but this is clearly not correct, since, for instance, Fe_3Al follows a curve of type 1. This erroneous generalization emerged from the theoretical work by

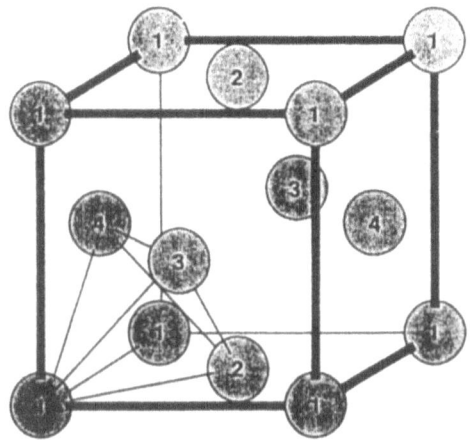

Fig. 7.6. A tetrahedral cluster of 4 atoms on distinct sublattices in a face-centred cubic unit cell, as used in CVM calculations (After *Inden* and *Pitsch* [7.20])

Bragg and Williams (see below), and was perpetuated in early surveys of LRO, for instance, in the generally excellent review by *Nix* and *Shockley* [7.14].

Even permanently ordered alloys (ones whose virtual critical temperature is above the melting-point) may lose measurable amounts of LRO before they melt. Ni_3Al, which has the same $L1_2$-type superlattice as Cu_3Au, is an important example. Electrical resistivity is sensitive to LRO and resistance measurements have been used to demonstrate reversible changes in LRO in Ni_3Al at temperatures as low as 590°C [7.15]. In this phase, which is unusual in a number of respects, it is also possible to drive T_c to just below the melting-point by going slightly off-stoichiometry, and this has been demonstrated both by dilatometry [7.16] and by high-temperature gamma-ray diffractometry [7.17]. The gamma-ray determination of S vs. temperature for an alloy of composition $Ni_{76}Al_{24}$ shows that the critical temperature here coincides approximately with the melting-point.

The Bragg-Williams parameter, S, has often been criticised by theoreticians as being hopelessly over-simple; nevertheless, it refuses to die. Perhaps this is because it is both conceptually and experimentally simple enough to be used by non-specialists in critical theory and statistical mechanics, who merely wish to achieve an approximate idea of how LRO order varies with temperature in different alloys. Many other, more complex forms of order parameter, applicable to partial LRO, have been proposed, some of them exceedingly difficult to relate to any experimentally measurable diffraction data. The so-called *Cowley–Warren short-range order parameters* [7.18] can be applied to the characterization of imperfect *long-range* order, in spite of their name, and are relatively readily related to diffraction measurements. More recently, *de Fontaine* [7.19] proposed a complex form of single-site average. The matter has, however, really become elaborate with the development of the concept of pair and cluster correlation functions. Here we are concerned with the statistics of defined *clusters*, e.g., the group of atoms 1–4 in Fig. 7.6. As explained, for instance, by *Inden* and *Pitsch* [7.20] in a clear and thorough recent exposition, one must consider occupation probabilities for single sites, pairs, triplets and quartets in this kind of tetrahedral cluster and combine these probabilities into linear combinations to generate order parameters which will both define degrees of order and specify the order *types* (i.e., distinguish different super lattice types). This approach is particularly useful for theorists who wish to predict stability ranges for different kinds of superlattice.

7.1.3 Critical Phenomena

This matter of the proper definition of order parameters is an instance of the divergence between atomic order as perceived by metallurgists and materials scientists on the one hand, and by theoretical solid-state physicists and theorists of critical phenomena on the other.

I have just mentioned the theory of *critical phenomena*. This approach to collective phenomena in physics has really come to notice since one of its chief exponents, K. Wilson, introduced the mathematical concept of the *renormalisation group* into the field in 1971 and won the Nobel Prize in consequence. In this approach (there is no proper name for it . . . one might attempt to speak of *criticality*) large groups of atoms are considered collectively, rather than according to the pair interaction model which sufficed for Bragg and Williams, and it applies equally to atomic order and ferromagnetism, and to many other phenomena as well. Thus, an order parameter, Q, (not necessarily specifying atomic order), as deduced from criticality theory, will obey a temperature law of the form

$$Q \approx (T_c - T)^\beta , \tag{7.2}$$

where β is a *critical exponent*. Such an equation applies, according to the theory, only when one gets rather close to T_c. The value of the critical exponent tells one something about the dimensionality of the physical situation, e.g., in the case of ferromagnetism, β will be different according as magnetic moments are constrained to lie along a line, in a plane or are free to lie in any direction.

The critical exponent concept has also proved useful in the study of atomic ordering. Criticality concepts are only rigorously applicable to transitions, such as order-disorder transitions, which are second-order in nature. A second-order transition is one in which the free energy and its first derivative are continuous at T_c but the second derivative is discontinuous. There is no latent heat. For such a transition, there is an infinitesimally small difference between the structures of the ordered and disordered phases across the critical temperature. In general, if the order parameter changes *continuously* (without any jump) to zero at T_c, then the transformation is presumed to be second-order. Cu–Zn (β-brass) orders via a second-order transition, *Als-Nielsen* [7.21]. We shall be encountering such transitions and other criticality concepts later in this chapter.

Yeomans [7.22] has written an elementary introduction to the basic concepts of criticality theory, while a very up-to-date account, accessible to modern materials scientists, has just been published by *Binder* [7.23], as part of a broader discussion of statistical mechanics.

7.2 Antiphase Domains

Just over half a century ago, a British physicist, Charles Sykes, made an unexpected discovery [7.24, 25]. He took X-ray diffraction (Debye–Scherrer) photographs of Cu_3Au, initially quenched from above T_c, at various stages of annealing below T_c. As he expected, he found that the intensity of the super-lattice lines gradually increased, as order built up, but he also saw that in the early stages, the superlattice lines were much broader than the fundamental lines (those which are unaffected by the state of order). Indeed, it took rather longer for this broadening to disappear completely than for the order parameter, S, to reach its equilibrium value. This differential broadening indicated that the regions in which the state of order was homogeneous were very much smaller

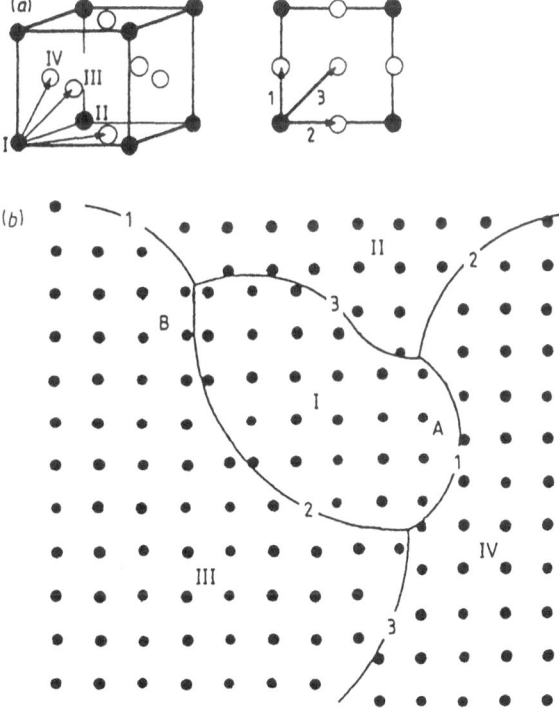

Fig. 7.7. (a) The $L1_2$-type (X_3Y) structure, together with a projection along [0 0 1]. The three types of displacement vector at antiphase domain boundaries, labelled 1, 2, 3, together with the 4 sublattices, I, II, III, IV, are indicated. (b) A simplified projection along [0 0 1] of an $L1_2$ superlattice, showing different kinds of APDB. A (displacement vector 2) and B (displacement vector 1) correspond respectively to locally conservative and non-conservative APDB [7.115]. In (b), only gold atoms are shown

than the crystal grains of his (polycrystalline) alloy, at least in the early stages of anneal. These homogeneous regions have come to be called *antiphase domains*.

The reason for this name appears from Fig. 7.7. Sykes recognized that LRO is nucleated independently at many sites within a single disordered grain; there are 4 independent sublattices (one for each distinct atom site in the cubic unit cell, Fig. 7.2a), and any of these can be where the gold atoms finish up. If two ordered domains nucleate with gold on distinct sublattices, as most often they will, then when they meet each other there will be a hickup, or antiphase mismatch, in the location of the gold atoms, of the kind shown in the sketch. Once the grain is entirely filled by ordered domains, they can progressively coarsen; some domains simply swallow up their neighbours. As Sir Lawrence Bragg recognized soon afterwards, in Cu_3Au a metastable *foam structure* can exist, because there are four sublattices and a three-dimensional array of 4 kinds of domains is topologically feasible. In β-brass (Fig. 7.3), where there are only two sublattices, no topologically metastable foam structure can form and the domains coarsen extremely rapidly; no broadened superlattice lines are seen.

More recently, antiphase domain boundaries (APDBs) have been much studied by TEM. If a dark-field image is produced with the aid of a superlattice reflection then a proportion of the domain boundaries will be in contrast (Fig. 7.8) and the size and morphology of the APD structure can be appreciated. In some alloys, APDBs prefer particular lattice planes, e.g., $\{1\,0\,0\}$ in Cu_3Au, as can be seen in Fig. 7.8, but the alignment is never perfect.

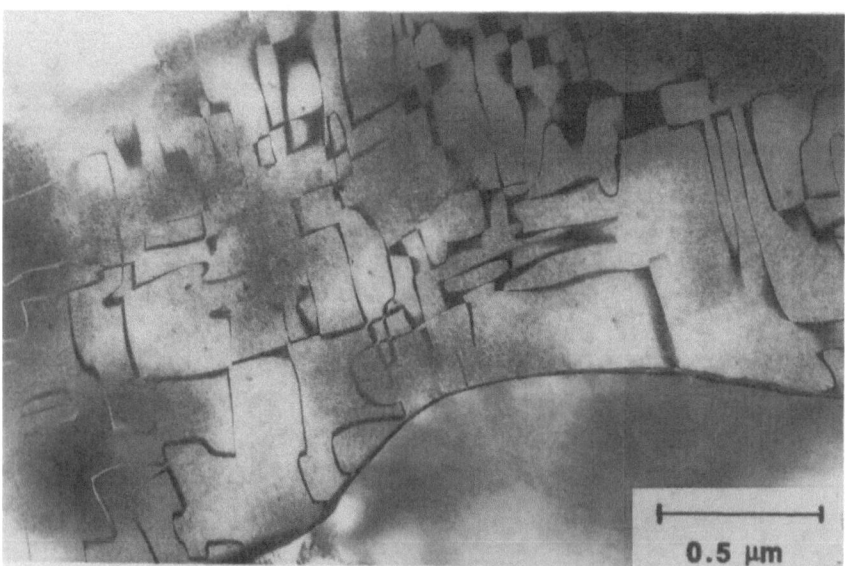

Fig. 7.8. Antiphase domains in cold-rolled and recrystallized Cu_3Au, specimen annealed for 8 days at 362°C, 18°C below T_c. (K.H. Westmacott, R.W. Cahn, unpublished research)

It is often not recognized that APD size and the degree of order as measured by S can be independently controlled by appropriate heat-treatment, within limits (it is not possible to have an ultrafine domain size in association with perfect order), and thus the effects of these two features on physical and mechanical properties can be independently examined.

Another point which is often not appreciated is that APDs can only form freely when the alloy freezes in the disordered form and orders on cooling in the solid state. Permanently ordered intermetallics which remain ordered until they melt have no or very exiguous APDs (and those few that form seem to be associated with single dislocations). This correlation between critical ordering temperature and APD formation was made very clear in a study of offstoichiometric Ni–Al alloys where, as we have seen, T_c plays hide-and-seek with T_m. Figure 7.9 shows a micrograph of a $Ni_{78}Al_{22}$ alloy which has been rapidly frozen and thus has a dendritic structure, with microsegregation. Using analytical electron microscopy, the Al content proved to be above 23 at.% in the regions with very coarse domains and below 23 at.% in the regions with

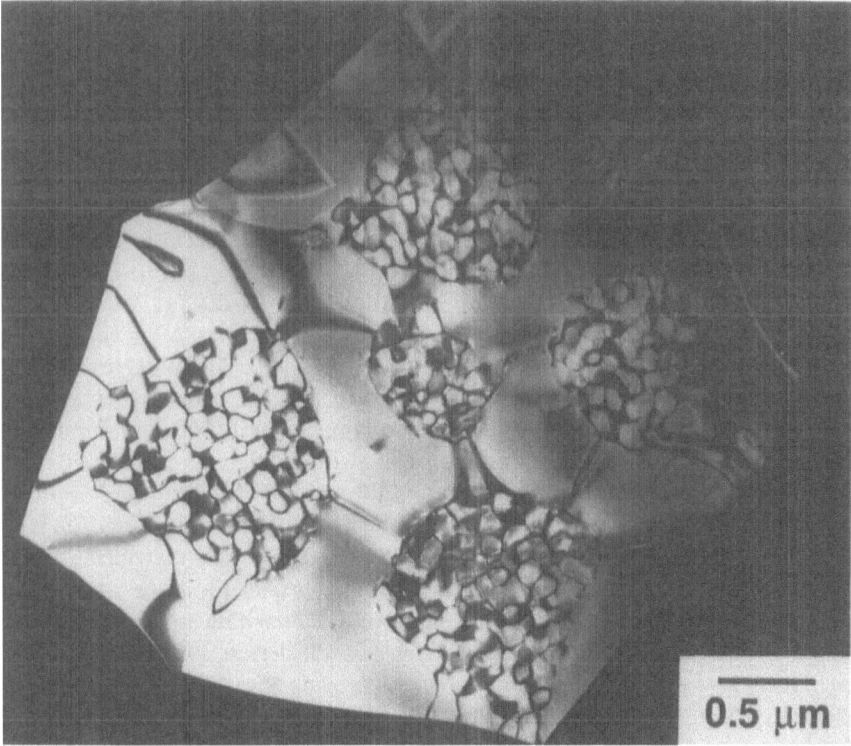

Fig. 7.9. Dendrites in rapidly cooled, microsegregated $Ni_{78}Al_{22}$, showing fine antiphase domains in Ni-rich regions and very coarse domains in Al-rich regions [7.26]

fine domains [7.26]; this corresponds to regions with $T_c > T_m$ and $T_c < T_m$, respectively.

Under some circumstances, domains can form when a deformed alloy recrystallizes below T_c to form a new population of grains. This has been observed in Cu_3Au, for example [7.27]; the APDBs seen in Fig. 7.8 were in fact obtained in this way. This observation implies the existence of a thin zone of disordered material near the grain boundary, in which domains are constantly being nucleated as an advancing grain boundary drags the disordered zone with it through otherwise ordered alloy.

7.2.1 Varieties of Domains

In addition to the kind of domain seen in Figs. 7.8, 9, there are several other types. One kind is generated by heterogeneous ordering generated behind an advancing grain boundary; this happens at temperatures too low for homogeneous ordering to be feasible, and has been observed, for instance, in FeCo [7.28] and in CuPt [7.29]; the heterogeneous domains in CuPt are coarse and can be seen by optical micrography, using polarised light (Fig. 7.10). Here, domains form heterogeneously at surface defects such as scratches, also. The morphology of heterogeneously formed domains is treated in a major survey of microstructures resulting from the ordering transformation, by *Tanner* and *Leamy* [7.30]. A discussion of the factors that decide whether an ordering transformation is homogeneous or heterogeneous will be found in a review by *Irani* [7.31].

Another kind of domain is associated with ordering reactions in which the crystal symmetry changes. In such reactions, a cubic crystal, for instance, will change into a tetragonal (CuAu), rhombohedral (CuPt) or orthorhombic (Ni_4Mo) one. To make the point clearer, a fcc disordered crystal of composition CuPt starts by having 4 threefold axes parallel to cube 'body diagonals'; after ordering, one of these will remain a threefold axis, the other three lose that characteristic, and the interaxial angle is no longer exactly 90°. This implies that there are four different kinds of domain, each with a different original cube body diagonal as its threefold axis. The domain boundaries between these represent differences in orientation, whereas the kind of APDB we saw in Fig. 7.8 is *not* associated with orientation differences. An ordered species which displays domain boundaries between differently oriented *variants* (these are not properly named 'antiphase') can in addition have true APDBs within any one orientation domain. The 'misorientation-type' domain boundaries are sometimes distinguished from real APDBs by calling them *twin boundaries*, which is acceptable if a wide interpretation of *twin* is acceptable. This kind of twin domain boundary has been exhaustively studied by the school of Amelinckx in Belgium, over many years, in alloys such as Au_3Mn, Au_4Mn, Ni_4Mo, Pt_8Ti, and others. Their geometry is very complex; an excellent overview of the use of high-resolution

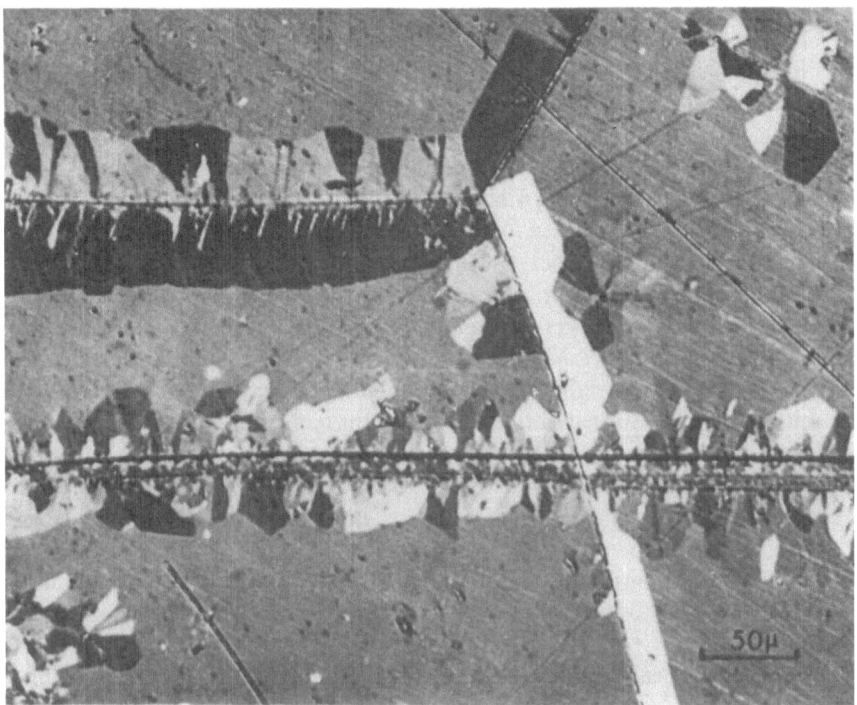

Fig. 7.10. Antiphase domains heterogeneously nucleated at grain boundaries and scratches, in CuPt annealed 85 h at 550°C below T_c. Photographed by polarised light [7.29]. Domains nucleated at grain boundaries, a scratch and surface impressions

TEM in the study of domains in such alloys has recently been published by a member of that team, *van Tendeloo* [7.32].

The scale and morphology of twin domains in superlattices that form with change of symmetry is controlled by the elastic stresses which arise from the change of shape of unit cell as a result of ordering. This aspect of domain structure in tetragonal CuAu (formed from a cubic disordered form) was first addressed in 1944 in a classic paper by *Harker* [7.33], and is also discussed by *Irani* [7.31]. This has led on to a whole series of theoretical and experimental studies of modifications to phase equilibria in coherent two-phase systems as a result of self-stressing by transformation products (see, e.g., *Johnson* et al. [7.34], also an overview by *Cahn* [7.35]). This kind of theory has by now become very sophisticated [7.36].

As we shall see below, proper antiphase domains (the first kind of boundary discussed above) have substantial effects on the properties and transformations of ordered phases; an excellent review of this aspect of APDBs, as well as their morphology and energetics, has just been prepared by *Morris* [7.37].

7.3 Theory of Ordering

7.3.1 The Ordering Energy

For many years, the formation of superlattices was interpreted in terms of two linked concepts – *pairwise interaction*, and the *ordering energy*. Pairwise interaction (better, *pair approximation*) implies the simplifying hypothesis that the free energy of a crystalline solid solution contains a term which can be expressed as the sum of the interaction energies of atoms taken two at a time, without taking any account of larger groupings (hence the alternative name for this approach, *quasi-chemical theory*); commonly, only the interactions between nearest neighbours were considered, although it is perfectly possible to take into account second-nearest neighbours and even more distant pairs as well. (Without doing this, no sense can be made of certain superlattices such as CuPt and Fe_3Al; CuPt, for instance, has the same number of first-nearest unlike neighbours whether it is ordered or disordered, and only the number of second-nearest Cu–Pt pairs is altered by ordering.

On the pair approximation, and taking account of nearest neighbours only, the ordering energy is defined as

$$V = \tfrac{1}{2}(V_{AA} + V_{BB}) - V_{AB} , \tag{7.3}$$

where V_{AA}, V_{BB} and V_{AB} are the energies associated with AA, BB and AB atom pairs, respectively. Clearly, if V is negative, then the internal energy (enthalpy) is reduced by converting a random assemblage into an ordered one.

Making use of this simple approach, Bragg and Williams, in the classic paper already cited [7.10], were the first to work out the statistical mechanics of the ordering reaction for simple systems without change of symmetry, (AB systems such as CuZn and A_3B systems such as Cu_3Au). Their approach was made familiar among a wider audience of metallurgists by *Cottrell* [7.38]. Bragg and Williams were able to make sense of the different forms of the S vs. T curves for these two types of alloys (see Fig. 7.5, above) in statistical mechanics terms. While the internal energy is always lowered by ordering when V is negative (and Bragg and Williams only considered this situation), the configurational entropy is of course much greater for random alloy, because of the larger number of microstates corresponding to one macrostate. The increasingly steep decrease of S as the temperature rises towards T_c is due to the fact that in an almost perfect state of order, each displacement of an atom to a "wrong" location costs a great deal of energy, but when the order is already distinctly imperfect, another wrong atom can be achieved at a much slighter cost in energy. This is a perfect example of the statistical mechanics of a *cooperative* process. The discontinuous drop in S for Cu_3Au also comes naturally out of this simple theory.

A by-product of the Bragg–Williams theory is the relation

$$T_c = -zV/2k , \tag{7.4}$$

where z is the coordination number. Direct measurement of the 'latent heat' associated with ordering in CuZn gave a value in fair agreement with that calculated from the observed value of T_c.

Adopting a more elaborate definition of order parameter (he used a distinct parameter, S_i, for each simple cubic sublattice of the fcc lattice), Shockley (who later went on to play a major part in the invention of the transistor) was able to rationalise the CuAu superlattice with its change of symmetry, and also deduced the first simple phase diagram for an ordered system with two-phase fields [7.39, 14].

It is important to be clear that the ordering energy, even in a very strongly ordered alloy that remains up to its melting-point, is merely a small perturbation on top of the basic cohesive energy of the alloy, which is there whether or not the atoms are ordered among themselves. De Fontaine (private communication) has likened the ordering energy to a little bird on the back of an elephant.

7.3.2 The Cluster Variation Model

In recent years, as already briefly indicated in Sect. 7.1.2, the pair approximation model has given way to the *cluster variation model* (CVM), which has proved to be a most powerful approach. The frequency of occurrence of clusters of specific kinds of atoms, typically a 4-fold cluster, on a specified group of lattice sites (as exemplified in Fig. 7.6, above) are calculated; each cluster has its own characteristic internal energy and the configurational entropy can be calculated in terms of cluster frequencies.

The CVM approach was originally proposed by *Kikuchi* [7.40] and for many years developed by him. An interim report was published by *de Fontaine* [7.41] as part of a monumental overview of the configurational thermo-dynamics of solid solutions. Later, it was shown [7.42] that "CVM actually provides a completely general and optimal way of describing partial order"; these words are borrowed from de Fontaine's excellent survey of the use of CVM for calculating alloy phase diagrams [7.43]. The matter is also discussed, at a high level of mathematical abstraction, by Inden and Pitsch in the chapter on atomic order already cited [7.20]. T_c can be calculated on the basis of a CVM approach; pair interactions and the concept of an ordering energy still apply, but now in groups applying to the various pairs in a cluster. According to *Sigli* and *Sanchez* [7.44], the CVM approach gives quite a different relationship between T_c and V than does the Bragg–Williams approach. *De Fontaine* [7.41] compares the theoretical reliability of CVM and Bragg–Williams, by comparing T_c and "latent heats" of ordering as calculated by means of the critical physics (Ising) approach and by these two methods. He claims that CVM gives results within typically 7% of "critical" calculations, but Bragg–Williams only comes within $\approx 60\%$.

Another approach for describing the gradual establishment of LRO is due to Khachaturyan, and is dealt with in detail in his book [7.45], and also briefly by *de Fontaine* [7.41]. *Static concentration waves*, alternatively *ordering waves*, are formed with a fixed wavelength of the order of the lattice parameter but with progressively increasing amplitude; any such wave represents a variation in the concentration of one of the constituent atoms, which is equivalent to progressively moving such atoms to the nodes of a specific sublattice. This approach has computational advantages.

7.3.3 "Criticality Physics"

One of the problems faced by theorists has been to integrate ideas such as CVM and concentration waves with the "criticality physics" approach, already briefly mentioned. CVM and concentration wave theories are "classical" theories, which are incompatible with the key theories of the criticalists. In 1974, *Tanner* and *Leamy* [7.30] called theories of ordering based on critical physics, the "modern theories"; this will hardly do nowadays, since several of the newer theoretical approaches are more modern still! The key critical model is the Ising model in two or three dimensions, with Onsager's exact solution for two dimensions and very good approximations for three dimensions. The application of these ideas to solid state transitions, including ordering or disordering close to T_c (critical theories only apply close to T_c) has been reviewed, for instance, by *Heller* [7.46] and by *Lines* [7.47]. The form of the transition, when critical physics is applicable, is insensitive to the form of the interatomic potential, and thus to parameters such as the ordering energy [7.48]. As we saw in the introduction, the crucial variable that emerges from a critical physics analysis of an ordering reaction is the index β in eq. (7.2). A development that may well help to link the "criticalists" to the classicists is due to *Suzuki* [7.48a]: he introduced the *coherent anomaly method*, which takes results from CVM in different orders of approximation to extrapolate better critical exponents; this approach is now being carefully examined.

Critical physics is only rigorously applicable to second-order transformations, sometimes also called *continuous* transformations. The usual test whether or not a particular ordering reaction in a particular alloy can be true second-order is to apply the Landau–Lifshitz symmetry rules (necessary but not sufficient rules), which apply to the space group of the ordered phase and to the wave vectors of the static concentration waves. These complicated rules are described in some detail by *Khachaturyan* [7.45], *de Fontaine* [7.41], and *Binder* [7.23]. All ordering transitions to the B2 (β-brass) superlattice satisfy the L–L rules, and thus any such superlattice *can* be second-order. The test is how closely the critical exponents satisfy Ising theory. For β-brass, for instance, the test is passed and the ordering reaction in β-brass is certainly second-order [7.21, 46]. It would be interesting to know whether some other ordering transitions of this type, e.g., in FeCo, are second-order or not. (Certain other superlattices, for instance the $L1_2$ or Cu_3Au type, can never be second-order, according to the L–L rules.) Once a transformation is identified as truly second-order, a number of aspects of the transformation are predetermined: for instance, no two-phase (ordered plus disordered) field can appear in the phase diagram in respect of that transformation. These matters are very fully discussed in an up-to-date way by *Binder* [7.23]. I give a wide berth here to tricky concepts such as "nearly second-order" transitions or a "degree of second-orderness"! I shall return to critical concepts, however, in Sect. 7.5 in connection with ordering kinetics.

A particular variant of critical statistical mechanics has been developed to interpret the occurrence of long-period superlattices, met in Sect. 7.1.1. Following years of unsuccessful attempts to understand these structures properly, and in particular, to interpret the varying periodicities in one and two dimensions, the ANNNI (axial next-nearest neighbour Ising) theory has now been developed which goes some way to interpreting the known facts, including some about highly variable long-period superlattices and those with a periodicity which is incommensurate with the underlying lattice. The "axis" referred to is the normal to the repeating antiphase boundaries. An accessible account of the ANNNI theory has been published by *Selke* [7.49].

Another complication in understanding some ordering reactions arises from the possibility that spinodal decomposition (a kind of anti-ordering) is combined with ordering, as sequential processes. The spinodal decomposition can produce long waves of changing composition and then ordering happens in certain parts of the waves. This arcane topic is fully discussed in a major paper by *Soffa* and *Laughlin* [7.50], with illustrations from several real systems. I shall not discuss it further here, but will return to it briefly in Sects. 7.5, 6.

7.3.4 Prediction of Phase Diagrams

One vitally important theoretical issue remains to be discussed: what is the basis of the *prediction* of the form of phase diagrams involving ordering transformations, and also how can the stable crystal structures of particular ordered compounds be predicted? Both relate to an even more basic question: what is the physical nature of the ordering "force", or in other words, what determines whether V in (7.3) is negative or not?

This is a question which has become very much the province of the theoretical physicist, and a metallurgist enters the fray at his peril! I shall say very little here, but refer the reader to some remarkable recent treatments. Traditionally (see for instance *Barrett* [7.6]) three qualitative "ordering forces" have been proposed: (1) the quasi-chemical or pairwise approach, in which bonds are regarded in the same way as bonds in chemical molecules; (2) the strain-relaxation theory, in which long-range order reduces the short-range elastic strains resulting from the coexistence of atoms of different sizes; and (3) the Brillouin-zone approach in which the collective energy of the conduction electrons is altered by a change in structure type and/or unit cell size and shape. As remarked at the outset of this chapter, "explanation" is at the best of times an uncertain concept in science, and the quasi-chemical approach is in effect a *description* of what is observed rather than an explanation.

The other two theories, however, have real physical content. The Brillouin-zone approach has developed gradually from Hume-Rothery's original empirical generalization that phase stability in Cu, Ag and Au-based alloys is closely correlated with the ratio of the numbers of conduction electrons to numbers of atoms... the famous e/A ratio, and the early interpretation of this generalization by Mott and Jones; both these date from the 1930s. The demonstration by "experimental fermiologists", led by Pippard, in the 1950s, that the Fermi surface in pure copper was not spherical as had been assumed by *Mott* and *Jones* [7.51] cast a pall over attempts to interpret the Hume–Rothery rules, of which the e/A criterion is one. Nevertheless, ideas developed gradually; e.g., *Heine* and *Samson* [7.52], working in the same laboratory as Pippard, argued that transition-metal alloys will order if the Fermi energy approximately half fills the d band. Very recently, *Pinski* et al. [7.53] have shown that this latest rule is firmly broken by an alloy of composition NiPt. This alloy has an almost full d band but orders vigorously. Without going into any details of the complex arguments, it is still worth pointing out that the authors, from purely electron-band-based arguments, are able to interpret the ordering and relate it to each of Hume–Rothery's three criteria . . . e/A ratio, the need to have constituent atoms not differing by more than 15% in radius, and a condition relating to the electronegativity of the atoms. The atom-size (alias strain-effect) criterion has received more than its share of contempt from subtle theorists, and it is satisfying, if mystifying also, to see estimates [7.53] of the effective sizes of Ni and Pt atoms emerge from electronic arguments! The important findings in this paper are clearly explained in a commentary in *Nature* by the editor, *Maddox* [7.54]. The paper can be regarded as the beginning of a rapprochement between the metallurgist's approach represented by Hume–Rothery's celebrated rules and the rigorous approach of the theoretical physicist.

A fuller and more general account of the kind of arguments concerning the electronic basis of ordering advanced by Pinski et al. [7.53] was published at about the same time by almost the same group of authors [7.55].

The application of arguments based purely on band theory and those based primarily on CVM to the prediction of ordering phase diagrams has made considerable progress recently; for details, the papers by *de Fontaine* [7.43] (already cited) and by Ducastelle [7.56] should be consulted. These methods, especially the CVM approach, have succeeded both in determining the *ground-state diagrams* (i.e., those relevant to 0 K, based on enthalpy arguments only) and the true phase diagrams, in which entropy arguments are adduced to convert the ground-state calculation into one allowing for changes of temperature.

7.3.5 Prediction of Crystal Structures

This still leaves the issue of predicting the crystal structures of different intermetallic compounds. This is a field where rapid progress has been made in the last few years. Such prediction falls into two categories: a semi-empirical approach, and a first-principles ("absolute") approach. The semi-empirical approach has developed gradually over the years, always based on the notion of a *structure map*. The idea here is that two variables characteristic of a compound are plotted along the Cartesian axes of a graph or map. These have been: average principal quantum number and electronegativity difference between the constituent atoms; atomic size difference and electronegativity difference; and other combinations. Some aspect of electronic behaviour, direct or indirect, always enters into such maps. The hope then is that different domains in the map will correspond to different structure types. The most recent is the *Pettifor map*, in which an imaginary string is run through the periodic table along a particular route and then stretched out, so that each element now has a serial or *Mendeleev number* to identify it. Light elements such as B and C are omitted from consideration. The two axes plot Mendeleev numbers for the two constituent atomic species, for a particular stoichiometry (say, AB_2) and it turns out that different fields do indeed correspond to particular ordered structure types. The discovery was empirical although subsequently quantum-mechanical arguments have been advanced in explanation.

One extension of such maps is to predict the effect of ternary additions on the structure of particular compounds: to achieve this, a third dimension is added to make a three-dimensional "map", which is then depicted in successive sections, each section representing a different concentration of a range of alternative ternary elements (Pettifor called this a *phenomenological structure map*). Figure 7.11 shows such a series of sections for the "vicinity" of NiAl. Any help in predicting, even tentatively, the effect of ternary alloying elements is of great practical importance in the effort to improve industrial intermetallics. The history, present status, underlying theory, use and limitations of structure maps have been set out by *Pettifor* [7.57–59].

Another semi-empirical approach which has proved widely useful is the macroscopic atom method pioneered by Miedema, as a means of estimating the enthalpy of formation of intermetallic compounds, especially those incorporating a transition metal. The name comes from the notion that the effect of putting one solute atom into a pure metal can be approximated by the effect of bringing two free surfaces of the metals together. The "electron density" at the boundaries of the Wigner–Seitz cells is an important parameter in this approach. All sorts of measured property, such as heats of vaporization or surface energies, can be used as empirical input to determine the enthalpies of formation of ordered intermetallics, among other alloy phases, and the Miedema approach has been very widely used to estimate enthalpies where no direct measurements are available. A concise account of the method and its theoretical foundations, as well as the results of numerous calculations, have been published in a recent book [7.60].

7.3.6 First-Principles Calculations

The ultimate objective, towards which the above procedures are waystations, is the calculation of phase *stabilities and phase diagrams from first principles*,

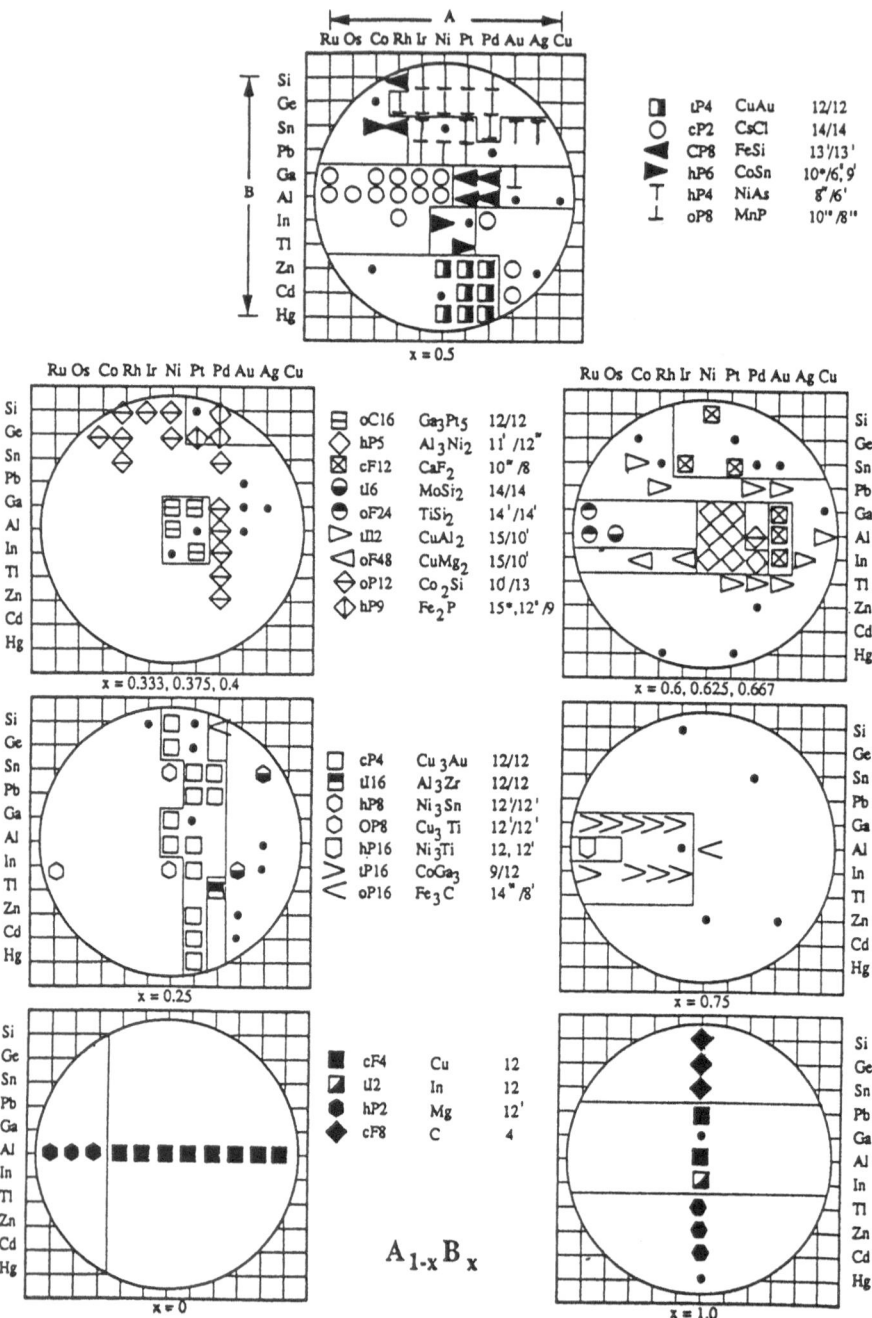

Fig. 7.11. "Neighbourhood map" for NiAl. The successive sections represent different values of x in $A_{1-x}B_x$, where A and B represent either Ni, Al, respectively, or their near neighbours in the periodic table. The two bottom figures represent alloys in which only Ni or only Al is replaced by nearby atom species, the other figures represent cases where both are replaced. The hope is that this kind of map will indicate the effect of moderate ternary alloying on NiAl [7.58, 59]

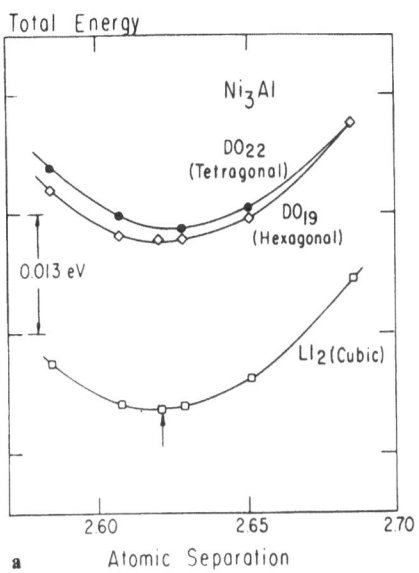

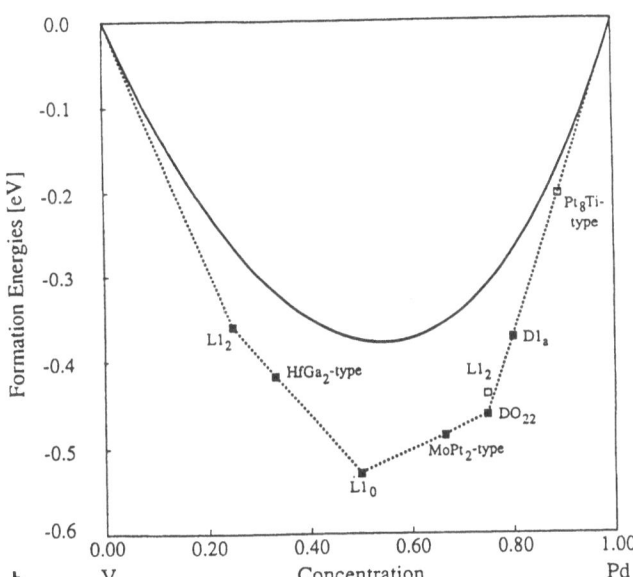

Fig. 7.12. (a) The total energy as a function of the Wigner–Seitz radius and its structure dependence for Ni_3Al. Open and closed circles, and diamonds indicate $L1_2$, DO_{22}, and DO_{19}, respectively [7.61]. (b) Calculated formation energies for Pd–V fcc ground states as a function of Pd concentration. [7.63]

without any adjustable parameters. A sine qua non for any such procedure is the availability of high-speed supercomputers. A number of physicists have made striking progress towards this objective with the help, in particular, of Cray computers. Perhaps the leading practitioner at present is Freeman, in America, and he has published two concise and very accessible overviews of his recent work [7.61, 62]. *Pettifor* [7.59] has also discussed very recent advances in first-principles calculation of phase stabilities. The key approach used by Freeman and several others is *local density functional* theory, which is based on two observables, the total electron (or charge) density and the total energy of the system (involving both kinetic energy of the electrons and electrostatically determined potential energy). By this theory, Freeman and his collaborators have calculated diagrams such as Fig. 7.12a, for Ni_3Al, where the relative ground state energies of three alternative crystal structures are compared as a function of interatomic separation. From the minimum, the lattice constant is calculated, and the elastic bulk modulus, antiphase domain boundary energy and heat of formation can also be determined; all four agree well with experiment [7.61]. It is also possible to calculate a density-of-electronic-states diagram and to find out in detail how this relates to electronic stabilization of a particular structure.

Freeman has also recently begun to combine statistical mechanics with his electron energy approach. (Statistical mechanics are of course essential if one is to go in due course from the ground state to high-temperature stabilities.) This approach, which is the non plus ultra of first-principles calculation, has been taken to considerable lengths by the group headed by *de Fontaine* [7.63]. Using a development of CVM (of the kind explained by *Inden* and *Pitsch* [7.20]) for the Pd–V system, in which the "embedding medium" is assumed to be a 50/50 solid solution for all actual compositions, the ground state diagram of Fig. 7.12b was obtained, with no adjustable or experimentally derived parameters. As de Fontaine points out, there are still unavoidable approximations in the quantum-mechanical input. He also asserts that "predicting, for a binary system, which intermetallic structures will have the lowest energy, for all concentrations, at zero Kelvin, is an impossible task." He means that one cannot compare energies for all conceivable crystal structures! Here, his team did calculations only for the familiar derivatives from the disordered fcc structure.

One advantage of this kind of calculation is that an approximate value of the *ordering energy* can be deduced, by comparing individual calculated points in Fig. 7.12 with the ideal energy of the corresponding disordered alloy, taken from the solid curve.

7.4 Special Experimental Methods

Both long-range and short-range forms of order are characterized by diffraction methods; most commonly, X-rays are used. The most detailed determination of the state of partly long-range ordered alloys and also of short-range ordered alloys requires the use of a single crystal, which then allows a proper study of regions of the reciprocal lattice between lattice points. This approach, which falls beyond the scope of this overview, is most commonly used for the study of SRO; an excellent example is the detailed study of SRO in the much-examined α-phase Cu–Al alloy [7.64]. The state of LRO in intermetallics is most commonly examined by polycrystal diffractometry, however.

When a simple superlattice such as CuZn or Cu_3Au is examined by diffraction, the structure factors of the superlattice and fundamental lines are in the ratio $(f_A - f_B)/(f_A + f_B)$, the fs being the atomic scattering factors. The intensities are roughly in a ratio given by the squares of these quantities (geometrical factors enter as well). This poses no problem with a phase such as Cu_3Au, where the two scattering factors are very different, but for CuZn or Ni_3Fe where the constituent atoms are close in the periodic table, the super-lattice lines may be too weak to be measurable above background, especially if the domain size is small and the superlattice lines are therefore broadened. Reducing the background by monochromatisation and/or by using a wave-length-discriminating detector helps, but sometimes more drastic measures are needed. CuZn was originally studied [7.65] by using an X-ray wavelength ($ZnK\alpha$) close to the absorption edge of one of the constituent atoms, which generates *anomalous dispersion* and greatly enhances the difference ($f_A - f_B$). Nowadays, the same objective is achieved more efficiently by using synchrotron X-radiation in conjunction with a double-crystal wavelength selector.

Recently, synchrotron radiation has been used to determine the locations of ternary alloying additions to binary ordered phases. *Marty* et al. [7.66] exam-ined the γ phase (based on Ni_3Al) in Ni–Al–X alloys, where X was Ti or Cr, at levels of 3.5–10 at. %, and X-ray wavelengths near the absorption edges of the constituent elements were used. This gave sufficient data for the third atom sites to be deduced. Ti goes largely to the Al sites, Cr is distributed between Ni and Al sites. These experimental findings agree well with predictions made independ-ently by a CVM calculation [7.67]. A similar study was made of the B2 phase, FeCo, with 2 at. % of V added; this is done industrially to enhance the mechanical properties of this useful magnetic alloy. *Williams* et al. [7.68] used synchrotron radiation, with wavelengths (a) remote from all absorption edges, and (b) close to each of the three absorption edges in turn, and in addition neutron diffraction was used. This gave five linearly independent structure factors for the ternary alloy; using Rietveld refinement (as normally employed in crystal structure determination) it was possible to demonstrate that V occupies Co sites preferentially, contrary to earlier studies. It is clearly very difficult to locate 2 at. % of a constituent; this is why the use of multiple wavelengths, with the additional information obtained thereby, was important.

CuZn has also been studied by means of neutron diffraction, since the neutron scattering cross-sections for Cu and Zn differ by more than the X-ray scattering factors. This is what *Als-Nielsen* and *Dietrich* [7.69, 21] used to verify that the ordering transition in CuZn is of strictly second-order nature. Diffuse neutron scattering has been extensively applied, especially by Kostorz, to the study of SRO in various alloys. Thus *Klaiber* et al. [7.70] studied SRO in a Ni-10 at. % Al alloy and found a preference for localised configurations resembling the $L1_2$ superstructure type. *Kostorz* [7.71] was able by neutron scattering to identify strong SRO in Cu-30 at. % Zn (containing ^{65}Cu isotope which scatters neutrons efficiently) aged for 21 d at 160°C, slightly below the presumed T_c of this very sluggishly ordering alloy. Figure 7.13 shows some of his results: the

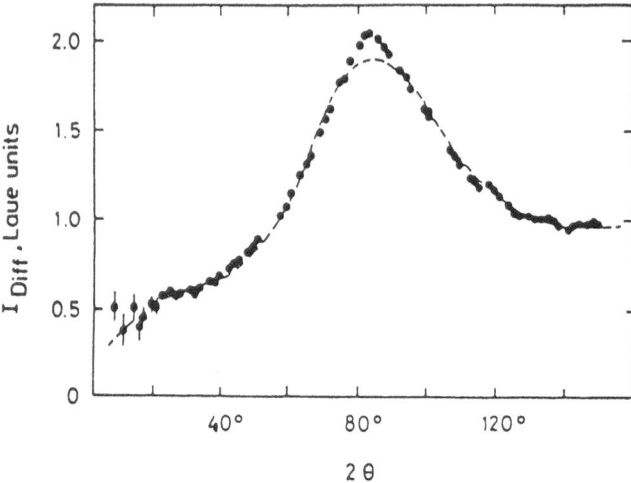

Fig. 7.13. Diffuse neutron scattering intensity, I_{Diff}, as a function of scattering angle , 2θ, for ^{65}Cu-30 at. % Zn aged 21 days at 160°C [7.71]

diffuse scattering peak is not quite at the position corresponding to a super-lattice line, as is quite commonly found with SRO.

Field-ion microscopy is increasingly being used to study the fine detail of the process of ordering in appropriate alloys. A good example of what can be done is a study by *Kuwano* et al. [7.72] of ordering of fcc CuPt to form the $L1_1$ superlattice, by a first-order process. These investigators observed the nucleation of extremely small APDs (1–2 nm across), followed by their growth and a simultaneous gradual increase of order within the domains, while the matrix remained obstinately disordered.

Mössbauer spectrometry, the subject-matter of the last chapter of this book, is finding application as a means to sense the changing local environment of atoms as order builds up, for instance in the work by Fultz to be mentioned in the next section, and a study by *Yoshida* et al. [7.73] in which SRO in Au–Fe was studied, as a function of temperature, by high-temperature Mössbauer spectrometry. This technique is, of course, limited to a few appropriate atomic species.

Another recently developed experimental method has shown potential for locating atoms in hitherto unfamiliar ordered phases. This is the ALCHEMI method (the acronym denotes Atom Location by CHanneling-Enhanced MIcro-analysis), introduced by Spence and Taftø [7.74] and more recently explained by *Banerjee* [7.75]. Here, characteristic X-rays are excited as usual by a beam of electrons, but the beam is "channeled" in a specific direction relative to the unit cell of a single crystal or crystal grain. In a ternary $A_xB_yC_z$ alloy of known composition but unknown crystal structure, the relative intensities of X-rays characteristic of A, B and C depends on the channeling direction of the electron beam, and from these intensities, if the structure is not too complicated, the locations of the different atomic species can be determined. The method was used to very good effect to determine the structure of a previously unknown orthorhombic ordered phase, based on the ideal composition Ti_2AlNb, in the Ti–Al–Nb system [7.76]. It has also been used by the same investigators to identify the site substitution behaviour of quaternary additions to Ti_2AlNb.

7.5 Ordering Kinetics and Disorder Trapping

A reversibly ordered alloy – one with a critical temperature below its melting-point – can often be obtained metastably in a substantially disordered form by rapid quenching from a temperature at which it is disordered in equilibrium. This is readily done, for instance, with Cu_3Au, FeCo, Ni_4Mo, CuPt, Co_3V and a number of others. If, however, T_c is very high, as for instance for Ti_3Al ($T_c = 1150°C$), then it is generally impossible to quench out disorder, since ordering during the quench is unavoidable. In that case, it is of course not possible to examine ordering kinetics, since no disordered starting material is available.

At the other extreme, if an alloy has a low T_c, say 150–250°C, (like, for instance, $CuAu_3$, with $T_c = 240°C$), then it may be difficult or even impossible to produce substantial order because of sluggish ordering kinetics.

7.5.1 Disorder Trapping

Special interest has attached to attempts to quench out disorder in Ni_3Al, because of its practical importance, so that physical and mechanical properties might be examined. This alloy is permanently ordered, with $T_m < T_c$, so the only way disorder can be "trapped" is by crystallizing so fast that the superlattice has no time to form during quenching. Recently, the general problem has been treated theoretically, by an extension [7.77] of Aziz's theory of "solute trapping" during rapid solidification. Solute trapping, in a given alloy, may be easier or more difficult than disorder trapping. *Boettinger* and *Aziz* [7.77] calculate that for TiAl and Al_3Nb, the critical crystal growth rate to trap disorder should be ≈ 60 and 200 cm/s, respectively; in fact, TiAl has been quenched in disordered form, Al_3Nb has not. The authors point out that Al_3Nb (but not TiAl) is a "line compound", i.e., has no measurable homogeneity range; this is normally an indication of a very steep-sided free energy minimum at stoichiometry, which in turn implies very strong ordering. Boettinger and Aziz did not make any calculation for Ni_3Al.

Various attempts have been made to trap disorder in Ni_3Al. Melt-spinning, an effective form of rapid solidification, does not work for the binary alloy, but has been found to succeed in disordering a less strongly ordered

$Ni_3Al + 13$ at. % Fe ternary alloy [7.78]. *Cahn* et al. [7.79] used spark-erosion as a means of quenching from the melt, and found that the same $Ni_3Al + Fe$ alloy could be only partly disordered by this method. However, quite recently, binary Ni_3Al has been completely disordered by evaporating the alloy on to a cold substrate [7.80] or by transient melting of a film by pulsed laser [7.81]; both of these give very fast effective quenching. It is also possible to disorder Ni_3Al completely by ball-milling [7.82]; the reordering kinetics of the milled powder was then examined. Details will be found in a general survey of ordering kinetics by *Cahn* [7.83]. Reordering of the disordered films just mentioned has not yet been studied in detail.

Special interest attaches to the quenchability of disorder in alloys which undergo a second-order phase transformation. B2 CuZn has quite a low T_c of $\approx 460°C$, yet it is impossible to quench out disorder. However rapidly it is cooled, CuZn orders fully. (However, no quenching from the melt appears to have been attempted.) Some students of this alloy have asserted that the impossibility of preventing ordering is a direct consequence of the second-order nature of the transition, in the sense that no activation barrier needs to be overcome to create order, but I cannot find any proper discussion of this issue. In fact, the ordering of rapidly quenched CuZn can be explained purely in terms of anomalously fast diffusion. Another alloy of the same crystallographic kind, FeCo, which some have asserted to be second-order in its transition (but I have seen no proper check on this assertion, by measurement of critical exponents), *can* be disordered by water-quenching from above T_c, which here is much higher at 720°C. If it turns out that FeCo is *not* second-order in its transition, then the comparison of this alloy with CuZn would suggests that in alloys ordering by a second-order mechanism disorder cannot be quenched out. This is an issue worthy of closer attention than it has had hitherto.

It is a curious aspect of critical behaviour in a phase that orders by a second-order mechanism that, very close to T_c, the rate of ordering (or, just above T_c) the rate of disordering is very sharply *reduced*. This is called the *critical slowing down*, and is interpreted according to the Ising critical model. Although according to Ising theory, this should be wholly restricted to second-order transformations, it also appears clearly in first-order transitions, for example in

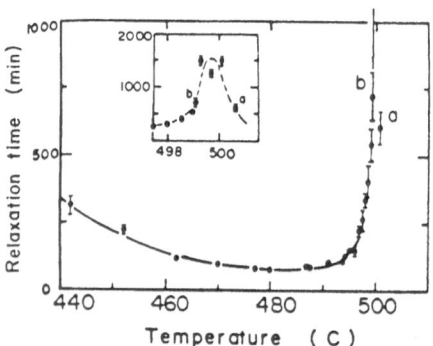

Fig. 7.14. Relaxation time of the Bragg order parameter, S, of Ni_3Mn ($L1_2$), held isothermally at various temperatures, after sudden change of temperature from a temperature below T_c [7.84]

Ni$_3$Mn (a L1$_2$ transition which cannot be second-order according to the L–L rules). Figure 7.14, taken from a classic study by *Collins* and *Teh* [7.84] shows this. Cu$_3$Au shows similar behaviour (see survey by *Cahn* [7.85]). At the time this quasi-critical behaviour by first-order superlattices was discovered, there were many protests that it was improper to describe this behaviour as being critical. This objection was answered by *Bolton* and *Leng* [7.86], who applied the Bragg–Williams phenomenological approach to show theoretically that A$_3$B alloys, even though Ising theory cannot apply to them, nevertheless should give a relaxation time $\tau = (T_c - T)^{-4}$ near T_c; this is the form characteristic of Ising critical slowing down. In other words, the superlattices forming by first-order transformations show *quasi-critical* behaviour. What this shows is that the Bragg–Williams model refuses to die, and that the battle of words between criticalists and classical physicists will not die either!

7.5.2 Phases with Low Critical Temperatures

There are two alloys which have, or are believed to have L1$_2$ superlattices but with very low critical temperatures, around 150°C, namely, Cu$_3$Zn and Cu$_3$Al. The slow incipient ordering in these alloys has been studied mainly by calorimetry; an X-ray study [7.64] and a neutron study [7.70] have been mentioned above. The most interesting alloy is FeNi, with an estimated T_c of 321°C; this temperature was identified by *Paulevé* et al. [7.87], who annealed the disordered alloy in an intense neutron flux. This enhanced the vacancy population and thus the diffusivity, and permitted the L1$_0$ (CuAu type) superlattice to form in spite of the low temperature. Later, *Gros* and *Paulevé* [7.88] used Mössbauer spectrometry to confirm the formation of the superlattice. The same superlattice has been observed in large meteorites, estimated to have cooled over millions of years in space [7.89]; here, presumably, order was induced without benefit of neutrons.

The details of ordering kinetics of normal reversibly ordered alloys, such as Cu$_3$Au and Ni$_3$Fe, would take us too far afield here. Suffice it to refer the reader to two reviews by the author [7.83, 85] and to a classical study of ordering kinetics in Cu$_3$Au, using electrical resistivity measurements, by *Burns* and *Quimby* [7.90]. *Cahn* [7.85] has attempted to relate ordering kinetics of L1$_2$ alloys to diffusivities in the various alloys, and has also surveyed the successive theoretical treatments of ordering kinetics. The most recent of these, not covered in Cahn's review, is Fultz's approach to ordering kinetics, which is particularly interesting in connection with alloys such as Fe$_3$Al in which ordering can pass through a transient metastable form of order (imperfect B2 order) before moving on to the stable form, the DO$_3$ Fe$_3$Al-type superlattice. In such alloys, Fultz shows that the alloy can follow alternative, quite distinct, ordering paths; the path actually chosen, which can be depicted on a form of *ordering map*, depends on the temperature. Thus, in Fe$_3$Al the formation of B2-like order tends to predominate over DO$_3$-like order as the temperature of isothermal ordering is lowered. Hyperfine magnetic field distributions deduced from Mössbauer spectrometry are used to obtain the necessary experimental information. The reader is referred to a recent review by *Fultz* [7.91]. (The uses of Mössbauer spectrometry with regard to NiFe and Fe_3Al indicate that this technique has an increasing role to play in the study of ordering, because when suitable atomic species are invoked, it can identify the local environment of particular kinds of atom.)

7.5.3 Rapidly Ordering Phases

It is particularly difficult to measure ordering rates when the whole process is so fast as to be complete within a few seconds. Recently, the use of *in-situ time-resolved X-ray diffraction* has been applied to deal with this problem [7.92]. The very intense X-ray beam from a synchrotron source was monochromated and the diffracted beam then measured, typically every 100 ms, by means of a position–sensitive detector. The technique was applied to measure the increase

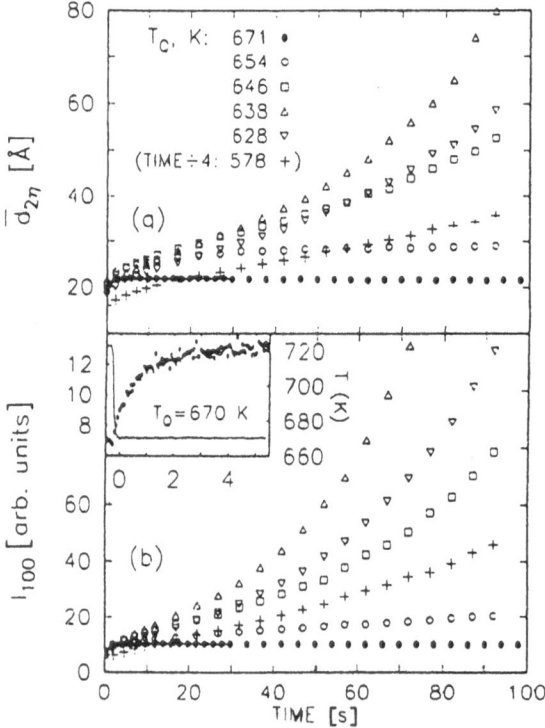

Fig. 7.15. (a) Average domain size, $d_{2\eta}$, and (b) (1 0 0) superlattice line intensity, I_{100} as a function of holding time for various T_Q. Inset: Typical temperature profile (solid line) and intial exponential relaxation of I_{100} (points) (above T_c) [7.92]

of (100) superlattice line intensity, and the growth of antiphase domains, in Cu_3Au quenched from above T_c to various temperatures, T_Q, not far below T_c. Figure 7.15 shows some results. The results were analysed in relation to the classical theory of spinodal decomposition followed by "continuous" ordering [7.93, 94, 50], and it was possible from results such as those in Fig. 7.13 to identify a spinodal limiting temperature at 356°C, 34 K below the critical temperature of 390°C. Above the spinodal limiting temperature, ordering proceeds by nucleation and growth, below, the ordering that follows the creation of spinodal waves is of a "continuous", i.e., homogeneous character. It would seem that these results are not altogether consistent with the earlier observations of pseudo-critical slowing down mentioned in the preceding section! The use of time-resolved XRD certainly will have much to contribute to the study of ordering, and disordering, reactions.

In addition to the many studies of ordering kinetics, attention has also been paid to the kinetics of antiphase domain growth; a $t^{1/2}$ law has been confirmed in a thorough study by *Allen* and *Cahn* [7.95].

7.6 Computer Simulation of Ordering and Disordering and Related Features

The use of Monte Carlo simulation to illuminate various aspects of the order–disorder transformation has become almost routine and here I have space only to cite a very few examples to illustrate what can be done.

A very early example of Monte Carlo simulation in this field is the work by *Beeler* [7.96], in 1965. He was interested in a problem which has excited a great deal of interest over the years, see recent reviews by *Bakker* [7.97, 98], the mechanism of atomic diffusion in a superlattice. The central problem here is that, in some forms of superlattice (notably in B2, to which most theorising has been applied) any exchange of an atom with a neighbouring vacancy is bound to create several wrong bonds and the activation energy for jumps thus becomes very high. Moreover, successive jumps may enhance the consequential disorder. This has led to complicated notions such as the six-jump mechanism – a sequence of jumps which minimises consequential disorder. Beeler was concerned specifically with the behaviour of individual vacancies is such a diffusion process. He discovered the *contraction effect*, a tendency for a migrating vacancy in a superlattice to visit a few sites repeatedly and thus to "contract" into a small migration volume. He also was the first to deduce, from such simulation, for both B2 and $L1_2$ superlattices, how rapidly "wrong" bonds are replaced by "right" bonds during simulated annealing of an initially disordered alloy . . . the central issue in the interrelation of diffusion and ordering, a problem discussed in detail by *Cahn* [7.85].

Experimentally, it is known that in the same B2 alloy, compared in the ordered and disordered forms, the activation energy for diffusion is much higher when there is a superlattice. This was verified for CuZn in a famous study [7.99]. For the $L1_2$ superlattice, there appears to be no change in activation energy, because this superlattice in any case includes a number of like neighbours and the energy penalty of replacing an unlike bond by a like bond is much slighter [7.85].

More recently, some attention has turned to diffusion behaviour in phases such as NiAl and CoGa which contain *constitutional vacancies*, i.e., large vacancy concentrations which do not depend on temperature. Thus *Bakker* et al. [7.100] performed a Monte Carlo simulation of diffusion in CoGa. They paid special attention to the *correlation factor*, a key concept in the theory of diffusion. They were able to interpret the change in activation energy at T_c, which had also been observed experimentally.

Computer simulation, in place of analytical calculation, has also been used to examine the ordering transformation in all its complexities. Perhaps the most impressive use of this approach is a very recent one by *Chen* and *Khachaturyan* [7.101, 102]. They simulated the precipitation of an ordered intermetallic phase from a disordered solid solution in a model binary alloy. The motive was to understand the precipitation of an $L1_2$ phase, Al_3Li, from a disordered supersaturated Al–Li solid solution, a system of great current practical importance. They also simulated a similar decomposition in Ni–Al alloys. They were able to demonstrate a primary process of congruent but metastable ordering, i.e., the formation of a superlattice with the same composition as the disordered starting material, which is not the composition of the stable ordered phase. This metastable superlattice then decomposes spinodally into an ordered phase with the right composition plus a disordered phase, which in fact nucleated at the antiphase domain boundaries of the metastable ordered phase. I return to this kind of heterogeneous nucleation of disorder in the next section. All this would no doubt be too complex a sequence to disentangle by purely analytical means, but the results of the simulation agree with experimental observations on the two alloys. This work shows the power of computer simulation when effectively applied.

7.7 Ordering and Disordering at Free Surfaces, Interfaces and at Antiphase Domain Boundaries

7.7.1 Free Surfaces

It has been known for some years that the order–disorder transition may not be the same at a free surface as it is in the bulk of an alloy. The first recognition of this came in 1974 from a LEED study of a (1 0 0) surface of Cu_3Au [7.103]: it was found that very close to the surface, the Bragg order parameter diminished *continuously* to zero at T_c, instead of showing the usual discontinuous drop as seen in Fig. 7.5. A later study by *Buck* et al. [7.104], with modern refinements,

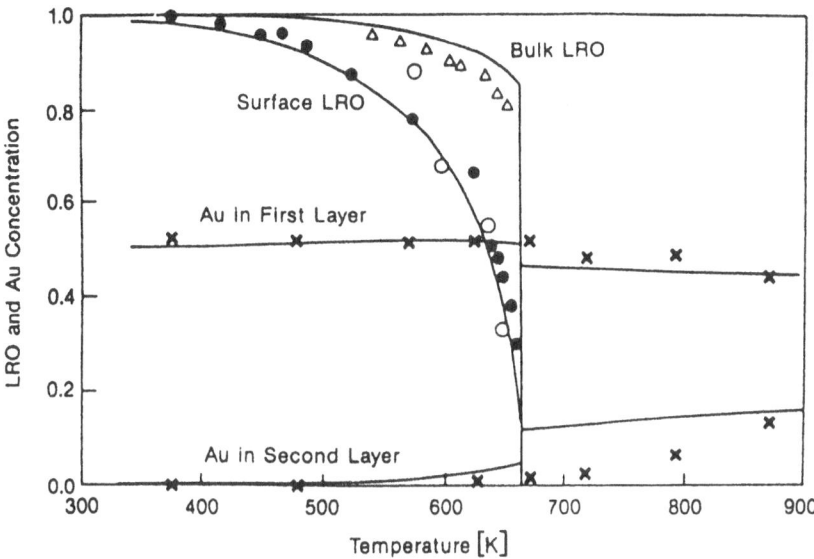

Fig. 7.16. Comparison between calculated (full lines) and experimental values (points) for Cu_3Au. Open and closed circles, the LRO parameter at an (001) surface. Crosses, the Au concentration in the first two layers [7.105]

confirmed this finding. The mean composition is also changed in the surface layer. The form of the variation of S with T is quite different at the surface and in the bulk, but T_c has the same value for both. A classical statistical mechanics treatment by *Sanchez* and *Morán-López* [7.105] gave results in close agreement with these experimental findings, as is shown in Fig. 7.16. The same problem was handled in terms of critical physics by *Lipowsky* [7.106, 107]: he concluded that a phase behaving classically and undergoing a first-order (discontinuous) phase transition in the bulk converts to a critical (second-order, continuous) mode of behaviour at the surface . . . which is what is observed. He also showed [7.108] that the preferentially *disordered* layer at the surface is in fact "macroscopic", i.e., stretches over several atomic layers. (A more accessibly written overview of Lipowsky's ideas can be found in *Ferroelectrics* [7.109].) One reflection induced by this group of studies is that, once again, classical and critical physics approaches to the same problem give similar quality of agreement with experiment, in spite of the spurts of hostility between the adherents of the two approaches! (This is why I have decided not to follow, in this chapter, the instruction I have received from one theorist to use different terms, *critical temperature* (T_c) and *transition temperature* (T_{tr}), for the temperature at which LRO disappears for second-order (critical) and first-order (classical) transformations, respectively.)

Other studies of anomalous ordering or disordering near surfaces have begun to appear, for instance, a Japanese study by field-ion microscopy of the surface of Ni_4Mo. This extensive work was reviewed by the investigators [7.110]. On curved surfaces, at lower temperatures they found *preferential ordering* on particular lattice planes, (1 0 0) and (1 1 1) of the disordered fcc structure, while near T_c, they again found preferential surface disordering.

A general survey of the results of electron-microscopic examination of "pre-transition" effects in alloys, including a number of order–disorder transitions, has been published recently by *van Tendeloo* et al. [7.111].

7.7.2 Interfaces

Recently, this kind of study has been extended from free surfaces to *interfaces*. Where grain boundaries are concerned, the studies have been purely experimental and all concerned with Ni_3Al, a permanently ordered phase. By TEM, evidence has been reported of the stable existence of a very thin disordered layer at grain boundaries, e.g., [7.112]. These findings have been hotly disputed and asserted by some to be electron microscopic artefacts. The latest situation, as reported by *Kung* et al. [7.113], is that disordered zones (≈ 1.5 nm thick) are found only in large-angle general boundaries, and then only when the alloy is doped with boron (as it generally is for practical use, because this ductilises the normally brittle polycrystalline alloy). As we saw above, the fact that APDBs are formed during the recrystallization, below T_c, of deformed Cu_3Au suggests that in this alloy, most grain boundaries, as they migrate, carry thin disordered zones along with them. No attempts at theoretical interpretation have been published.

Another category of experiment has referred to the role of antiphase domain boundaries in disordering, and here there is both better agreement as to the facts and better theoretical understanding. Some years ago, TEM studies of ternary alloys based on Ni_3Al indicated that in alloys which, under heat-treatment, should change from a single-phase $L1_2$ superlattice into a two-phase (ordered + disordered) structure with the two phases of different compositions, the disordered phase nucleates exclusively at APDBs in the initial ordered phase, e.g., see [7.26]. Even before this, a theoretical analysis in Russia [7.114] predicted a reduction in order close to an APDB, in a way similar to what is seen in Fig. 7.16; this work is reviewed by Morris in the paper referred to earlier [7.37].

Both experimental and theoretical aspects of this issue have now been taken to a rigorous level by some excellent work done in France [7.115, 116]. Experimentally [7.115], the $L1_2 \rightarrow$ Al transformation was studied by both TEM and HREM in a $Co_{30}Pt_{70}$ alloy. It was observed that, at temperatures starting about 40 K below T_c, APDBs began to thicken and did so increasingly as the temperature was brought closer to T_c. (The technical name for this in critical physics theory is "wetting of the APDBs by the ordered phase".) Detailed study showed that there were in fact two competing disordering processes: wetting of the APDBs, and nucleation and growth of the disordered phase inside ordered

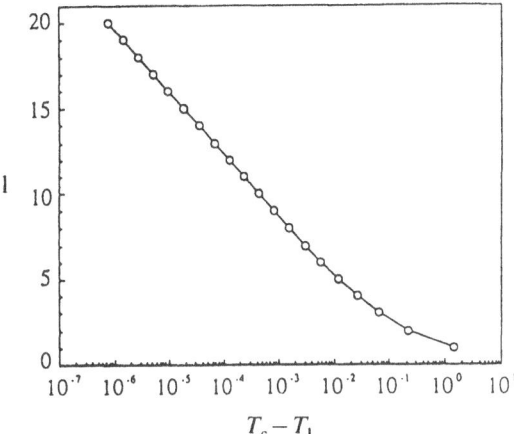

Fig. **7.17.** Half-thickness of the wetting layer (disordered zone), l, vs $\log_{10}(T_c - T)$. Logarithmic behaviour, as predicted from critical theory, is observed over more than 4 decades [7.116]

domains. The heating rate has a major effect on this competition. Thickening of APDBs near T_c had been observed by various earlier investigators, but this latest work has established the process in detail. One aspect which is not made very explicit in this work, but seems to match what is reported, is that the catalytic effect of APDBs is restricted to *non-conservative* APDBs (see the distinction explained in Fig. 7.7, above).

The theoretical counterpart of this work [7.116] (following an earlier treatment by *Sanchez et al.* [7.117]) applies CVM methods to the interpretation of the behaviour of a non-conservative (1 0 0) APDB as T_c is approached. This very interesting paper shows, among other things, that as the temperature rises, there is an infinite series of first-order transitions associated with entropy jumps, each corresponding to the disordering of an additional plane of atoms adjacent to the APDB; this in fact assures complete "wetting" of the APDB by the time T_c is reached. Figure 7.17 shows the half-thickness of the wetted layer as a function of $(T_c - T)$. The critical exponent of the straight line is -1, which is known to be characteristic for wetting phenomena in general.

The nucleation of disorder at an APDB and the gradual thickening of the disordered zone as T_c is approached can also be seen in the images of computer simulation of disordering associated with spinodal decomposition, by *Chen* and *Khachaturyan* [7.101, 102], mentioned above.

7.8 Magnetic and Atomic Order

Magnetic spins and atomic sites behave in ways which, in statistical mechanics terms, are very similar. The plot of the magnetization of a ferromagnet as a function of temperature, up to the Curie (critical) temperature is similar to that of the Brass LRO parameter in CuZn (Fig. 7.5); similarly, short-range magnetic order in the paramagnetic domain has points of similarity with SRO above the critical temperature for the disappearance of LRO. It is therefore not surprising that magnetic interactions can influence atomic ordering, and differential interactions between like and unlike nearest neighbours can modify the normal dependence of magnetic order on temperature. The phenomena are varied and the literature is large.

The first survey of these two-way interrelations was in 1969, by *Kouvel* [7.118], who first considered the effect of local environment on the atomic moments of individual atoms. *Miodownik* [7.119] discussed the effect of magnetic interactions on the form of phase diagrams, while *Inden* [7.120] examined the behaviour of the Fe–Al, Fe–Si and Co–Pt systems in detail. A very complete survey, taking in both experimental and theoretical aspects, and with over 200 citations to the literature, was published recently by *Cadeville* and *Morán-López* [7.121]. The topic has become too large for me to do more than cite a few examples of what has been done.

Perhaps the most detailed experimental study of the influence of magnetic interactions on the order–disorder transition is by *van Deen* and *van der Woude* [7.122], who examined Ni_3Fe (a $L1_2$ superlattice). They were able to show that there is a range of a few degrees just below T_c in which magnetic interactions govern the stability of the superlattice. In such cases, an applied field can affect the stability of order.

A good deal of information about the state of order can be obtained by measurements of electrical resistivity, making use of the accurate theory of resistivity in ordered alloys due to *Rossiter* [7.123]. *Leroux* et al. [7.124] made use of Rossiter's theoretical expression for resistivity, with the addition of a term taking account of spin-disorder scattering, to interpret the behaviour of $Co_{30}Pt_{70}$ (an off-stoichiometric $L1_2$ alloy) in the paramagnetic range. Here magnetic order is enhanced by increased chemical disorder. The magnitude of the effect was identified by comparison with $Ni_{30}Pt_{70}$, in which spin-disorder scattering does not play a role. The effect of the state of order (equilibrium or metastable) on Curie temperatures and the reverse influence of magnetic interactions on the phase diagram were measured and theoretically interpreted by *Sanchez* et al. [7.125]. Figure 7.18 shows both experiment and theory for both these features.

7.8.1 Directional Order

A quite different kind of influence of LRO on magnetic behaviour arises in connection with *induced magnetic anisotropy*. When a ferromagnetic solid solution alloy is annealed either in a magnetic field or under applied stress in the elastic range, the alloy finishes up with a magnetic anisotropy induced by the applied influence. (The high temperature is needed to allow atoms to move about.) In the 1950s, this phenomenon was shown (independently by Néel and by Taniguchi) to be due to the creation of a state of *directional order*, in the sense

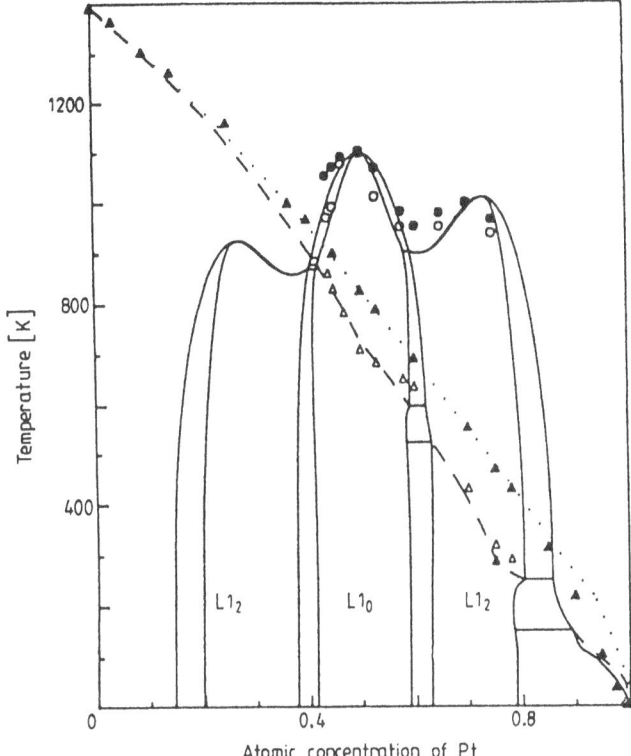

Fig. 7.18. The phase diagram for $Co_{1-x}Pt_x$. The calculated order–disorder diagram is shown by full lines and the experimental results, by full and open circles. The calculated magnetic Curie temperatures are shown by dotted (metastable, disordered) and broken (equilibrium) lines. The corresponding experimental results are shown by full and open triangles, respectively [7.125]

that the number of unlike nearest atom neighbours differ slightly parallel to the field (or stress) and perpendicular to it. The effect is highly sensitive to small order anisotropies. (Néel and Taniguchi's treatments were theoretical; until recently, the actual existence of directional short-range order had not been demonstrated, but recently, its reality has been demonstrated by very precise X-ray diffraction in various azimuths of a metallic glass with induced magnetic anisotropy [7.126, 127].) Many years ago, *Birkenbeil* and *Cahn* [7.128] studied the creation of magnetic anisotropy in Fe_3Al induced by an applied stress, and proved that when this alloy was long-range ordered, the induced anisotropy was negligible, while alloys deviating somewhat from stoichiometry (where LRO cannot be perfect) showed substantial induced anisotropy. Here, there was clearly interference between (isotropic) LRO and anisotropic induced directional SRO.

7.9 Ordering in Semiconductors and Other Non-Metals

In addition to the great mass of research on long-range ordering in metallic solid solutions, there are separate subcultures of atomic (spatial) ordering in minerals and in compound semiconductors. I have space only to give a brief summary of these two subcultures, both of which have become quite extensive in recent years.

7.9.1 Minerals

In minerals, attention has been paid mainly to feldspars. A precocious early study, in 1952, was due to Laves, who studied albite, $NaAlSiO_8$.. In this mineral, Na and Al ions occupy distinct sublattices. *Laves* [7.129] showed that this mineral could be turned by heat-treatment into analbite, in which the distribution of Na and Al atoms has been made random. Laves further showed that under shear, analbite can form deformation twins, albite cannot. The presence of LRO implies that a shear deformation would represent a martensitic phase transformation instead of a reorientation without change of phase, and this implies too large an energy penalty. Other feldspars, such as leucite, $K(Al, Si_2)O_6$, undergo more complicated kinds of order–disorder transition (though here also, some disordering of Al and Si is believed to happen). Many of the ordered crystals show a special kind of "superelasticity" which causes them to be called *ferroelastic* crystals, by analogy with ferroelectrics. It would take us too far to go into details here; the reader is referred to a very up-to-date text by *Salje* [7.130].

7.9.2 Semiconductors

Where semiconductors are concerned, the literature on LRO, although it is already quite extensive, is very recent. The observations stem from research on semiconductor strained-layer superlattices, a rather grandiose name for devices consisting of periodic arrangements of epitaxial layers of two kinds of semiconductor; these are often ternary or quaternary because this permits light emission at tuned wavelengths. In some such superlattices, the ternary or quaternary layers have recently been shown to possess LRO. Such ordering has been observed, for instance, in InGaAs, $GaInP_2$, (Ga, Al)InP, and Ga_2AsSb layers in superlattices, complete with domains. The ordering is always of CuPt ($L1_1$) type; in this superlattice, successive close-packed layers are occupied wholly by Cu, then wholly by Pt, atoms, and the cubic fcc lattice becomes rhombohedral, with only one triad axis instead of four. Adjacent domains have differently oriented triad axes. Figure 7.19, from an up-to-date overview of semiconductor ordering by *Mahajan* and *Philips* [7.131], shows how CuPt-type

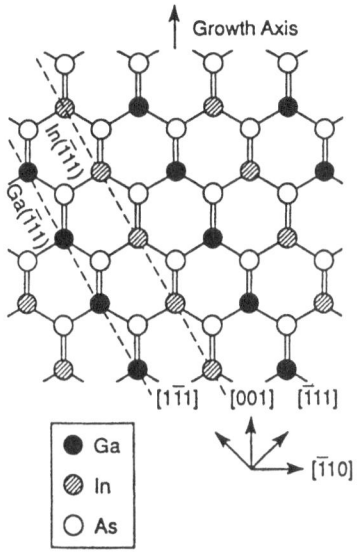

Fig. 7.19. Schematic drawing showing the atomic arrangement in an InGaAs layer that has undergone CuPt-type ordering. One orientational variant only is shown [7.131]

ordering translates into an InGaAs layer. Whether or not a superlattice forms depends on the mode of deposition; thus, molecular beam deposition seems more favourable than some vapour deposition methods. The ordering is metastable: it turns out to be induced by the strains inherent in multilayer formation (although the details are still in dispute), and if an ordered layer induced by such interfacial strain is heated above the critical temperature and cooled again, the superlattice does not return. The metastability is consistent with the fact that calculation shows that some superlattices other than CuPt-type (in particular, CuAu-type) should have lower free energies than CuPt type; the preferred interpretation of this awkward fact is that CuPt ordering represents the stable state at a layer surface or interface but not in its bulk.

It now appears that the degree of ordering, at least in some semiconductor systems, can be controlled by heat treatment, and that by this means the bandgap can be fine-tuned, useful for? for photovoltaic applications [7.158]. This very recent paper also cites many other related papers.

These findings have led to considerable activity in the theory of semiconductor structures. Thus. *Wei* et al. [7.132] have published a first-principles calculation of phase diagrams for semiconductor "alloys", including the LRO phenomenon, and this in turn has led the same group on to apply the same theoretical methods to a new kind of first-principles statistical mechanics of the structural stability of *intermetallic* phases [7.133]. They succeeded in deriving the correct ground states for compounds such as NiAl, CuPt and CuAu out of $\approx 65\,000$ possible configurations... an achievement which de Fontaine recently thought to be impossible! This represents an excellent example of cross-fertilisation between distinct but related scientific subcultures.

7.9.3 Superconductors

In the 1980s, it was established that the superconducting properties of Nb- and V-based A15 compounds, including the superconducting transition temperature, are sensitive to the perfection of the normal state of LRO in these compounds. LRO can be reduced or even eliminated by irradiation; in some phases, order can be somewhat reduced by quenching from a high temperature. Thus, in V_3Ga, the transition temperature varies with the quenching temperature as shown in Fig. 7.20 [7.134]; in the figure, measured values of the Bragg order parameter are included. Reordering kinetics were then measured by making use of the associated changes in the transition temperature [7.135]. Later, the same approach was applied to obtain more detailed information about the form of partial disorder in another A15 compound superconductor, $Ca_3Rh_4Sn_{13}$ [7.136].

A particularly impressive example of the interplay of theory and experiment is found in the recent theoretical prediction, by a CVM approach with first-principles parameters [7.137], of the two-dimensional long-range ordering of oxygen atoms and vacancies in the "basal" plane of the $YBa_2Cu_3O_x$ crystal structure – the standard high-temperature superconductor. Very long O–Cu–O and $\square$–Cu–$\square$ chains (where $\square$ is an oxygen vacancy) are predicted, the details varying with the oxygen concentration, x, and are found to agree very well with electron microscopic studies.

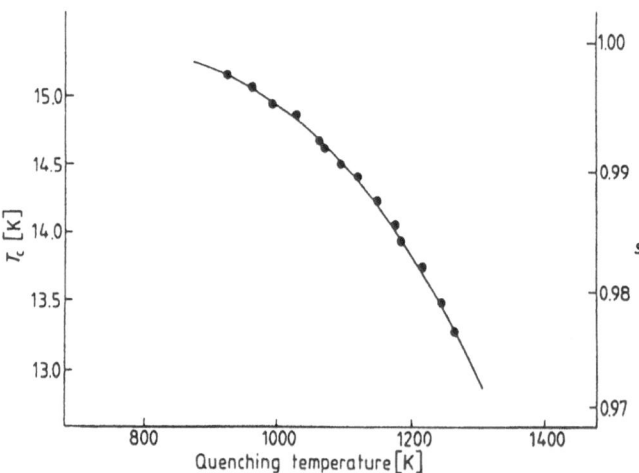

Fig. 7.20. Superconducting transition temperature of V_3Ga as a function of quenching temperature. The curve drawn is a fit to a theoretical model. The axis on the right shows the values of the Bragg order parameter, S, corresponding to different T_c [7.134]

7.9.4 Constitutional Vacancies

Not only combinations of oxygen atoms and vacancies, but vacancies alone, can undergo long-range order. This applies to compounds, which are numerous, in which large concentrations of vacancies appear "constitutionally", i.e., as a means of accommodating non-stoichiometry; the concentrations of such vacancies depend little on temperature. Constitutional vacancies were first discovered in Al-rich NiAl, and later it was discovered that these vacancies can order [7.138], in a series of different patterns. *Chattopadhyay* et al. [7.139] have discovered a relation between this kind of phase and quasiperiodic phases; as the periodicity of crystalline vacancy ordering becomes longer, they claim that the resulting crystal becomes progressively closer to a quasicrystal. Many other compounds, for example, FeS, show vacancy ordering, and the whole topic was recently reviewed by *Amelinckx* [7.140]

7.9.5 Plastic Crystals

There is a large family of compounds, inorganic and organic, in which ordering is orientational rather than positional, in the sense that bonds between particular atom pairs can have different orientations; alternatively, complete radicals can lose their anchoring to the cage of cations in which they sit and begin freely rotating or librating in the crystal structures – the so-called *rotator phases* or *plastic crystals*. As the temperature changes, such crystals undergo order–disorder transitions, but not in the positional sense. This large branch of chemistry can only be mentioned here; it is well surveyed in a book by *Parsonage* and *Staveley* [7.141]. A recent example of such a transition was discovered in gallium-lanthanide binary alloys (66–80 at. % Ga), in which lanthanide atoms are replaced by *pairs* of gallium atoms. The vector joining the Ga–Ga pair is variable in *both* position and orientation [7.142].

The question of atomic (positional) ordering in quasicrystals is still very disputed, because of the great difficulty of obtaining reliable information about the positions of atoms in such materials. Recently, some progress has been made in the interpretation of diffraction patterns to obtain such information [7.143], but it is too soon to draw any conclusions.

7.10 Order and Mechanical Properties

This chapter is concerned with the nature of long-range order, the reasons for its existence, and the intricacies of the order–disorder transition, quite a large enough composite subject in its own right. Accordingly, the correlation between the state of order on the one hand, and physical and mechanical properties on the other, is really beyond the scope of a short survey like this. This section, then, has been kept very brief. This is certainly not to suggest that the topic is unimportant; on the contrary, the section is included at all only because it lies at the basis of current applications of ordered alloys.

Metallurgists the world over are currently investigating the mechanical behaviour of ordered alloys – the term *intermetallics* is now invariably used – with a view to their widespread use as high-temperature load-bearing materials, particularly in aero-engines. Most of the very extensive research and development of intermetallics in recent years has been concentrated on permanently ordered alloys, those ordered up to their melting-points, whereas this chapter is largely devoted to order–disorder *transformations*. This means that those who investigate transformations and those who examine intermetallics constitute, on the whole, separate communities. Nevertheless, there are overlaps: in particular, first-principle predictions of ordered structures spans the divide between these two communities, and so do some conferences on ordered alloys, for instance, a recent NATO Advanced Research Workshop [7.144].

Mechanical behaviour of intermetallics has been studied since the 1950s; in the early days, attention was largely devoted to what would now be called model materials, because they disorder at too low temperatures to be of interest for high-temperature applications ... alloy phases such as Cu_3Au, $CuZn$, Fe_3Al and FeCo. The emphasis in all these researches was to compare the characteristics of the same alloy in wholly or partially ordered states and in the disordered state. These findings were drawn together for the first time in a review by *Stoloff* and *Davies* [7.145]. A quarter of a century later, another review in the same periodical (*Yamaguchi* and *Umakoshi* [7.146]) shows how interests have changed: the emphasis now is on the very complex minutiae of dislocation structures and mechanical behaviour of permanently ordered intermetallics such as the Ni, Ti and Fe aluminides, Ni_2AlTi, $MoSi_2$ and the like. A recent major conference proceedings [7.147] covers the same group of topics in over 1000 pages. An excellent, full survey by Sauthoff is in the press [7.158].

Intermetallics in general share the following features: the superlattice confers high work-hardening rates; the flow stress tends to be high if the ordering energy is high, and peaks for partially ordered alloys with $S \approx 0.5$; dislocations move in pairs linked by strips of APDB ... *superdislocations*, which begin to separate when $S \approx 0.5$; properties can vary drastically with quite small deviations from stoichiometry; creep resistance is much enhanced by the presence of strong order, but is also correlated with diffusivity, being lower when diffusivity is higher – moreover, as a rule strong order depresses diffusivity; grain boundary mobility is depressed by order, thus recrystallization of cold-worked alloys is retarded and consequential softening is slowed as well. This drastic simplification of a mass of findings indicates why interest in intermetallics is now so intense, but it is necessary to refer to the Achilles' heel of the whole family: the great majority of intermetallics is brittle, or nearly so, near ambient temperature, and the more so, the higher the ordering energy, especially in polycrystalline form. However, the discovery, in 1979, in a classic study [7.148], of the possibility of ductilizing polycrystalline Ni_3Al by doping with boron, which accumulates at grain boundaries, gave a strong impetus to intermetallics research.

This is not the place to detail the growth-points of current applied research or to assess the problems and promise of load-bearing intermetallics, but the reader is directed to a few recent overview articles. The most authoritative is one by Fleischer et al. [7.149]. Two very substantial groups of metallurgically inclined overviews were published in the proceedings of two successive conferences on aluminides [7.150]. A concise overview was published by *Cahn* [7.151]. The brittleness problem remains the overriding concern, but ways forward are visible. A consensus is beginning to appear that the most useful intermetallics will be multiphase alloys in which, typically, a stronger, more strongly ordered but less ductile phase is combined with a somewhat weaker, less strongly ordered but more ductile one. If circumstances are right (and this may include the existence of coherent or semicoherent interfaces between the phases) then glissile dislocations can be "injected" into the intrinsically brittle phase from the more ductile one [7.152, 153]. A further family of multiphase intermetallics which are used at ambient rather than high temperatures is the extensive range of gold-based alloys used in dentistry [7.154]. These depend greatly on age-hardening associated with progressive ordering of certain phases in these

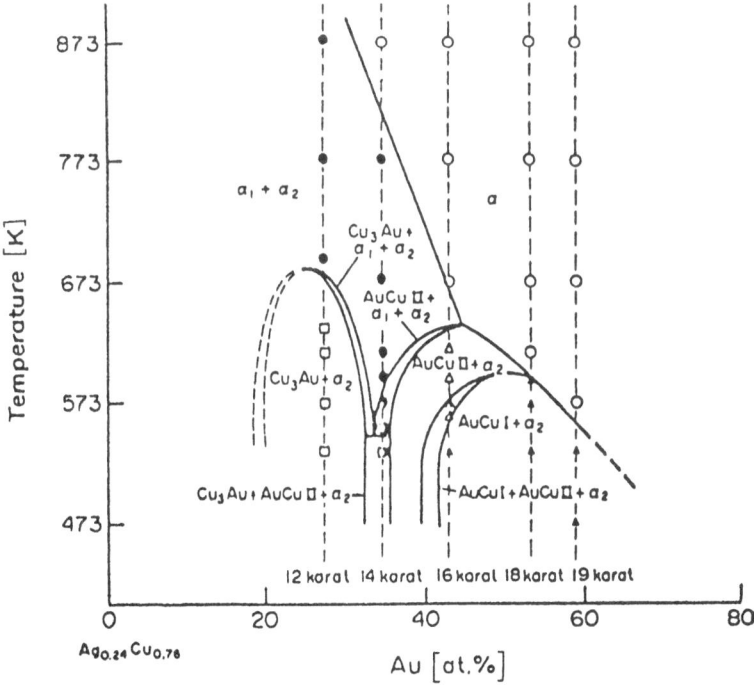

Fig. 7.21. Experimental "coherent" phase diagram of the $Au_x(Ag_{0.24}Cu_{0.76})_{1-x}$ pseudobinary phase diagram [7.156]

complex alloys [7.155]; Fig. 7.21 shows a characteristic pseudo-binary phase diagram section from one such group of alloys, the Au–Ag–Cu alloys, taken from a study by *Nakagawa* and *Yasuda* [7.156], in which several distinct superlattice types are present.

7.11 Conclusion

The order–disorder transformation in alloys, and concerns associated with it, present a paradigm of modern physical metallurgy. Approaches to the field range from severely practical metallurgical research aimed at creating creep-resistant high-temperature alloys, via phenomenological experimental work on superlattice types, phase diagrams and ordering kinetics, via research on associated themes such as novel experimental probes and related properties such as diffusion measurements, to sophisticated experimental and theoretical physics on such subsidiary topics as first-principles crystal structure prediction, interpretation of second-order transitions and nucleation of disorder at anti-phase domain boundaries. There are plenty of scholarly disputes of the kind that characterize active fields, for instance, the disputes between "critical" and "classical" physicists. For many years, the pure physicists and the horny-handed metallurgists had virtually no contact, but this is rapidly changing. In particular, theorists interested in crystal structure prediction are working in close coopera-tion with metallurgists seeking guidelines for developing improved ternary and quaternary alloys. The healthy growth of intermetallics research depends on the healthy growth of further forms of cooperation of this kind.

7.12 Addendum

Recently, there has been renewed interest in the behaviour of lattice vacancies in certain ordered phases. The classic problem here is that of *constitutional va-cancies*, first discovered in B2-NiAl by *Bradley* and *Taylor* in 1937 [7.159]. NiAl has a broad homogeneity range, and *Bradley* and *Taylor* discovered, by a critical comparison of densities and lattice parameters, that the Al-rich phase ($Ni_{50-x}Al_{50+x}$) accommodates its non-stoichiometry by introducing 'constitu-tional' vacancies on the Ni sublattice; the adjective denotes that the vacancy concentration is determined by composition, not by temperature, and that the concentration may reach several percent. Recent experimental research has confirmed the old findings in great detail [7.160]. Early attempts to explain constitutional vacancies in terms of the electron-to-atom ratio failed, and their origin (confirmed in a number of other phases, such as CuGa) has remained a mystery. Older attempts at interpretation were surveyed by *Cahn* [7.161]. Now *Cottrell* [7.162] has provided what appears to be an entirely satisfactory expla-nation, in terms of the modern theory of cohesion in metals. He compared the total cohesive energy (in terms of numbers and energies of different kinds of nearest-neighbour bonds) for a non-stoichiometric alloy with and without constitutional vacancies, and found that the vacancies lower the total free energy. A little later [7.163], he extended his theory to Al-Ni-Cu alloys, the constitutional vacancy content of which was first studied experimentally as a function of composition in 1938, and was able to make sense of these old findings.

Another aspect of vacancies in ordered alloys has arisen from the greatly enhanced interest, for technological reasons, in ordered iron-aluminum alloys, especially those near the FeAl (B2-type) composition. In an important paper, *Chang* et al. [7.164] established by lattice parameter measurements that aL-rich FeAl contains constitutional vacancies proportional to the deviation from stoichiometry, and also that the microhardness increases linearly with the square root of the vacancy concentration (this is a large effect); in fact, this plot applies to both Fe-rich and Al-rich alloys. In Fe-rich FeAl, a range of research (outlined by *Munroe* [7.165]) going back to the 1970s, shows that exceptionally high concentrations of *thermal* vacancies are present, which can be modified by heat-treatment following quenching, with concomitant large reductions in the hardness. As *Munroe* showed in his cited paper, the effect of heat-treatment on hardness can be moderated by the addition of ternary additives, notably nickel. Research on vacancies in FeAl is burgeoning.

Another field which has greatly expanded in the last few years is the study of mechanical disordering of intermetallics, notably by milling in ball-mills. A series of specialised conferences [7.166, 167] has been devoted to this recondite topic and, very recently, an excellent survey article by *Bakker* et al. has appeared [7.168]. This survey describes some unexpected findings, notably the creation of magnetic spin-glasses in certain milled intermetallics such as near-stoichiometric AuFe and CuMn. Recent research by *Le Caër* in Nancy [7.169] on ferromagnetic ($L2_1$) Heusler phases (Ni_2MnSn and Co_2MnSn) shows that quenching from high temperatures partly disorders them without destroying ferromagnetism, but ball-milling disorders them in a different way with, apparently, the replacement of ferromagnetic order by a spin-glass structure. This is one of a growing number of cases where thermal and mechanical disordering appears to lead to different structural changes.

Another ternary compound with long-range order which has attracted much attention recently because of its possible practical use as a high-temperature structural material is Ti_2AlNb, also referred to as the "O-phase". There are two slightly different structure variants (both orthorhombic) and the phase can be thermally disordered. It was discovered in Hyderabad, India, by a team of metallurgists led by D. Banerjee; its structure has been studied by X-ray, electron and neutron diffraction and also by the ALCHEMI technique, alias channeling-enhanced microanalysis in the electron microscope. *Banerjee* has recently surveyed what is known about the structure and mechanical properties of this intriguing phase [7.170].

Creep of ordered intermetallic compounds is of ever-increasing practical concern, and this topic has been discussed in subtle detail in a fine book by

Nabarro and *de Villiers* [7.171]. Much of the book is concerned with creep in simple metals or solid solutions, but three chapters are devoted to creep in the important phase Ni_3Al, in superalloys and in "possible new high-temperature creep-resistant materials". As the present author has remarked in a review of this book, "the standard of critical discussion, of both experimental facts and theory, is very high throughout and the authors rigorously eschew handwaving."

Most intermetallic compounds are ordered up to the melting-temperature, and their ordering energies (the difference between the free energy of the ordered form and the corresponding disordered form) therefore cannot be estimated from a measured disordering temperature. The only way of estimating the ordering energy (which affects such variables as the creep resistance) is by theory. Until now, no comprehensive discussion of the various methods of tackling this problem, with their pros and cons, has been available. Now, chapter 7 of a new book on CALPHAD (Calculation of PHAse Diagrams) admirably fills this gap [7.172].

Finally, it is important here to record the publication of an outstanding 2-volume overview of the entire field of ordered intermetallic compounds; the first, larger, volume is devoted to basic science, the second to engineering applications. Between the two volumes, 101 contributors have covered the whole field in exemplary fashion. These volumes, edited by *Westbrook* and *Fleischer* [7.173], will be the standard reference for many years to come.

References

7.1   N. Kurnakow, S. Zemczuzny, M. Zasedatelev: J. Inst. Met. **15**, 305 (1916)
7.2   C.H. Johansson, J.O. Linde: Ann. Physik **78**, 305 (1925)
7.3   W. Hume-Rothery: J. Inst. Metals **35**, 309 (1926)
7.4   C.H. Desch: *Intermetallic Compounds* (Longmans Green, London 1914)
7.5   C.H. Desch: *The Chemistry of Solids* (Cornell Univ. Press, Ithaca 1934) pp. 153–160
7.6   C.S. Barrett: *The Structure of Metals* (McGraw-Hill, New York 1943)
7.7   E.C. Bain: Trans. AIME **68**, 625 (1923)
7.8   A. Loiseau, G. van Tendeloo, R. Portier, F. Ducastelle: J. Physique **46**, 595 (1985)
7.9   A. Loiseau, J. Planes, F. Ducastelle: In Alloy Phase Stability, G.M. Stocks, A. Gonis (eds) (Kluwer, Dordrecht 1989) pp. 101–106
7.10   W.L. Bragg, E.J. Williams: Proc. Roy. Soc. (London) A **145**, 699 (1934)
7.11   D.T. Keating, B.E. Warren: J. Appl. Phys. **22**, 286 (1951)
7.12   N.S. Stoloff, R.G. Davies: Acta Metall **12**, 473 (1964)
7.13   A. Lawley, R.W. Cahn: J. Phys. Chem. Solids **20**, 204 (1961)
7.14   F.C. Nix, W. Shockley: Rev. Mod. Phys. **10**, 1 (1938)
7.15   R. Kozubski, M.C. Cadeville: J. Phys. F **18**, 2569 (1988)
7.16   R.W. Cahn, P.A. Siemers, J.E. Geiger, P. Bardhan: Acta Metall. **35**, 2737 (1987)
7.17   F.J. Bremer, M. Beyss, H. Wenzl: Phys. stat. sol. (a) **110**, 77 (1988)
7.18   J.M. Cowley: Phys. Rev. **77**, 669 (1950)
7.19   D. de Fontaine: J. Phys. Chem. Solids **34**, 1285 (1973)

7.20 G. Inden, W. Pitsch: In *Phase Transformations in Materials*, ed. by P. Haasen (VCH Verlag, Weinheim 1991) pp. 497–552

7.21 J. Als-Nielsen: In *Phase Transitions and Critical Phenomena*, Vol. 5A, ed. by C. Domb, M.S. Green (Academic, New York 1976) p. 88

7.22 J. Yeomans: In *Solid State Theory – Past, Present and Predicted*, ed. by D.L. Weaire, C.G. Windsor (Hilger, Bristol 1987) pp. 223–236

7.23 K. Binder: In *Phase Transformations in Materials*, ed. by P. Haasen (VCH Verlag, Weinheim 1991) pp. 143–212

7.24 C. Sykes, H. Evans: J. Inst. Metals **58**, 255 (1936)

7.25 C. Sykes, F.W. Jones: Proc. Roy. Soc. **157**, 213 (1936)

7.26 R.W. Cahn, P.A. Siemers, E.L. Hall: Acta Metall. **35**, 2753 (1987)

7.27 R.W. Cahn: In *High Temperature Aluminides and Intermetallics*, ed. by S.H. Whang et al. (TMS, Warrendale, PA 1990) pp. 245–270

7.28 M. Rajkovic, R.A. Buckley: Metal Science **15**, 21 (1981)

7.29 R.S. Irani, R.W. Cahn: J. Mater. Sci. **8**, 1453 (1973)

7.30 L.E. Tanner, H.J. Leamy: In *Order-Disorder Transformations in Solids*, ed. by H. Warlimont (Springer, Berlin Heidelberg 1974) pp. 180–239

7.31 R.S. Irani: Contemp. Phys. **13**, 559 (1972)

7.32 G. van Tendeloo: In *Alloy Phase Stability*, ed. by G.M. Stocks, A. Gonis (Kluwer, Dordrecht 1989) pp. 75–100

7.33 D. Harker: Trans. Amer. Soc. Metals **32**, 210 (1944)

7.34 W.C. Johnson, M.B. Berkenpas, D.E. Laughlin: Acta Metall. **37**, 3149 (1988)

7.35 R.W. Cahn: Adv. Mater. (part 8), 300 (1989)

7.36 M.J. Pfeifer, P.W. Voorhees: Met. Trans. A, in press

7.37 D.G. Morris: In *Ordered Intermetallics – Physical Metallurgy and Mechanical Properties*, ed. by C.T. Liu, R.W. Cahn, G. Sauthoff (Kluwer, Dordrecht 1992) pp. 123–142

7.38 A.H. Cottrell: *Theoretical Structural Metallurgy* (Edward Arnold, London 1948)

7.39 W. Shockley: J. Chem. Phys. **6**, 130 (1938)

7.40 R. Kikuchi: Phys. Rev. **81**, 988 (1951)

7.41 D. de Fontaine: Solid State Physics, **34** (Academic, New York 1979) pp. 74–294

7.42 J.M. Sanchez, F. Ducastelle, D. Gratias: Physica **128A**, 334 (1984)

7.43 D. de Fontaine: In *Alloy Phase Stability* ed. by G.M. Stocks, A. Gonis (Kluwer, Dordrecht 1989) pp. 177–203

7.44 C. Sigli, J.M. Sanchez: Acta Metall. **33**, 1097 (1985)

7.45 A.G. Khachaturyan: *The Theory of Structural Transformations in Solids* (Wiley, New York 1983)

7.46 P. Heller: Rep. Progr. Phys. **30**, 731 (1967)

7.47 M.E. Lines: Ann. Rev. Mater. Sci. **3**, (1973)

7.48 B. Widom: Science **157**, 375 (1967)
      M. Suzuki: J. Phys. Soc. Jpn. **55**, 4205 (1986)

7.49 W. Selke: In *Alloy Phase Stability*, ed. by G.M. Stocks, A. Gonis (Kluwer, Dordrecht 1989) pp. 205–232

7.50 W.A. Soffa, D.E. Laughlin: Acta Metall. **37**, 3019 (1989)

7.51 D. Shoenberg: In *Solid State Science, Past, Present and Predicted*, ed. by D.L. Weaire, C.G. Windsor (Hilger, Bristol 1987) pp. 109–142

7.52 V. Heine, J. Samson: J. Phys. F **13**, 2155 (1983)

7.53 F.J. Pinski, B. Ginatempo, D.D. Johnson, J.B. Staunton, G.M. Stocks, B.L. Gyorffy: Phys. Rev. Lett. **66**, 766 (1991)

7.54 J. Maddox: Nature **349**, 649 (1991)

7.55 B.L. Gyorffy, G.M. Stocks, B. Ginatempo, D.D. Johnson, D.M. Nicholson, F.J. Pinski, J.B. Staunton, H. Winter: Phil. Trans. Roy. Soc. (London) A **334**, 515 (1991)

7.56 F. Ducastelle: In *Alloy Phase Stability*, ed. by G.M. Stocks, A. Gonis (Kluwer, Dordrecht 1989) pp. 293–327

7.57 D.G. Pettifor: In *Second Supplementary Volume of the Encylopedia of Materials Science and Engineering*, ed. by R.W. Cahn (Pergamon, Oxford 1988) pp. 51–59

7.58   D.G. Pettifor: Mater. Sci., Technol. **4**, 2480 (1988)
7.59   D.G. Pettifor: In *Intermetallic Compounds – Structure and Mechanical Properties*, ed. by O. Izumi, (Japan Inst. of Metals, Sendai 1991) pp. 149–156
7.60   F.R. de Boer, R. Boom, W.C.M. Mattens, A.R. Miedema, A.K. Niessenm: *Cohesion in Metals: Transition Metal Alloys* (North-Holland, Amsterdam 1988)
7.61   A.J. Freeman: In *Alloy Phase Stability*, ed. by G.M. Stocks, A. Gonis (Kluwer, Dordrecht, 1989) pp. 365–375
7.62   A.J. Freeman: In *Ordered Intermetallics – Physical Metallurgy and Mechanical Behaviour*, ed. by C.T. Liu, R.W. Cahn, G. Sauthoff (Kluwer, Dordrecht 1992) pp. 1–14
7.63   D. de Fontaine, C. Wolverton, G. Ceder, H. Dreysse: In *Ordered Intermetallics – Physical Mettallurgy and Mechanical Properties*, ed. by C.T. Liu, R.W. Cahn, G. Sauthoff (Kluwer, Dordrecht 1992) pp. 61–72
7.64   J.E. Epperson, P. Fürnrohr, C. Ortiz: Acta Cryst. Sect. A**34**, 667 (1978)
7.65   F.W. Jones, C. Sykes: Proc. Roy. Soc. (London) A **161**, 440 (1937)
7.66   A. Marty, M. Bessière, F. Bley, Y. Calvayrac, S. Lefebvre: Acta Metall. et Mater. **38**, 345 (1990)
7.67   M. Enomoto, H. Harada: Met. Trans. A **20**, 649 (1990)
7.68   A. Williams, G.H. Kwei, A.T. Ortiz, M. Karnowski, W.K. Warburton: J. Mater. Res. **5**, 1197 (1990)
7.69   J. Als-Nielsen, O.W. Dietrich: Phys. Rev. **153**, 706, 711 (1967)
7.70   F. Klaiber, B. Schönfeld, G Kostorz: Acta Cryst. A **43**, 525 (1987)
7.71   G. Kostorz: In: *Dynamics of Ordering Processes in Condensed Matter* (Plenum, New York 1987)
7.72   N. Kuwano, M. Umeda, A. Mukai, N. Kawahara, K. Oki: J. de Physique, Colloque C6, suppl. 11 **48**, C6-391 (1987)
7.73   Y. Yoshida, F. Langmayr, P. Fratzl, G. Vogl: Phys. Rev. B **39**, 6395 (1989)
7.74   J.C.H. Spence, J. Taftø: J. Microscopy **130**, 147 (1983)
7.75   D. Banerjee: In *Second Supplementary Volume of the Encyclopedia of Materials Science and Engineering*, ed. by R.W. Cahn (Pergamon, Oxford 1990) pp. 785–789
7.76   D. Banerjee, A.K. Gogia, T.K. Nandi, V.A. Joshi: Acta Metall. **36**, 871 (1988)
7.77   W.J. Boettinger, M.J. Aziz: Acta Metall. **37**, 3379 (1989)
7.78   A.R. Yavari, B. Bochu: Phil. Mag. A **59**, 697 (1989)
7.79   R.W. Cahn, J.L. Walter, D.W. Marsh: Mat. Sci. Eng. **98**, 33 (1988)
7.80   S.R. Harris, D.H. Pearson, C.M. Garland, B. Fultz: J. Mater. Res. **6**, 2019 (1991)
7.81   J.A. West, J.T. Manos, M.J. Aziz: Mat. Res. Soc. Symp. Proc. **213**, 859 (1991)
7.82   S. Gialanella, S.B. Newcomb, R.W. Cahn: In *Ordering and Disordering in Alloys*, ed. by A.R. Yavari (Elsevier, Barking 1992) pp. 67–78
7.83   R.W. Cahn: In *Ordered Intermetallics – Physical Metallurgy and Mechanical Properties*, ed. by C.T. Liu, R.W. Cahn, G. Sauthoff (Kluwer, Dordrecht 1992) pp. 511–524
        H. Chen, V.K. Vasudevan (eds.): *Kinetics of Ordering Transformations in Metals* (The Min. Met. Mater. Soc., Warrendale, PA 1992)
7.84   M.R. Collins, H.C. Teh: Phys. Rev. Lett. **30**, 781 (1973)
7.85   R.W. Cahn: Mater. Res. Soc. Symp. Proc. **57**, 385 (1987)
7.86   H.C. Bolton, C.A. Leng: Phys. Rev. B **11**, 2069 (1975)
7.87   J. Paulevé, D. Dautreppe, J. Laugier, L. Néel: Compt. Rend. Acad. Sci. (Paris) **254**, 965 (1962)
7.88   Y. Gros, J. Paulevé: J. Physique **31**, 459 (1970)
7.89   J.F. Peterson, A. Aydin, J.M. Knudsen: Phys. Lett. A **62**, 192 (1977)
7.90   F.P. Burns, S.L. Quimby: Phys. Rev. **97**, 1567 (1955)
7.91   B. Fultz: *Ordering and Disordering in Alloys*, ed. by A.R. Yavari (Elsevier, Berking 1992) pp. 31–42
7.92   K.F. Ludwig, Jr., G.B. Stephenson, J.L. Jorda-Sweet, J. Mainville, Y.S. Yang, M. Sutton: Phys. Rev. Lett. **61**, 1859 (1988)
7.93   H.E. Cook, D. de Fontaine, J.E. Hilliard: Acta Metall. **17**, 765 (1969)
7.94   H. Chen, J.B. Cohen, R. Ghosh: J Phys. Chem. Solids **38**, 855 (1977)
7.95   S.M. Allen, J.W. Cahn: Acta Metall. **27**, 1085 (1979)

7.96   J.R. Beeler, Jr.: Phys. Rev. A **138**, 1259 (1965)
7.97   H. Bakker: In *Diffusion in Crystalline Solids*, ed. by G.E. Murch, A.S. Nowick (Academic, New York 1984) pp. 189–256
7.98   H. Bakker, D.M.R LoCascio, L.M. Di: In *Ordered Intermetallics – Physical Metallurgy and Mechanical Properties*, ed. by C.T. Liu, R.W. Cahn, G. Sauthoff (Kluwer, Dordrecht 1992) pp. 433–448
7.99   A.B. Kuper, D. Lazarus, J.R. Manning, C.T. Tomizuka: Phys. Rev. **104**, 1536 (1956)
7.100 H. Bakker, N.A. Stolwijk, L.P. van der Meij, T.J. Zuurendonk: Nuclear Metallurgy **20**, 96 (1976)
7.101 L.-Q. Chen, A.G. Khachaturyan: Scr. Metall. et Mater. **25**, 61, 67 (1991)
7.102 L.-Q. Chen, A.G. Khachaturyan: Acta Metall. et Mater. **39**, 2553 (1991)
7.103 V.S. Sundaram, R.S. Alben, W.D. Robertson: Surf. Sci. **46**, 653 (1974)
7.104 T.M. Buck, G.H. Wheatley, L. Marchut: Phys. Rev. Lett. **51**, 43 (1983)
7.105 J.M. Sanchez, J.L. Morán-López: Surf. Sci. **157**, L297 (1985)
7.106 R. Lipowsky, W. Speth: Phys. Rev. B **28**, 3983 (1983)
7.107 R. Lipowsky: Z. Phys. B **51**, 165 (1983)
7.108 R. Lipowsky: Phys. Rev. Lett. **49**, 1575 (1982)
7.109 R. Lipowsky: Ferroelectrics **73**, 69 (1987)
7.110 M. Yamamoto, S. Nenno: Surf. Sci. **213**, 502 (1989)
7.111 G. van Tendeloo, D. Broddin, C. Leroux; D. Schryvers; L.E. Tanner, J. van Landuyt, S. Amelinckx: Phase Transitions **27**, 61 (1990)
7.112 I. Baker, E.M. Schulson, J.R. Michael: Phil. Mag. B **57**, 379 (1988)
7.113 H. Kung, D.R. Rasmussen, S.L. Sass: In *Intermetallic Compounds – Structure and Mechanical Properties*, ed. by S. Izumi (Japan Inst. of Metals, Sendai 1991) pp. 347–354
7.114 L.E. Popov, E.V. Kozlov, N.S. Golosov: Phys. Stat. Sol. **13**, 569 (1966)
7.115 C. Leroux, A. Loiseau, M.C. Cadeville, D. Broddin, G. van Tendeloo: J. Phys: Condens. Matt. **2**, 3479 (1990)
7.116 V. Finel, V. Mazauric, F. Ducastelle: Phys. Rev. Lett. **65**, 1016 (1990)
7.117 J.M. Sanchez, S. Eng, J.K. Tien: Mat. Res. Soc., Symp. Proc. **81**, 57 (1987)
7.118 J.S. Kouvel: In *Magnetism and Metallurgy*, ed. by A.E. Berkowitz, E. Kneller (Academic, New York 19969)
7.119 A.P. Miodownik: Bull. Alloy Phase Diagr. **2**, 406 (1982)
7.120 G. Inden: Bull. Alloy Phase Diagr. **2**, 412 (1982)
7.121 M.C. Cadeville, J.L. Morán-López: Phys. Rep. **153**, 331 (1987)
7.122 J.K. van Deen, F. van der Woude: Acta Metall. **29**, 1255 (1981)
7.123 P.L. Rossiter: J. Phys. F **10**, 1459 (1980)
7.124 C. Leroux, M.C. Cadeville, R. Kozubski: J. Phys., Condensed Matter, **1**, 6403 (1989)
7.125 J.M. Sanchez, J.L. Morán-Lóez, C. Leroux, M. C. Cadeville: J Phys. C **21**, L1091 (1988)
7.126 Y. Suzuki, J. Haimovich, T. Egami: Phys. Rev. B **35**, 2162 (1987)
7.127 T. Tomida, T. Egami: Mater. Sci. Eng. A **133**, 931 (1991)
7.128 H.J. Birkenbeil, R.W. Cahn: Proc. Phys. Soc. **79**, 381 (1961)
7.129 F. Laves: Naturwiss. **30**, 546 (1952)
7.130 E.K.H. Salje: *Phase Transitions in Ferroelastic and Co-elastic Crystals* (Cambridge University Press, Cambridge 1990)
7.131 S. Mahajan, B.A. Philips: In *Ordered Intermetallics – Physical Metallurgy and Mechanical Properties*, ed. by C.T. Liu, R.W. Cahn, G. Sauthoff (Kluwer, Dordrecht 1992) pp. 93–106
7.132 S.-H. Wei, L.G. Ferreira, Alex Zunger: Phys. Rev. B **41**, 8240 (1990)
7.133 Z.W. Lu, S.-H. Wei, Alex Zunger, S. Frota-Pessoa, L.G. Ferreira: Phys. Rev. B **44**, 512 (1991)
7.134 A. van Winkel, A.W. Weeber, H. Bakker: J Phys. F **14**, 2631 (1984)
7.135 A. van Winkel, H. Bakker: J. Phys. F **15**, 1565 (1985)
7.136 J.P.A. Westerveldt, D.M.R. Lo Cascio, H. Bakker, B.O. Loopstra, K. Goubitz: J. Phys. Condensed Matter **1**, 5689 (1989)
7.137 D. de Fontaine, G. Ceder, M. Asta: Nature **343**, 544 (1990)
7.138 S.S. Lu, T. Chang: Acta Phys. Sinica **13**, 150 (1957)

7.139 K. Chattopadhyay, S. Lele, N. Thangaraj, S. Ranganatha: Acta Metall. **35**, 727 (1987)

7.140 S. Amelinckx: In *First Supplementary Volume of the Encyclopedia of Materials Science and Engineering*, ed. by R.W. Cahn (Pergamon, Oxford 1990) pp. 77–85

7.141 N.G. Parsonage, L.A.K. Staveley: *Disorder in Crystals* (Clarendon, Oxford 1979)

7.142 G. Kimmel, W.D. Kaplan: Scr. Metall. et Mater. **25**, 571 (1991)

7.143 S. Song, E.R. Ryba, A. Ramani, P.R. Howell: J. Mater Sci. Lett. **10**, 237 (1991)

7.144 C.T. Liu, R.W. Cahn, G. Sauthoff: In *Ordered Intermetallic – Physical Metallurgy and Mechanical Properties* (Kluwer, Dordrecht 1992)

7.145 N.S. Stoloff, R.G. Davies: Progr. Mates. Sci. **13**, 1 (1966)

7.146 M. Yamaguchi, Y. Umakoshi: Progr. Mates. Sci. **34**, 1 (1990)

7.147 S. Izumi: *Intermetallic Compounds – Structure and Mechanical Properties (JIMIS-6)* (The Jpn. Inst. of Metals, Sendai 1991)

7.148 K. Aoki, S. Izumi: J. Jpn. Inst. Metals **43**, 1190 (1979)

7.149 R.L. Fleischer, D.M. Dimiduk, H.A. Lipsitt: Ann. Rev. Mat. Sci. **19**, 231 (1989)

7.150 S.H. Whang, C.T. Liu, D.P. Pope, J.O. Stiegler Eds.: *High Temperature Aluminides and Intermetallics I* (TMS-AIME, Warrendale 1990)
Proc. of High Temperature Aluminides and Intermetallics II, Mater. Sci. Eng. Vol. A **152** (1992)

7.151 R.W. Cahn: MRS Bulletin **16**/5, 18 (1991)

7.152 A. Misra, S. Hartfield-Wünsch, R. Gibala: In *Intermetallic Compounds – Structure and Mechanical Properties (JIMIS-6)* ed. by O. Izumi, (The Jpn. Inst. of Metals, Sendai 1991) pp. 597–607

7.153 R. Yang, J.A. Leake, R.W. Cahn: Mates. Sci. Eng. A **152**, 227 (1992) in press

7.154 K. Yasuda: In *Concise Encyclopedia of Dental Materials*, ed. by D. Williams, (Pergamon, Oxford 1990) pp. 197–205

7.155 K. Yasuda: Gold Bull. **20**, 90 (1987)

7.156 M. Nakagawa, K. Yasuda: J. Less-Comon Metals **138**, 95 (1988)

7.157 A. Zunger, S. Wagner, M. Petroff: J. Electron. Mater. **22**, 10 (1993)

7.158 G. Sauthoff: In *Structure and Properties of Non-Ferrous Alloys*, ed. by K.H. Matucha (VCH, Weinheim 1994)

7.159 A.J. Bradley, A. Taylor: Proc. Roy. Soc. (Lond.) A **159**, 56 (1937)

7.160 M. Kogachi, S. Minigawa, K. Nakahigashi: Acta Metall. & Mater. **40**, 1113 (1992)

7.161 R.W. Cahn: *Artifice and Artefacts* (IOP, Bristol 1992) p. 19

7.162 A.H. Cottrell: Intermetallics **3**, 341 (1995)

7.163 A.H. Cottrell: Intermetallics **4**, 1 (1996)

7.164 Y.A. Chang, L.M. Pike, C.T. Liu, A.R. Bilbrey, D.S. Stone: Intermetallics **1**, 107 (1993)

7.165 P.R. Munroe: Intermetallics **4**, 5 (1996)

7.166 A.R. Yavari (ed.): *Ordering and Disordering in Alloys* (Elsevier, London 1992)

7.167 A.R. Yavari (ed.): *Mechanically Alloyed and Nanocrystalline Materials,* Mater. Sci. Forum **179–181** (1995)

7.168 H. Bakker, G.F. Zhou, H. Yang: Progr. Mater. Sci. **39**, 159 (1995)

7.169 G. Le Caër, P. Delcroix, B. Malaman, R. Welter, B. Fultz, E. Ressouche: Mat. Sci. Forum **235–238**, 589–594 (1997)

7.170 D. Banerjee: Progr. Mater. Sci. **42**, 135 (1998)

7.171 F.R.N. Nabarro, H.L. de Villiers: *The Physics of Creep* (Taylor and Francis, London 1995)

7.172 N. Saunders, A.P. Miodownik: CALPHAD: A Comprehensive Treatment (Pergamon, Oxford 1998)

7.173 J.H. Westbrook, R.L. Fleischer: *Intermetallic Compounds – Principles and Practice* (Wiley, Chichester 1995) 2 volumes

8 Usefulness of Electron Microscopy

H. Fujita and N. Sumida

Electron microscopy is the most effective technique to obtain directly and dynamically topographic information in the atomic scale, and results in the sensation of both "seeing is believing" and "the materials are living". In the present paper, usefulness and applications of electron microscopy to materials science are discussed from the following two viewpoints, i.e., high-resolution electron microscopy and high-voltage electron microscopy.

8.1 Background

It is well known that the materials are very structure sensitive, and that their behaviour is always determined by the behaviour of lattice defects, such as vacancies and interstitials, dislocations, boundaries and interfaces, and solute atoms. The behaviour of these lattice defects always co-operates with each other so that the processes of phenomena are self-controlled to minimize the total free energy, as will be mentioned in a later section. Namely, "the materials are living", and their behaviour is always self-controlled by the co-operative inter-action among more than two factors just like biological materials. Therefore, the following three important pieces of information are necessary to be obtained for materials science research: a) What the phenomenon is; b) where it happens; and c) how it proceeds under certain conditions. The first two pieces of information are topographic, and details about these can be accurately obtained in the atomic scale by high-resolution electron microscopy (HREMy). In situ experiments with electron microscopes (EMs), especially with high-voltage electron microscopes (HVEMs), are an epoch-making method for getting the third information, and detailed behaviour of materials has been dynamically investigated in the atomic scale or the scale of lattice defects under necessary conditions, i.e., living conditions.

   Recently, not only the processes of phenomena, but also general rules for the formation of non-equilibrium phases, which may have a potential for adding new functional characteristics to materials, can be investigated directly in the atomic scale by HVEMy [8.1].

   The present paper is concerned with the advantages of EMy and their indispensable applications to materials science.

8.2 Principles of Image Formation

In this section basic theories of image formation are briefly described to understand the nature of image contrast and use an electron microscope effectively for characterizing the microstructures and the crystal defects of materials. The image contrast depends on how electrons leaving the bottom surface of a specimen are modified on passing through the microscope lenses to form the image. There are two important imaging mechanisms; diffraction contrast imaging and phase contrast imaging. The former is simply a high-magnification map of the intensity distribution across the transmitted beam or one of the diffracted beams, so-called the Bragg reflections, produced by diffraction phenomena in the specimen. In the latter case, the transmitted beam and some of the diffracted beams are recombined to form an image so that phase differences among them at the bottom surface are converted into the intensity differences in the image by interfering with each other.

Diffraction contrast imaging is the dominant technique for observing object details $\geq$ 1 nm in crystalline materials and is widely used for studies of crystal defects. On the other hand, phase contrast imaging is a very powerful method in lattice resolution studies.

8.2.1 Diffraction Contrast Imaging

Diffraction contrast is achieved in either of two imaging modes; a bright field (BF) image formed by the transmitted beam and a dark field (DF) image formed by only one diffracted beam. Figure 8.1 shows how to operate the BF and the DF imaging mode in the microscope. In the BF imaging mode, the objective aperture, which is positioned at the centre of the diffraction pattern formed on the back focal plane of the objective lens, excludes all other electron beams except the transmitted beam from contributing to the image. Alternatively, the DF image can be obtained by either shift of the objective aperture or deflection of the incident beam as shown in Fig. 8.1 b, c.

The latter of the DF imaging modes allows the particular diffracted beam of interest to pass along the optic axis, and thereby reduces the error arising from the spherical aberration of the objective lens. Therefore, we can observe the DF images at resolution approaching the highest capabilities of the microscope as will be discussed in Sect. 8.3.1.

The BF and the DF images are interpreted in terms of suitable electron diffraction theories (the kinematical and the dynamical theory of electron diffraction), which have been described in detail in a number of books, for example, *Heidenreich* [8.2], *Hirsch* et al. [8.3], *Amelinckx* et al. [8.4], *Valdré* et al. [8.5, 6], *Thomas* et al. [8.7] and *Cowely* [8.8].

The kinematical theory is applicable only to thin specimens and for far off-Bragg conditions, but may help electron microscopists to better understand

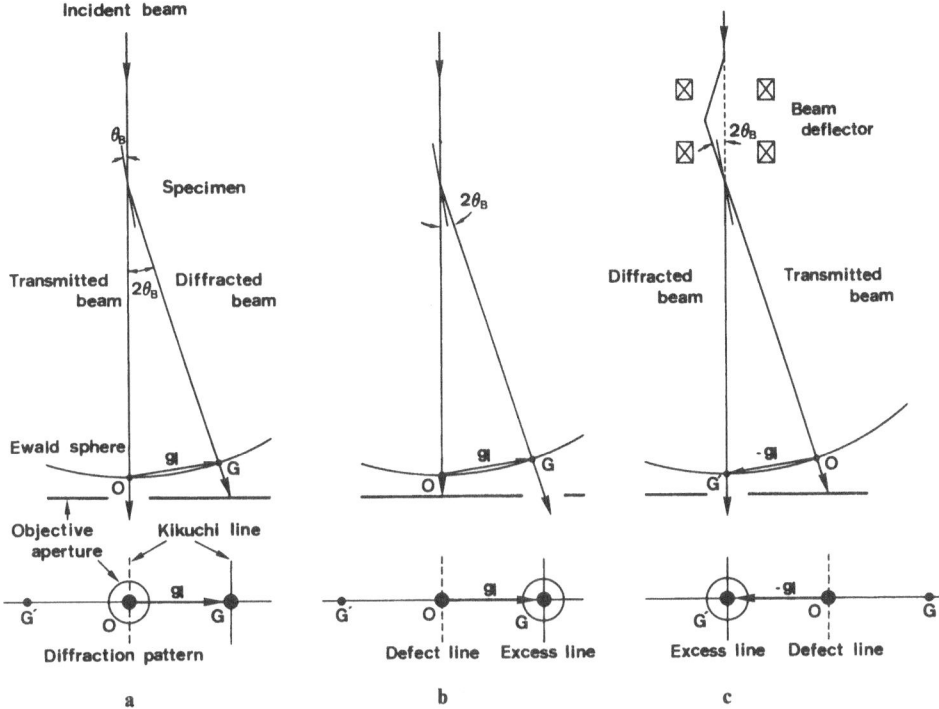

Fig. 8.1. Ewald sphere constructions and the corresponding diffraction patterns suggesting (a) the BF imaging mode, and (b) and (c) the DF imaging modes. The DF mode of (b) is obtained by the shift of the objective aperture from the position of the transmitted beam to that of the diffracted beam. In (c), so-called beam-tilted DF imaging mode, the deflection of incident beam tilted by an angle $2\theta_B$ allows the diffracted beam G′ to pass along the optic axis. It should be emphasized in this case that the direction of reflection vector in (c) is reversed with respect to that in (a)

the nature of image contrast. The dynamical theory more accurately describes the diffraction phenomena in materials and is important especially for the interpretations of high voltage electron microscope images, where large numbers of fairly strong Bragg reflections occur simultaneously. The formulation of the theory including several diffracted beams is summarised in the review articles by *Uyeda* [8.9] and *Howie* [8.10].

a) Kinematical Theory of Electron Diffraction

As mentioned above, the kinematical approximation can be applied when the Bragg reflection is weakly excited in a thin specimen, that is, the diffracted amplitude is negligibly small in comparison with the transmitted amplitude. Therefore, we use the first Born approximation to solve the intergral formulation of the Schrödinger equation. The wave function $\psi(r)$ for the scattered wave having wave vector k is

$$\psi(r) = \psi^{(0)}(r) + \pi\mu \int \{\exp(2\pi i k|r - r'|)/|r - r'|\} V(r')\psi(r')dr', \qquad (8.1)$$

where $\psi^{(0)}(r)$ represents the incident wave having wave vector k_0; $V(r)$, the scattering potential; $\mu = 2me/h^2$. We assume that we can replace $\psi(r')$ in the integral by the incident plane wave. As $|r|$ is large compared with the atomic dimensions r', one can write (8.1) as follows:

$$\psi(r) = \exp(2\pi i k_0 \cdot r) + \phi(q) \{\exp(2\pi i k r)/r\}, \qquad (8.2)$$

where

$$\phi(q) = \pi\mu \int V(r') \exp(-2\pi i q \cdot r')dr' \qquad (8.3)$$

and $q = k - k_0 = 2\sin\theta/\lambda$ is the scattering vector. The second term in (8.2) represents the scattered wave function. One should note that (8.3) is equivalent to the Fourier transform of the scattering potential $V(r)$ and it represents the scattering amplitude.

For an assembly of atoms at positions r_j, the electro-static potential distribution can be written

$$V(r) = \sum_j V_j(r) * \delta(r - r_j), \qquad (8.4)$$

where $V_j(r)$ represents the electron density of the atom at r_j, $\delta(r - r_j)$ is a delta function at $r = r_j$, and the symbol $*$ means a convolution operation. Fourier transform of (8.4) gives

$$\phi(q) = \sum_j f_j(q) \exp(-2\pi i q \cdot r_j), \qquad (8.5)$$

$$f_j(q) = \pi\mu \int V_j(r) \exp(-2\pi i q \cdot r)dr, \qquad (8.6)$$

where $f_j(q)$ is the atomic scattering amplitude for electrons.

Here, the scattering vector q is equal to $g + s$ at the off-Bragg condition as shown in Fig. 8.2a, where g and s represents the reciprocal vector of reflecting planes and the deviation vector from the Bragg condition, respectively. For a periodic object, we may rewrite (8.5) as follows:

$$\phi_g = \sum_n F_g \exp[-2\pi i(g + s) \cdot r_n], \qquad (8.7)$$

where F_g is the structure amplitude of the unit cell; r_n is equal to $n_1 a + n_2 b + n_3 c$ (a, b, c are the unit cell translations, n_1, n_2, n_3 are integers).

In the distorted crystal r_n should be replaced by $r_n + R$ in (8.7), where R is the displacement vector. Since $g \cdot r_n$ is an integer and $s \cdot R$ is negligibly small, (8.7) can be written as

$$\phi_g = \sum_n F_g \exp[-2\pi i(g + s) \cdot (r_n + R)]$$

$$= \sum_n F_g \exp[-2\pi i(s \cdot r_n + g \cdot R)]. \qquad (8.8)$$

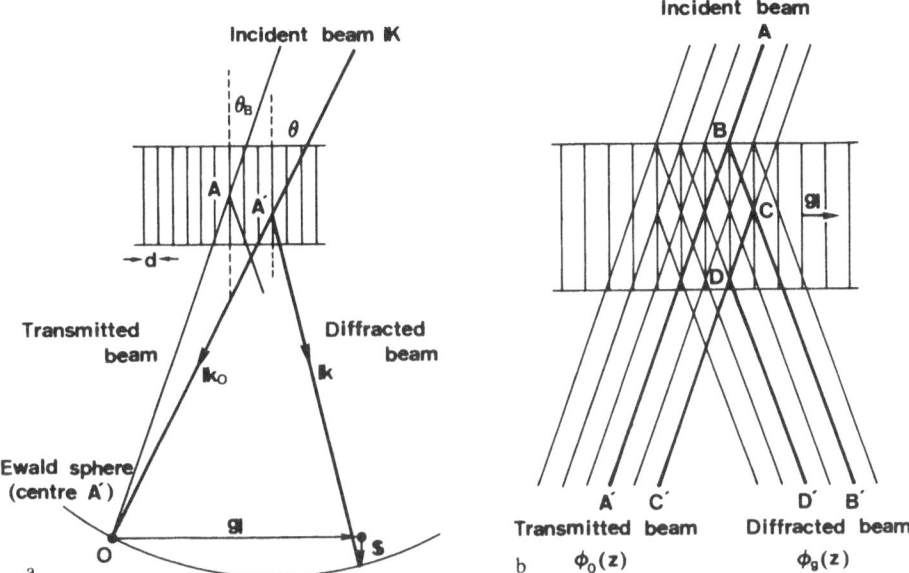

Fig. 8.2. (a) Ewald sphere construction illustrating the definition of deviation parameter s. (b) Illustrating the dynamical interaction between the incident and the diffracted beam at the exact Bragg condition

The amplitude of the diffracted beam at the bottom surface can be calculated by an integral of (8.8)

$$\phi_g = F_g \int_0^t \exp[-2\pi i(s_z z + g \cdot R)]\,dz, \tag{8.9}$$

where the product $s \cdot r_n$ has been substituted for $s_z z$ because s is very nearly parallel to z as seen in Fig. 8.2a; t, the specimen thickness. The intensity variation $|\phi_g|^2$ depending on both the specimen thickness t and the phase angle $2\pi g \cdot R$ due to lattice distortion shows the DF image contrast. The intensity variation $|\phi_g|^2$ at the bottom surface calculated in the perfect crystal, and the imperfect crystals containing stacking fault and dislocation has explained well the observed image contrast taken at the off-Bragg condition, but the interpretation of the image contrast at the Bragg diffraction condition should be treated by the dynamical theory.

b) Dynamical Theory of Electron Diffraction

When fast electrons enter the specimen at the Bragg condition and leave the bottom surface, the diffracted beam may be reflected by the lattice plane due to the strong interaction between electrons and materials, as shown in Fig. 8.2b. Therefore, the amplitude of the transmitted and the diffracted beam should be functions of the depth z in the specimen, $\phi_0(z)$ and $\phi_g(z)$, respectively. Consider only one set of reflecting planes in near the Bragg position. A total wave function of the electrons $\Psi(r)$ propagating in a column of crystal is given by

$$\Psi(r) = \phi_0(z)\exp(2\pi i \chi \cdot r) + \phi_g(z)\exp(2\pi i \chi' \cdot r), \tag{8.10}$$

where χ ($|\chi|$ is equal to $\sqrt{2meE}/h$; eE is the kinetic energy of electron) is the wave vector of the incident beam in field-free space; χ' denotes the wave vector of the diffracted wave written by

$$\chi' = \chi + g + s. \tag{8.11}$$

There are two formulations to treat the dynamical coupling between $\phi_0(z)$ and $\phi_g(z)$: (1) Wave-optical formulation which has been developed in a series of papers by *Howie* and *Whelan* [8.11], and *Kato* [8.12] and (2) wave-mechanical formulation which was originally given by *Bethe* [8.13] and further developed by *MacGillavry* [8.14], *Heidenreich* [8.15], *Kato* [8.16] and others. The first formulation yields a pair of differential equations known as *Howie-Whelan* equations [8.11] written by

$$\mathrm{d}\phi_0'/\mathrm{d}z = i\pi\phi_g'/\xi_g$$

$$\mathrm{d}\phi_g'/\mathrm{d}z = i\pi\phi_0'/\xi_g + \{2\pi i s_z + 2\pi i g \cdot (\mathrm{d}R/\mathrm{d}z)\}\phi_g', \tag{8.12}$$

where

$$\phi_0'(z) = \phi_0(z)\exp(-i\pi z/\xi_0),$$

$$\phi_g'(z) = \phi_g(z)\exp(2\pi i s_z - i\pi z/\xi_0 + 2\pi i g \cdot R). \tag{8.13}$$

These equations have been successfully applied to establish the dynamical diffraction contrast of the thickness and the bend extinction contours in perfect crystals.

Moreover, the amplitudes of $\phi_0(z)$ and $\phi_g(z)$ in crystals containing defects were calculated by the so-called column approximation which is a good approximation for electron diffraction, mainly because the Bragg angles are very small. The atomic displacement function in a narrow column is written by $R(z)$ as if it were of great lateral extent. Therefore, it is thought that the deformed crystal has been made up of a number of columns, which are perfect but displaced relative to each other according to $R(z)$. The contrast near defects is obtained by varying the positions x, y of the column where $R(z)$ will vary. The calculated image contrast of crystal defects such as stacking faults [8.17], dislocation [8.18], small precipitates with spherical symmetrical strain [8.19], by the column approximation agrees very well with the observed image contrast.

On the other hand, the wave-mechanical formulation becomes important for the treatment of contrast effect in the HVEM because many Bragg reflections inevitably occur simultaneously to a great extent depending on the strength of the crystal potential and the kinetic energy of incident electrons. Critical voltage effect [8.20, 21], Bloch wave channeling [8.22, 23] and multi-beam imaging [8.24], which were found by the HVEM, have been explained by the many beams approximation of the wave-mechanical formulation.

8.2.2 Phase Contrast Imaging

a) Moiré Patterns

When two optical gratings with a little different interplanar spacings and/or a small relative twist overlap, they form a beat pattern ('Moiré pattern') with some periodicity. We call this pattern formed by the two gratings which are parallel but different in spacing a 'parallel Moiré pattern' to distinguish it from the 'rotation Moiré pattern' formed by the two gratings which have identical spacing but a small relative twist angle.

The formation of Moiré patterns by two overlapping crystals can be easily considered as the optical analogy of two overlapping gratings as mentioned above. The superposition of two lattice planes whose diffraction takes place in each crystal causes double diffraction. It is recombined with the transmitted beam to form a Moiré pattern on the image plane. In the general case where the spacings of lattice planes are d_1 and d_2 and the small relative twist angle is β, the spacing and the direction of Moiré patterns can be interpreted in term of the double diffraction pattern as shown in Fig. 8.3. P and Q denote the diffraction spots from the lattice planes with the reciprocal lattice vectors of $\boldsymbol{g}_P(d_1 = 1/|\boldsymbol{g}_P|)$ and $\boldsymbol{g}_Q(d_2 = 1/|\boldsymbol{g}_Q|)$ in the upper and the lower crystals, respectively. The diffracted beam P can be diffracted by the lattice plane with the spacing of d_2 and produce the double diffraction beam spots R rotated by an angle ω about the reciprocal lattice vector $\boldsymbol{g}_P$. The direction of the Moiré fringes are perpendicular to the vector joining the transmitted beam spot O to the double diffraction beam spot R and the spacing D is equal to the inverse length of OR which is

$$D \sim d_1 d_2 / \{(d_1 - d_2)^2 + d_1 d_2 \beta^2\}^{1/2}. \tag{8.14}$$

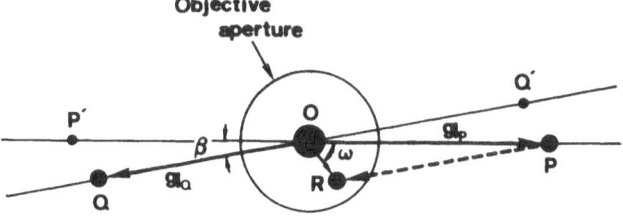

Fig. 8.3. The geometry of the double diffraction pattern caused by two overlapping crystals with a little different interplanar spacings, $1/|\boldsymbol{g}_P|$, $1/|\boldsymbol{g}_Q|$ and a small relative twist angle β. The Moiré fringes are perpendicular to the line OR joining the pair of the transmitted spot O and the double diffraction spot R

Electron microscope Moiré patterns were observed first by *Hashimoto* and *Uyeda* [8.25], and *Bassett* et al. [8.26]. *Hashimoto* et al. [8.27] explained the image contrast depending on the diffraction conditions and the specimen thicknesses in detail using the dynamical electron diffraction theory. The images do not resolve directly the actual lattice periodicities; they give us useful information about lattice defects as will be illustrated in Sect. 8.3.2 a.

b) Lattice Images

It is very instructive to explain the electron diffraction for lattice images because we can get the conception on the formation of multi-beam lattice images. Suppose two waves whose wave vectors are χ and $\chi + g$ exist in the free space. They interfere with each other and produce the interference fringes with the periodicity of $1/|g|$ as shown in the right-hand side of Fig. 8.4. The dynamical theory for the formation of the lattice image by *Hashimoto* et al. [8.28] allows the properties such as the position and the spacing of the fringes to depend on the specimen thickness and the Bragg condition. The outline is as follows:

When only one set of reflecting planes with the interplanar spacing of $d = 1/|g|$ is satisfied the exact Bragg condition, the electrons at the bottom surface of the crystal are described by a wave function $\Psi(r)$ in the same manner as (8.10).

$$\Psi(r) = \cos(\pi t/\xi_g)\exp(2\pi i\chi\cdot r)$$
$$+ i\cdot\sin(\pi t/\xi_g)\exp\{2\pi i(\chi + g)\cdot r\}, \tag{8.15}$$

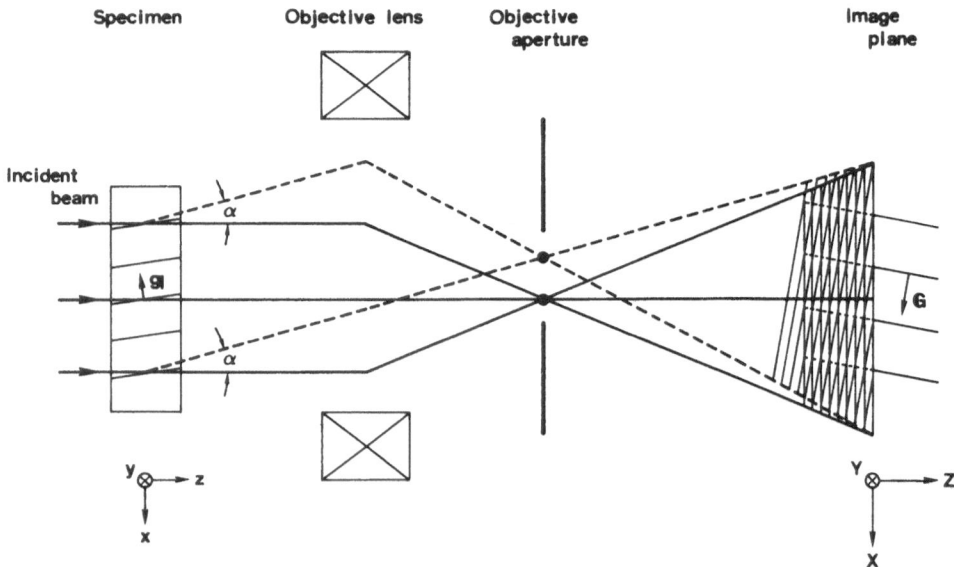

Fig. 8.4. Optical ray diagram showing the formation of lattice fringes

where t is the thickness of crystal; ξ_g, the extinction distance for the reflection g. Since the interference fringes are formed by recombining these two coherent electron waves (the transmitted and the diffracted beam), we obtain the total intensity distribution $I(R)$ on the image plane as

$$I(R) = \Psi(R) \cdot \Psi^*(R),$$
$$= 1 - \sin(2\pi t/\xi_g)\sin(2\pi G \cdot R), \tag{8.16}$$

where R and G are the position and the reciprocal lattice vectors in the image space, respectively. Let us assume that Z-axis is parallel to the incident beam and $G = (G_x, 0, G_z)$ as shown in Fig. 8.4, so that the equation can be rewritten as

$$I(R) = 1 - \sin(2\pi t/\xi_g)\sin(2\pi X/d + \pi \lambda Z/d^2). \tag{8.17}$$

Now for both a given thickness t except for thicknesses equal to multiples of $\xi_g/2$ and a focus position Z, (8.17) gives a modulated sinusoidal intensity in the X direction with the same periodicity equivalent to interplanar spacing d as

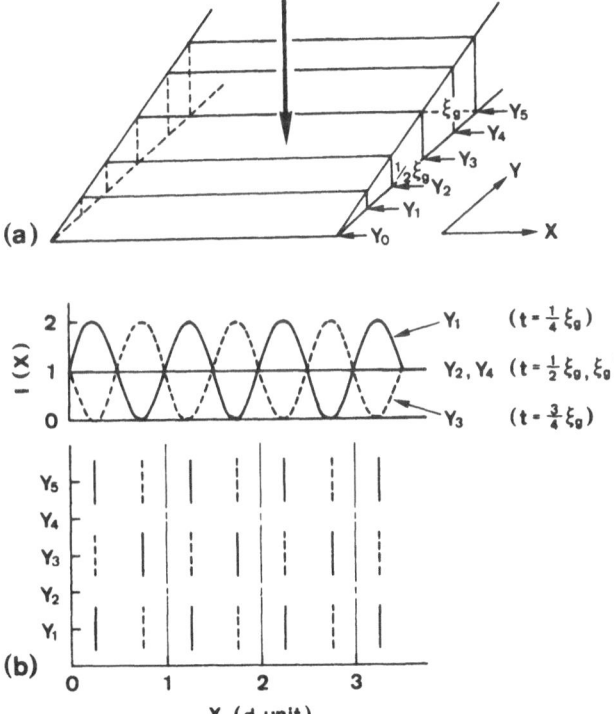

Fig. 8.5. (a) Illustrating the wedge shape specimen. (b) Intensity distribution of lattice fringes depending on both the positions X between the lattice planes, and the specimen thicknesses Y. Full and broken lines denote the maximum and the minimum intensity, respectively. Chain lines at $X = 1, 2$ and 3 represent the positions of lattice planes

shown in Fig. 8.5. Here, special attention should be paid to the relation between the positions of the maximum (or minimum) fringe contrast and the lattice plane. It is noted in Fig. 8.5 that the fringes do not coincide with the positions of lattice plane.

If many beams of systematic reflection in the case of the incidence parallel to the lattice planes, are allowed to contribute to the image, a few faint fringes appear within the periodicity of d, but the position of the main fringe approaches that of the lattice plane with increasing number of beams as studied by *Miyake* et al. [8.29].

c) Multi-Beam Lattice Images

When electrons fall along a zone axis with high symmetry in the crystalline specimen, large numbers of fairly strong diffracted beams are excited simultaneously because of the slowly curved Ewald sphere for high energy electron diffraction. Suppose the objective aperture allows such many electron beams to form the phase contrast image then according to the Abbe theory, the image process by the lens can be described in terms of two Fourier transforms as indicated in Fig. 8.6. All electrons scattered by an object having the transmission function $q(x, y)$ interfere on the back-focal plane to give the Fraunhofer diffraction pattern $Q(u, v)$ given by a Fourier transform function of $q(x, y)$; $Q(u, v) = \mathscr{F}q(x, y)$: Then, the wave from the back-focal plane forms an interference fringe pattern on the image plane. The amplitude distribution $\psi(x_i, y_i)$ is again given by the Fourier transform $\psi(x_i, y_i) = \mathscr{F}Q(u, v)$, so that the image amplitude given an ideal lens would be $q(-x_i, -y_i)$ and the intensity $I(x_i, y_i) = |q(-x_i, -y_i)|^2$.

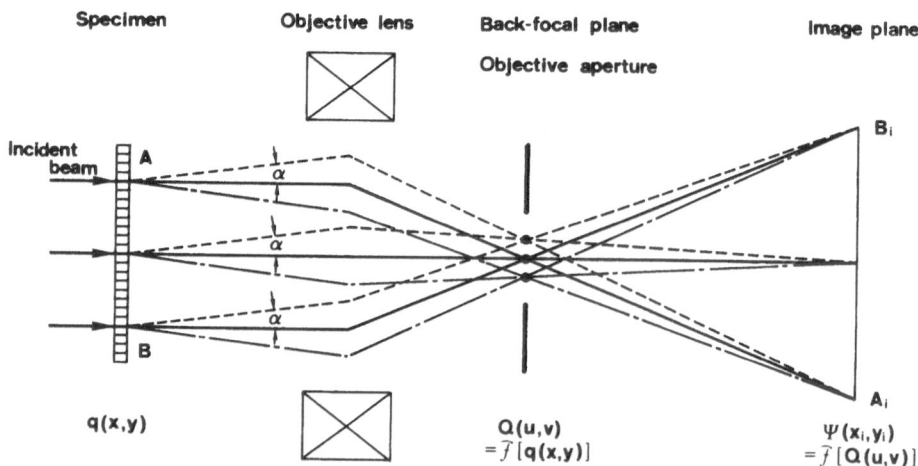

Fig. 8.6. Optical ray diagram illustrating the phase contrast imaging process in the Abbe theory

Since the intensity distribution of the image is proportional to the transmission function of the object, we may obtain the positions of lattice planes and/or atomic rows in the crystal from the multi-beam lattice images. In fact, we can sometimes observe beautiful dot arrangements which are likely to be the projected atomic rows. However, the relation between the multi-beam image and the arrangement of atomic rows is generally complex because of the dynamical effect of the electron, the lens aberration, the limited number of beams participating in forming the image and the capability of the resolving power of the electron microscope. This has considerable practical importance for understanding the formation of multi-beam lattice images which is given here briefly. More complete descriptions have been given by *Cowley* [8.8], *Spence* [8.30] and *Horiuchi* [8.31].

Multi-beam lattice images of a crystal are normally observed in a very thin specimen. For the thin objects, the transmission function on traversing the electro-static potential field $V(r)$ of the specimen is given by

$$q(x, y) = \exp[-i\sigma V_P(x, y)t], \tag{8.18}$$

where $V_p(x, y) = \int V(r)\,dz$ is the periodic function representing the projected electro-static potential (the atomic rows) and $\sigma = \pi/\lambda E$ is called the interaction constant. If the weak-phase object approximation can be applied, we assume $\sigma V_p(x, y)t \ll 1$ and (8.18) becomes

$$q(x, y) = 1 - i\sigma V_p(x, y)t. \tag{8.19}$$

If all waves leaving the weak phase object pass through an ideal objective lens to form the multi-beam image, the intensity distribution on the image plane reflect the projected potential $V_p(x, y)$, as stated above. But the aberration and the defocus of the lens prevent us from observing the projected potential. The effects of the spherical aberration C_s and the amount of defocus Δf of the lens are represented by introducing the phase change on the back-focal plane written by

$$\gamma(u, v) = (\pi/\lambda) \cdot [(u^2 + v^2)\lambda^2 \Delta f + (\tfrac{1}{2})C_s\lambda^4(u^2 + v^2)^2]. \tag{8.20}$$

Fraunhofer diffraction pattern modified by $\exp[-i\gamma(u,v)]$ is

$$\begin{aligned}
Q'(u, v) &= Q(u, v)\exp[-i\gamma(u,v)] \\
&= [\delta(u,v) - i\sigma\Phi(u,v)]\exp[-i\gamma(u,v)] \\
&= \delta(u,v) - \sigma\Phi(u,v)\sin[\gamma(u,v)] - i\sigma\Phi(u,v)\cos[\gamma(u,v)],
\end{aligned} \tag{8.21}$$

where $\Phi(u, v) = \mathscr{F}\{V_p(x, y)t\}$. The amplitude of the interference fringes on the image plane is calculated by the Fourier transform of $Q'(u, v)$ as

$$\begin{aligned}
\psi(x_i, y_i) &= 1 - \sigma\mathscr{F}\{\Phi(u,v)\sin[\gamma(u,v)]\} \\
&\quad - i\sigma\mathscr{F}\{\Phi(u,v)\cos[\gamma(u,v)]\}.
\end{aligned} \tag{8.22}$$

Neglecting the square terms, the intensity distribution of the image is expressed as

$$I(x_i, y_i) = 1 - 2\sigma \mathscr{F} \{\Phi(u, v) \sin[\gamma(u, v)]\}$$
$$= 1 - 2\sigma V_p(-x_i, -y_i)t * \mathscr{F} \{\sin[\gamma(u, v)]\}, \quad (8.23)$$

where $\sin[\gamma(u, v)]$ is called the *contrast transfer function* (CTF). If $\sin[\gamma(u, v)] = \pm 1$, the intensity is simplified as

$$I(x_i, y_i) = 1 \mp 2\sigma V_p(-x_i, -y_i)t. \quad (8.24)$$

Consequently, the interference image contrast reveals the projected potential exactly. Although the value of CTF is not generally equal to unity, $\sin[\gamma(u, v)]$ for the Scherzer optimum focus condition ($\Delta f = C_s^{1/2} \lambda^{1/2}$) is near unity for a wide range of scattering angles before it goes into wild oscillation at high angles. Figure 8.7 shows the example of CTF at 300 kV ($C_s = 0.96$ mm) which is near unity for the range of 2.0–5.2 nm^{-1}. This suggests that variation of potential 0.19–0.5 nm will be well represented and finer detail will be represented in a very confused manner.

However, in most of specimens of interest in materials science, the weak-phase object approximation can be hardly applied and the distances among the atomic rows are beyond the resolution limit of the microscope. Therefore, the observed images should be compared with the simulated images based on an assumed structure model using the dynamical electron diffraction theory. The image simulation commonly used is done by multi-slice formulation based on physical optics originally developed by *Cowley* and *Moodie* [8.32], and further developed practically by *Ishizuka* et al. [8.33] using fast Fourier transform algorithm. An alternative approach to the multi-slice method (real space method) by *Van Dyck* et al. [8.34] has also been developed using the formulation from the Schrödinger equation.

8.3 High-Resolution Electron Microscopy

The main purpose of high-resolution electron microscopy or the characterization of materials is to understand the microstructures, and the nature, the geometry and the shape of the crystal defects in fine scale. Multi-beam lattice imaging enables us to observe directly the different arrangements of atomic rows locally. However, we must pay particular attention to the interpretation of the images because the abrupt change of the strain field around the crystal defects affects the formation of phase contrast image significantly. Comparison of the observed images and the simulated images based on the assumed structure models is necessary for the analysis of defect structures.

Alternatively, the diffraction contrast images at high resolution, the so-called *weak-beam DF images* (WB–DF images), are often used to study strained crystal

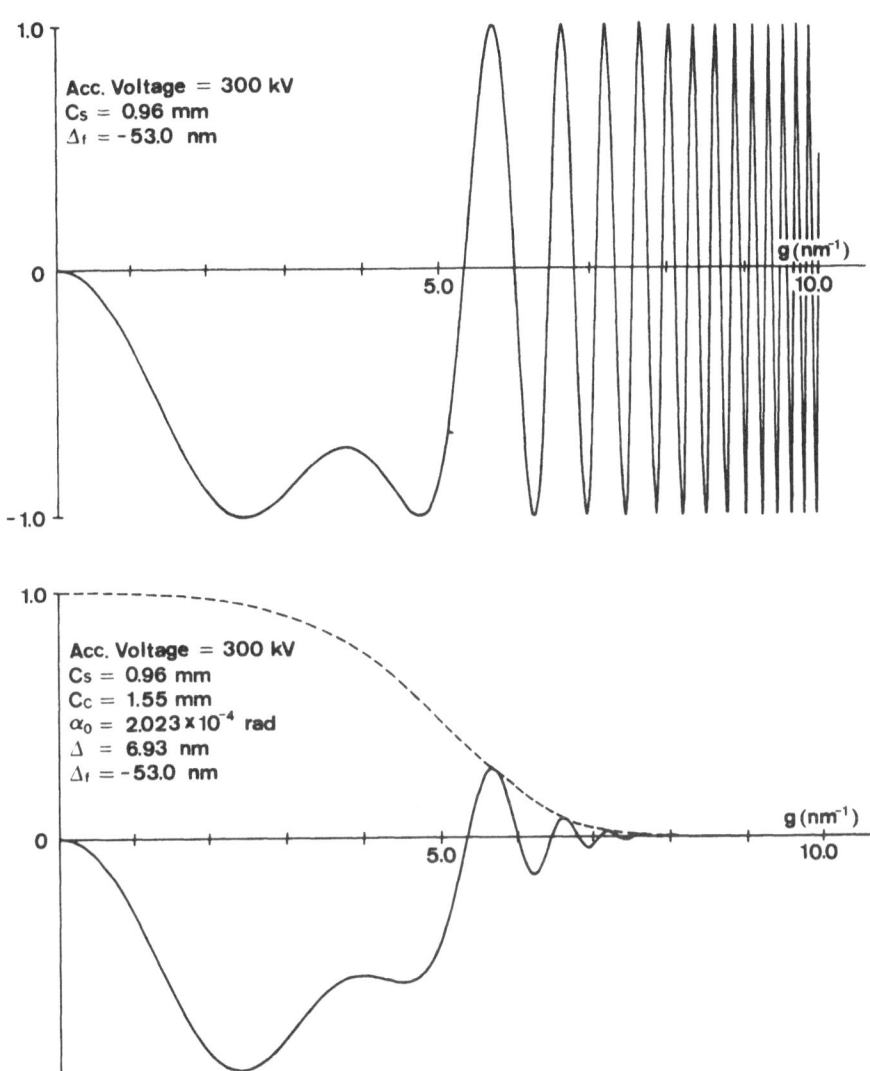

Fig. 8.7. Contrast transfer functions at 300 kV, $C_s = 0.96$ mm and the optimum focus $\Delta f = -53$ nm (a) without and (b) with the envelop function, where the chromatic aberration coefficient $C_c = 1.55$ mm

defects, even though the image resolution is not on the atomic scale. It gives us very useful information for the geometry and the nature of the individual defects. Principles of the weak-beam electron microscopy and some applications for defect analysis will be shown in the following section, and then some exciting examples of phase contrast imaging will also be shown in the Sect. 8.3.2.

8.3.1 Weak-Beam Electron Microscopy

Strain-related crystal defects such as dislocations and precipitates produce the wide image contrast in both the BF and the DF image. For example, the image peak width at half maximum of a dislocation for the dynamical two-beam condition is approximately between $\xi_g/3$ and $\xi_g/5$, where ξ_g is the extinction distance, i.e., typically ≥ 10 nm [8.18]. It follows that dislocations closer together than about 20 nm are not generally distinguished in the images. However, studies on the dissociation of dislocations, individual defects in the high density defects and a shape of small defects whose strain field overlap each other require their high resolution images. The poor resolution caused by the dynamical electron diffraction effect can be overcome, to great extent, by application of the WB–DF imaging technique which was pioneered by *Cockayne* et al. [8.35–37].

The effective extinction distance ξ_g^{eff} of the diffracted beam g at the far off-Bragg position is reduced as

$$\xi_g^{\text{eff}} = \xi_g / \{1 + (s_g \xi_g)^2\}^{1/2}, \tag{8.25}$$

where s_g represents the deviation parameter from the Bragg position. If ξ_{220} for Cu at 200 kV is 53 nm and s_g made about 0.2 nm^{-1}, $\xi_g^{\text{eff}} \sim 5$ nm and the image width of the dislocation is expected as about 1.3 nm. The detailed descriptions on the WB electron microscopy have been given experimentally and theoretically in the papers by *Cockayne* [8.36, 37]. His calculations show that provided the optimum diffraction conditions are met with $|s_g| \geq 0.2$ nm^{-1}, $|w_g| = |s_g \xi_g| \geq 5$ and no other reflections are strongly excited, the dislocation image behaves as follows: The image peak position is within ~ 2 nm of the core and the image intensity profile has sufficient narrow width of ~ 1.5 nm. The WB–DF images of dislocation enable its position to be accurately defined experimentally.

Since the diffracted beam is inclined to the optic axis at an angle of $2\theta_B$ ($= \lambda/d$) as shown in Fig. 8.1b, the spherical aberration of the objective lens results that all points on the DF image formed by shift of the objective aperture are elongated by an amount of $3C_s \lambda^2 \Delta\alpha/d^2$, where C_s is the spherical aberration coefficient; λ, the wave length of the electron beam; d, the interplanar spacing; $\Delta\alpha$, the divergence of the diffracted beam, typically $\sim 2 \times 10^{-3}$ for obtaining a reasonable image intensity. This elongation at 200 kV may be 2 nm for $d = 0.2$ nm and $C_s = 2$ mm. This affects the resolution of the DF image seriously because the resolution limit of the modern microscope is around 0.2 nm. Therefore, the diffracted beam of interest must be aligned accurately along the optic axis to get an ultimately good quality of the DF image (hereafter will be called the beam-tilted DF image for short), as shown in Fig. 8.1c.

The finer image contrast of crystal defects can be obtained in the beam-tilted DF image formed by a weakly excited diffracted beam. The range of applications for which the WB electron microscopy may be gainfully employed is quite extraordinary. Two main trends of defect analysis using the WB–DF image are demonstrated as follows:

The most common natural diamonds (classified as type Ia) contain numerous platelet defects, usually with mean diameters in the 5–100 nm range [8.38, 39]. The shape of platelet defects is a lath-like planar defect elongated along $\langle 1\,1\,0 \rangle$ directions and lying on $\{1\,0\,0\}$ planes [8.40, 41]. For studies of platelet the preferred specimen orientation is a foil parallel to $(1\,1\,0)$. Then, the sizes and shapes of platelets lying on both $(1\,0\,0)$ and $(0\,1\,0)$ making $45°$ with the specimen surface can be measured without tilting the specimen.

In Fig. 8.8a, the visibility of platelets on all three cube planes is good, but what sizes and shapes the platelets on $(1\,0\,0)$ and $(0\,1\,0)$ possess is highly uncertain. By contrast, the WB–DF image in Fig. 8.8b shows their dimensions clearly and the most common platelet diameter being about 20 nm. The images of $(0\,0\,1)$ platelets have reduced the visibility in Fig. 8.8b compared with that in Fig. 8.8a. In Fig. 8.8b, platelets on both cube planes can be seen equally well by the good visibility of their fault fringes.

The platelet population density can be measured by counting platelets in specimen volumes of accurately determined thickness. Combining the population density with the size measured by using WB–DF micrographs gives us the total platelet area per unit volume precisely. Measurement of the infrared absorption through the same thin specimen in which the platelet counting was performed allows the discussion on the relation between the nitrogen impurity and the platelet defect in type Ia diamond, and the problem of elucidating structures of nitrogen impurity aggregates [8.42].

The WB–DF image of platelets formed with only $[0\,1\,0]$ spike associated with a weakly excited Bragg reflection $(\bar{2}\,2\,0)$, one would expect to see only platelets parallel to $(0\,1\,0)$ [8.43, Fig. 8]. With weak-beam spike DF imaging of platelets, one would hope to identify the orientation of small platelets.

One of the most fruitful applications to materials science using WB electron microscopy is to observe the dissociation of dislocations. The stacking fault energies [8.44] and antiphase boundary energies [8.45, 46] were successfully estimated in a number of materials. Several kinds of interactions among dislocations during deformation were also examined to understand their mechanism. For example, the formation of faulted dipoles in silicon was studied by *Winter* et al. [8.47] and it was shown to distinguish vacancy and interstitial dipoles having intrinsic or extrinsic dissociation from their contrast behaviour.

It was very difficult to observe the partial dislocations in diamond because of its high stacking fault energy. A careful study on a type IIa diamond cut parallel to $(1\,1\,1)$ on which glide dislocations were lying was carried out to obtain the evidence for the dissociation [8.48]. The stacking fault contrast between two partial dislocations is clearly seen in Fig. 8.9a taken with $(1\,\bar{1}\,\bar{1})$ reflection normal to $[2\,1\,1]$. Moreover, Fig. 8.9b–d show images of an extended node of dislocation formed by the three $\{2\,2\,0\}$ reflections normal to $[1\,1\,1]$. Each image shows two arms of the node.

Two partial dislocations were clearly resolved for many dislocations with different line orientations, with separations being in the range of 3.6–4.8 nm. Although in this study WB electron microscopy is being used near limit of

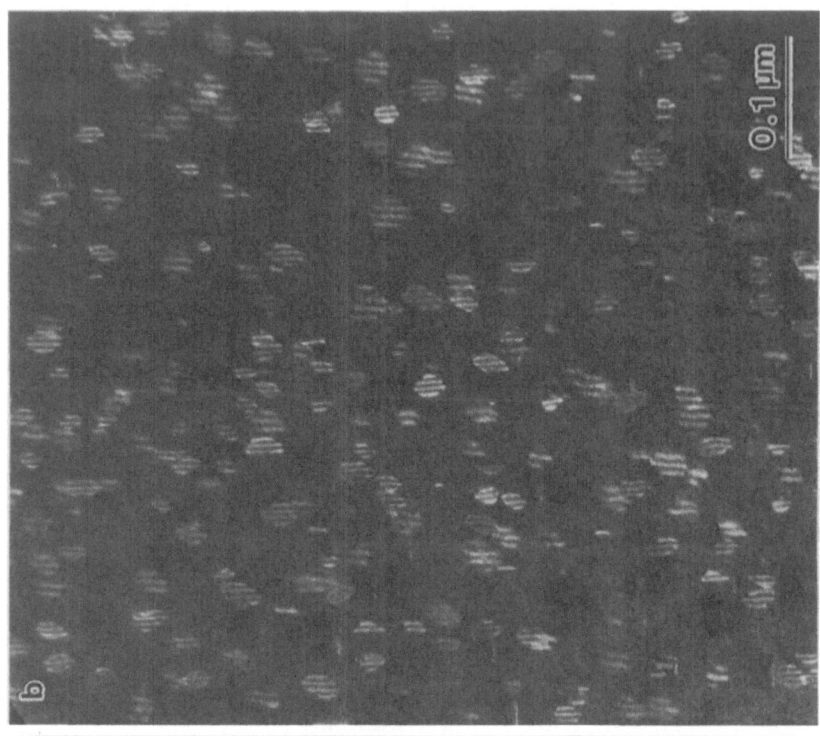

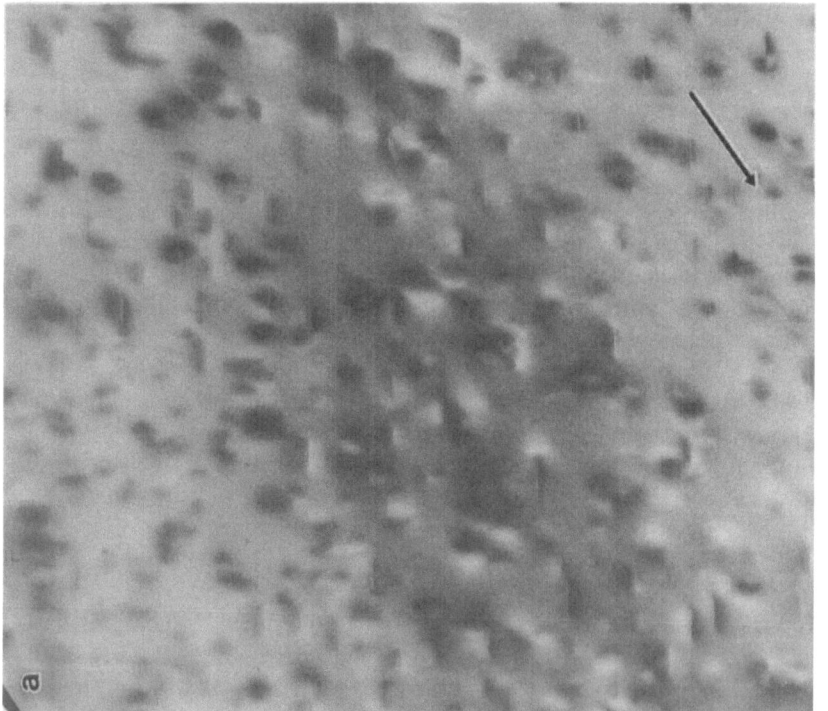

Fig. 8.8. Comparison of (a) the BF image and (b) the WB–DF image ($s_g = -0.16\ nm^{-1}$) of the same area in type Ia diamond taken with $(1\bar{1}\bar{1})$ reflection as indicated by a black arrow. An absence of contrast from the bounding partial dislocations is general observation in (b) because the visibility decreases with increasing $|s_g|$ in $\{111\}$ reflections [8.42]

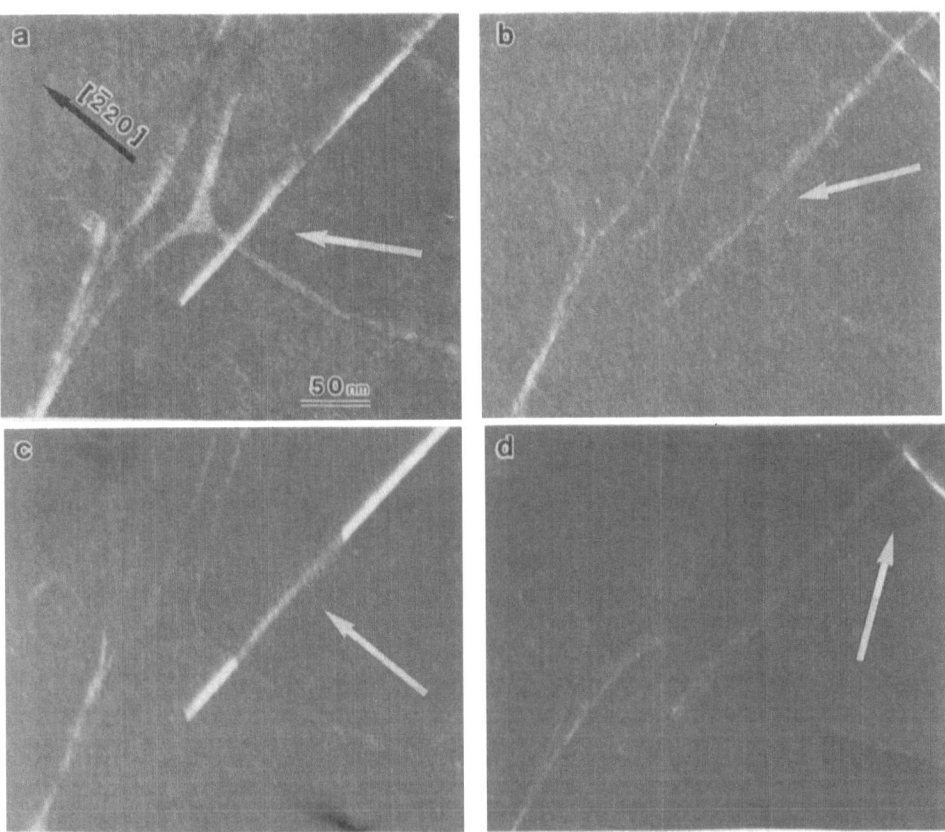

Fig. 8.9. WB–DF images of an extended node in type IIa diamond taken with (a) $(1\bar{1}\bar{1})$, (b) $(0\,2\,\bar{2})$, (c) $(\bar{2}\,2\,0)$, (d) $(\bar{2}\,0\,2)$ reflection, as indicated by white arrows. The stacking fault contrast from the extended node can be clearly seen in (a). The incident beam directions in (a) and (b)–(d) are nearly parallel to $[2\,1\,1]$ and $[1\,1\,1]$, respectively [8.48]

applicability, the estimates of the stacking fault energies from the distribution of the separations of the partial dislocations depending on the line orientations and from extended nodes are in good agreement with each other. It has been determined to be 279 ± 41 mJm^{-2}.

As described above, the WB electron microscopy is a very powerful technique to characterize lattice defects, especially dislocations.

8.3.2 High-Resolution Images with Phase Contrast

Phase contrast images in a very thin foil such as Moiré patterns, lattice images and multi-beam lattice images may be used to relate the specimen structure in atomic scale. But the care on the image interpretation must be beared in mind. Applications of phase contrast imaging will be illustrated in the following:

a) Moiré Patterns

Although Moiré patterns are indirectly resolved lattice images, the imaging technique has sometimes gained significant support in the periodic structure and the location of lattice imperfections of a crystal. For example, Moiré technique are used for measuring small displacements of one periodic object relative to another of similar periodicity as can be understood by (8.14).

When studying atomic-scale displacement due to dislocations, the interpretation can be straightforward if a dislocation-free crystal is superimposed on the dislocation-containing crystal. The number of Moiré fringes terminating at the dislocation outcrop is $g \cdot b$, where b is the Burgers vector of the dislocation [8.49, 50]. In the case of planar defect, the fringe displacement at the outcrop of the planar defect is $g \cdot f$, where f is the relative translation between crystal matrices on either side of the defect.

The Moiré pattern appears in the narrow strip of overlap in which the specimen laminae above and below were slightly rotated with respect to each other, as shown in Fig. 8.10 [8.51]. We can view the outcrop of $(0\,0\,1)$ edge-on platelets in the $(1\,1\,0)$ specimen. Moiré fringes at the edge-on platelet in the centre of the field looks exactly like the image of pure edge dislocation. The number of extra fringes there is estimated to lie between 1.40 and 1.45. Taking $g \cdot f$ for $(0\,0\,4)$ reflection to be the values between 1.40 and 1.45 and f parallel to

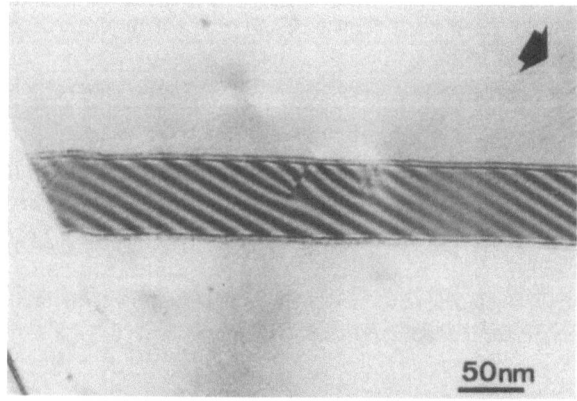

Fig. 8.10. A BF image taken with $(0\,0\,4)$ reflection as indicated by a black arrowhead. Rotation Moiré fringes appear where cleavage and overthrust have led to superposition of type Ia diamond. It is noted that the clear-cut displacement of the Moiré fringes at the $(0\,0\,1)$ edge-on platelet can be recognized within the overlap region. The specimen surface is parallel to $(1\,1\,0)$ [8.51]

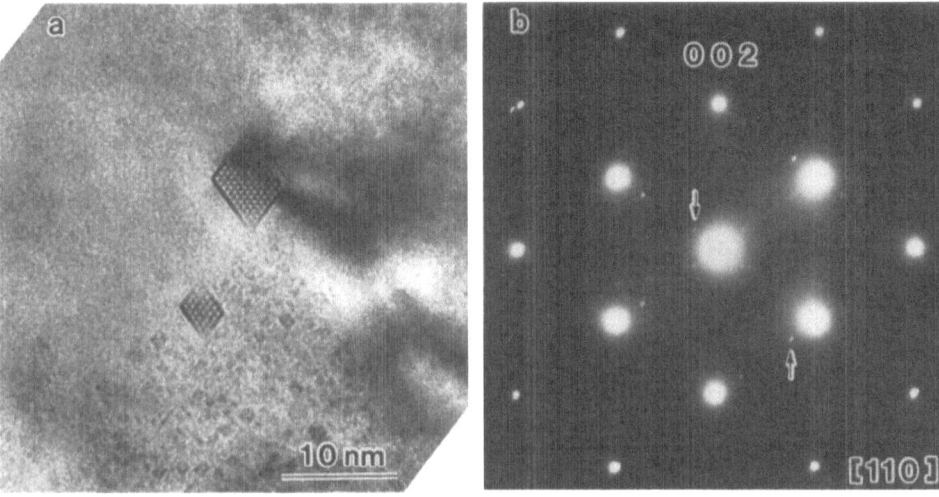

Fig. 8.11. (a) A BF image of a voidite-containing type Ia diamond and (b) the diffraction pattern of exact [1 1 0] zone axis. Note the satellite spots both around the transmitted beam and the diffracted beams due to double diffraction effect in (b), and the corresponding two dimensional Moiré fringes within the voidites in (a) [8.58, 61]

[0 0 1], the magnitudes of f lie between $0.35a_0$ and $0.362a_0$, a_0 being the lattice constant of diamond, 0.357 nm.

It is important to have a strictly accurate value for the lattice displacement produced by the platelet structure for their model-making. Multi-beam lattice images of platelets taken by *Barry* et al. [8.52] and *Humble* et al. [8.53] can be also used to estimate the displacement. However, as most of distances among atomic rows in these images are beyond the resolution limit of the present microscope, the platelet structure models derived from their images should be carefully compared with the models obtained by other methods. We would say that the present experiment using Moiré technique provides the best and most direct measurement obtained to date.

Small {1 1 1}-faceted void-like defects ('voidites') with diameters in the range 1–10 nm are seen in type IaB diamonds, generally in association with degraded {1 0 0} platelets [8.54–56]. Assuming that the platelet are composed of nitrogen, it is very likely that this nitrogen is going into the voidites when the platelet collapses [8.56, 57]. This has been demonstrated by means of energy-dispersive X-ray analysis by *Hirsch* [8.58]. On the other hand, the high resolution image in the (1 1 0) specimen exhibits two dimensional 'superlattice' contrast within the voidites [8.59] which was later interpreted as being the Moiré fringes from the overlapping diamond and a solid crystalline phase in the voidites [8.58, 60, 61]. This interpretation is established by the double diffraction spots around both the transmitted and the diamond diffraction spots in the diffraction pattern taken from the single voidite with the Moiré fringes, as shown in Fig. 8.11. The detailed analysis reveals that the crystalline phase is a face-centered cubic phase of solid ammonia at high pressure and the lattice parameter varies from voidite to voidite [8.58, 61]. However, no infrared absorptions which could possibly be attributable to stretching of N–H bonds could be detected in diamonds expected to contain voidites [8.62, 63]. The crystal phase of voidites is still uncertain at present, so that further experiments on Moiré patterns and the corresponding diffraction patterns will be necessary to solve this problems.

b) Multi-Beam Lattice Images

A few examples suggesting the usefulness of multi-beam lattice images for characterization of materials such as plane defects, microstructures and amorphous materials will be explained in this subsection.

Many growth faults ('lamellae') are seen in ζ_β-FeSi$_2$. It was concluded from the BF images and the diffraction patterns that the lamellae are a plane defect with [0 1 1] displacement vector lying on (1 0 0) plane [8.64]. Figure 8.12a shows the BF image of the edge-on lamellae and Fig. 8.12b–d shows the diffraction patterns taken from circular regions of A, B and C in Fig. 8.12a. Although these regions are in the same grain, the orientations of A and C are identified as [0 2 1] and [0 1 $\bar{2}$], respectively. The pattern from B can be got by the superposition of both patterns from A and C. The multi-beam lattice images of the corresponding regions in Fig. 8.12e–g can clearly solve the curious problem caused by introducing lamellae. In order to make the position of lamellae clear, a white line connecting white peaks are drawn in each micrograph. In Fig. 8.12e, the white line lies obliquely in the perfect lattice but it lies vertically at the lamellae. Increasing a number of lamellae as shown in Fig. 8.12f, the length of vertical line increases, and the very small region containing only a few unit cells changes their orientation to [0 1 $\bar{2}$] as shown in Fig. 8.12g. This observation gives a support to the structure model of lamellae derived from the comparison of the observed image and the simulated image [8.64].

A thin (0 0 1) foil specimen of non-stoichiometric β'-NiAl alloys annealed at 623 K shows a well modulated structure with about 10 nm width along nearly $\langle 1 0 0 \rangle$ directions in the BF image and complex intensity distribution around the fundamental and the superlattice spots in the diffraction pattern [8.65]. The diffraction pattern as shown in the inset of Fig. 8.13 is classified into three kinds of net patterns, which correspond to the patterns of [0 0 1] zone axis in β'-NiAl, [1 1 0] zone axis in tetragonal Ni$_3$Al$_2$ and an unidentified phase. The high resolution image also shows three characteristic images: The image with interplanar spacing of 0.20 nm (β'-NiAl); the image with rows of bright dots appeared every six planes nearly parallel to the $\langle 1 1 0 \rangle$ direction of β'-NiAl; and the granular image between them. According to the results of element analysis at the mixed region of β'-NiAl image and the granular image, enrichment of aluminium atoms occurs there. Taking both the multi-beam lattice image and the result of element analysis into account, the region showing the granular

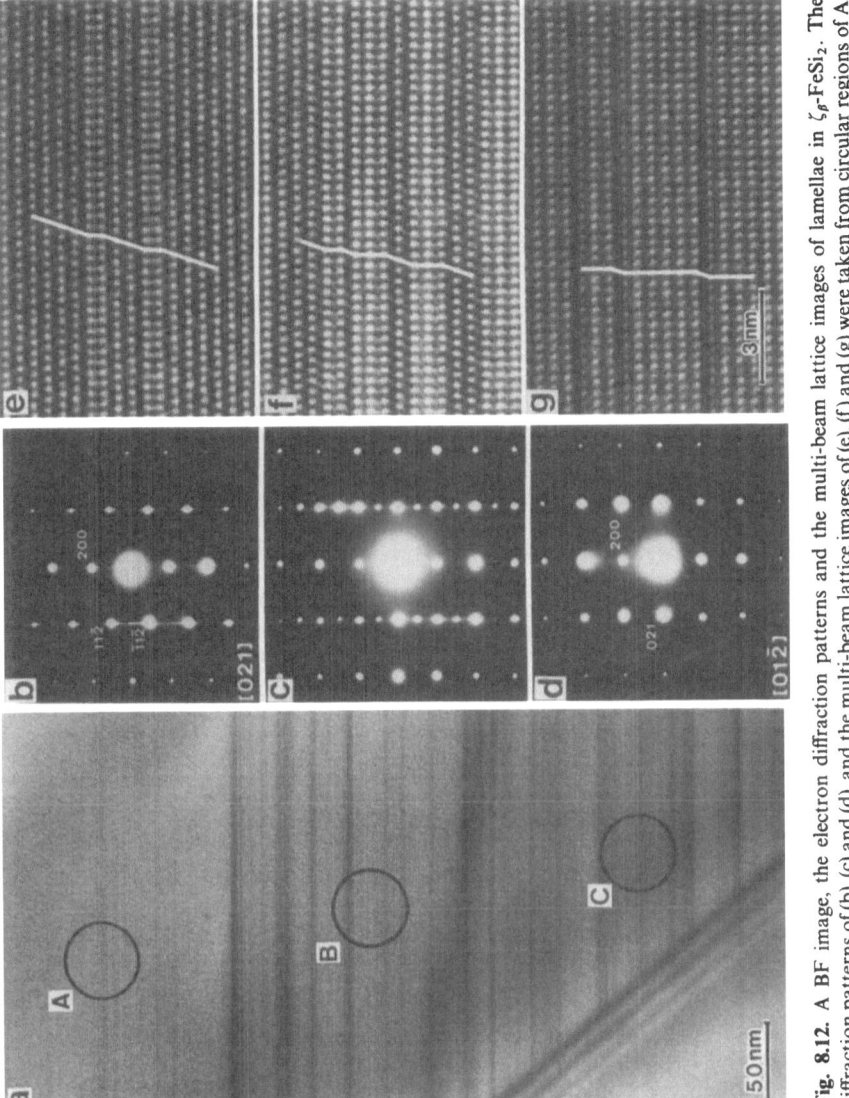

Fig. 8.12. A BF image, the electron diffraction patterns and the multi-beam lattice images of lamellae in ζ_β-FeSi$_2$. The diffraction patterns of (b), (c) and (d), and the multi-beam lattice images of (e), (f) and (g) were taken from circular regions of A, B and C in (a), respectively. It should be noted that the different diffraction patterns and images can be observed in the same grain [8.64]

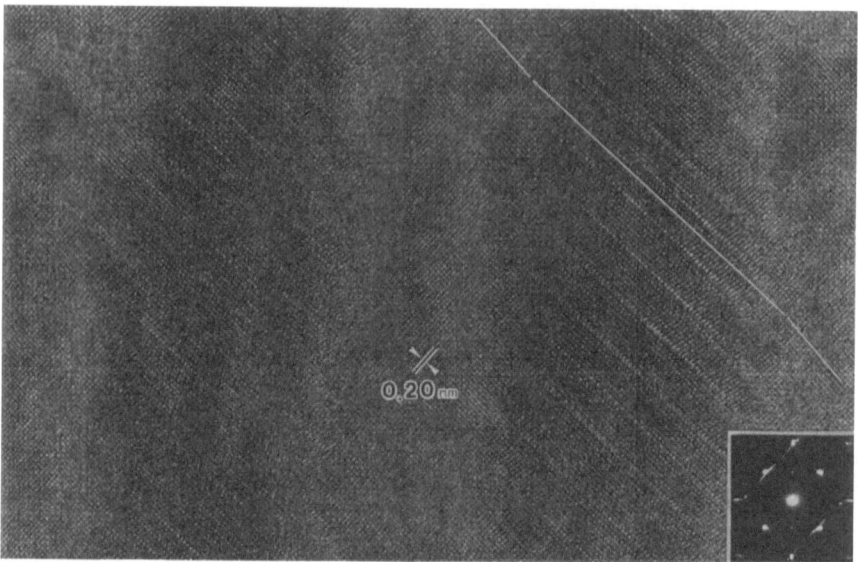

Fig. 8.13. High resolution image of β'-NiAl annealed at 623° K for 1 h. The thin specimen, whose surface is parallel to (0 0 1), finished for electron microscope observation was annealed. The crystal structure of β'-NiAl, a new periodic structure with rows of bright dots nearly parallel to the $\langle 1\,1\,0 \rangle$ directions of the matrix as indicated by a white line and granular contrast between them can be recognized [8.65]

image is likely to be Ni_2Al_3 of which the existence is confirmed by the diffraction pattern of the other zone axis. In this manner, we can distinguish the crystal structure within small regions less than 10 nm.

High resolution images of thin evaporated carbon film, which is an ideal amorphous phase, are normally used to estimate the performance of an electron microscope by their through focal series because their granular image contains many periodicities randomly [8.66]. On the other hand, high resolution images of some amorphous materials show often lattice-like fringe contrast locally, which suggests the existence of local order of atomic arrangement. The simulated images based on assumed small atomic assemblies by the kinematical approximation were also compared with the observed images [8.67, 68] to explain the atomic arrangements in the amorphous materials. They concluded that the lattice-like fringes arise from the medium range order regions of a few-nanometers in diameter, if the images were observed under careful experimental conditions.

Another approach to confirm the conclusion stated above was done by an elegant method [8.69]. High energy electron irradiation at low temperature produces the amorphous phase easily in NiTi alloy [8.70]. This amorphization method has a great advantage of preparing the specimen in which both the crystalline and the amorphous phase exist. The boundary between the irradiated

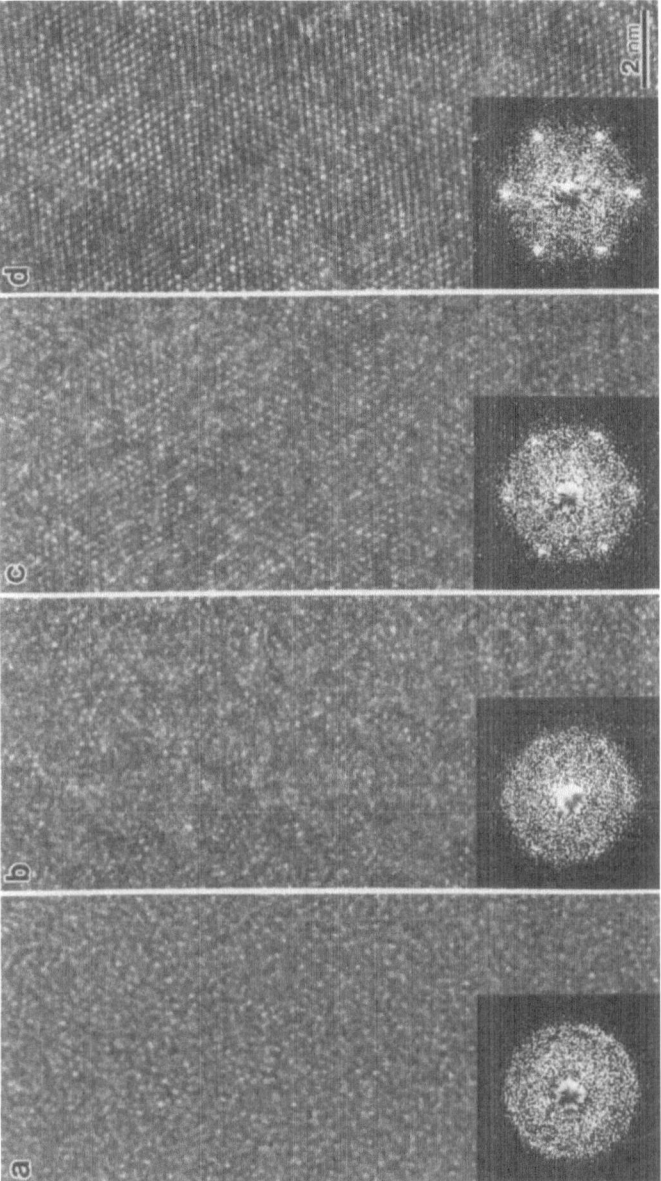

Fig. 8.14. High resolution images and their corresponding optical diffractograms taken from various parts of the irradiation-induced amorphous region in NiTi alloy. The micrograph of (a) shows a central part of the irradiated region. The micrographs of (b), (c) and (d) correspond to the parts of the broad boundary between the irradiated and the unirradiated region [8.69]

region (the amorphous phase) and the unirradiated region (the crystalline phase) is broad and obscure. This suggests that the loose cut-off intensity distribution of electron beam leads to incomplete amorphization on the boundary. High resolution micrographs of the boundary were taken so as to cover both the amorphous and the crystalline phase by one shot. The micrographs in Fig. 8.14a and b–d show a central part of the irradiated region and various parts of the boundary, respectively. The micrographs of Fig. 8.14b and c correspond to the parts closer to the amorphous and the crystalline phase, respectively. Typical granular contrast in Fig. 8.14a suggests that the amorphization occurs completely there. But it should be noted that the local order of bright dots within a few nanometer regions, can be still seen in Fig. 8.14b and c. The corresponding optical diffractograms reveal that the local order has the same lattice image as the original crystal. Namely, they are crystallites with the same orientation. Needless to say, the sizes of these crystallites become smaller as the region extends farther from the crystalline region. Therefore, using this observation as important clues, we will be able to understand the amorphization of the crystalline material in atomic level.

It is finally stressed here that in situ observation of high resolution images will open the door to new research fields for materials science. For example, *Iijima* et al. [8.71] recorded structural change and the coalescence of fine gold particles by video-tape recording system and shows new directions of high resolution electron microscopy. *Mitome* et al. [8.72] observed the formation process of gold particles in situ in ultra-vacuum and high-resolution electron microscope and suggested their instability depending on their size.

Irradiation effect can be positively used to induce the structural change of materials during observation. For example, formation process of point defect aggregates in aluminium [8.73] and the decomposition process of GP (I) zone in Al–4 at. % Cu alloy [8.74] were observed in situ in a high-resolution electron microscope operated at 300 kV.

Figure 8.15 shows another example on successive stages of the decomposition of (0 0 1) edge-on platelet in type Ia diamond [8.75]. The platelet image in Fig. 8.15a, which shows two rows of bright dots along (1 0 0) plane and the displacement of two sets of {1 1 1} lattice fringes at the platelet, disappears with increase in the irradiation time of 300 keV electrons. This evidently implies that the platelet decomposes when the radiation damage is introduced into the diamond. On the way to the decomposition, the bright spots on the platelet have been scattered around the matrix as seen in Fig. 8.15b. It is very difficult at present to identify the origin of the bright dots by only the simulated image. However, both the contrast change of platelet and their simulated image will help the establishment of the structure model of platelet in the near future.

High-resolution electron microscopy of a wide-range study such as surface structures at a monoatom layer level, grain boundary, interface and artificially prepared periodic superlattice structure is also a highly promising technique for understanding the structures and/or the behaviours directly at atomic scale.

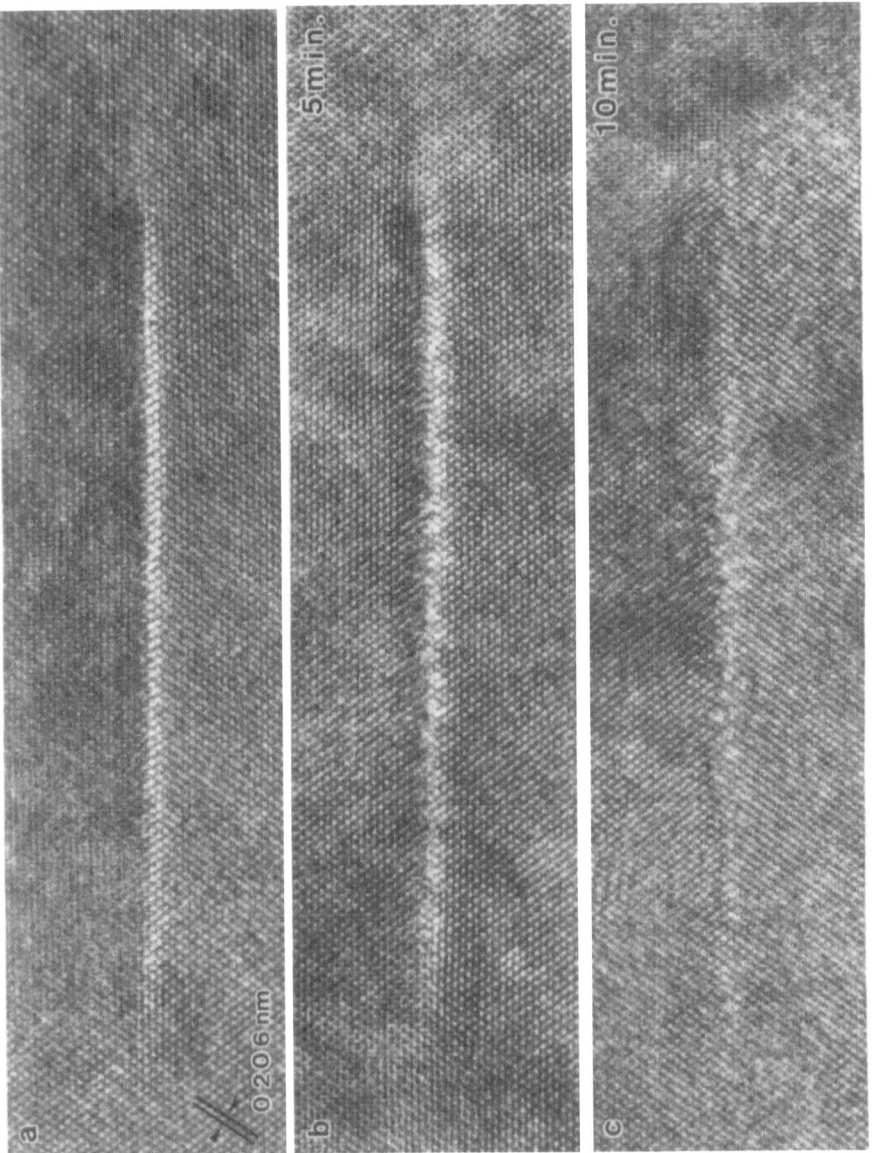

Fig. 8.15. Successive stages of the decomposition of edge-on platelet in type Ia diamond during observation with HREM operated at 300 kV [8.75]

8.4 Indispensable Applications of HVEMy

8.4.1 Quick Response of Lattice Defects to Applied Conditions

In situ experiment with HVEM's shows epoch-making advantages for studying natural science, especially materials science. It is well known, however, that in order to observe the behaviour of lattice defects as occurs in bulk materials the specimen must be thicker than a critical thickness determined by the mean free path of lattice defects that contribute to each individual phenomenon [8.76–80]. Even in the case of point defects, their mean free paths are frequently determined by the interaction with such sinks as surfaces and dislocations, the strain field of which is generally large. The mean free path of lattice defects is $\sim 3 \, \mu m$ in general, and thus the specimen thickness should be larger than 3 μm for the in situ experiment. Erroneous information might result from thin specimens.

To carry out the in situ experiment, several types of specimen-treatment devices have been developed [8.81]. All of these devices are mounted on a universal goniometer stage, and most of these specimen holders can be easily exchanged using an air-lock system.

Using these devices, dynamic observation of various phenomena in materials has been carried out under/in such conditions as stressing, various atmospheres, electron irradiation, magnetic field, and combinations of the foregoings with a wide range of temperatures, from the temperature of liquid helium to higher than 2300°K.

By the in situ experiment, not only mechanisms of various phenomena but also the origin of microstructures, such as secondary point defects, atom clusters etc., can be verified in detail [8.76–82]. For example, even when the size of structures is extremely small, their characteristics and structures can be investigated after they grow to a sufficiently large size for characterization by in situ specimen treatments. Furthermore, the following indispensable information can be also obtained in local regions by the in situ experiment: 1) Interaction among the lattice defects and between the lattice defect and other microstructures, 2) position and shape of the lattice defects under experimental conditions, e.g., under applied stress, 3) motion of each individual lattice defect, and others. The above information is a typical example showing "the materials are living". As a result, reliability of image resolution by EMy is very much improved by these in situ experiments.

By the in situ experiment satisfying the necessary condition mentioned above, indispensable information on behaviour of the lattice defects has been obtained under various conditions. Based on the results, mechanisms of most mechanical behaviour of materials, such as the work hardening, the fracture, the mechanical twinning, the martensitic transformation etc., including fatigue and creep deformation, have been made qualitatively clear [8.81, 82].

Figures 8.16 and 8.17 are some examples of the in situ experiment on the mechanical behaviour of metals [8.83, 84] and ceramics [8.84]. It is recognized

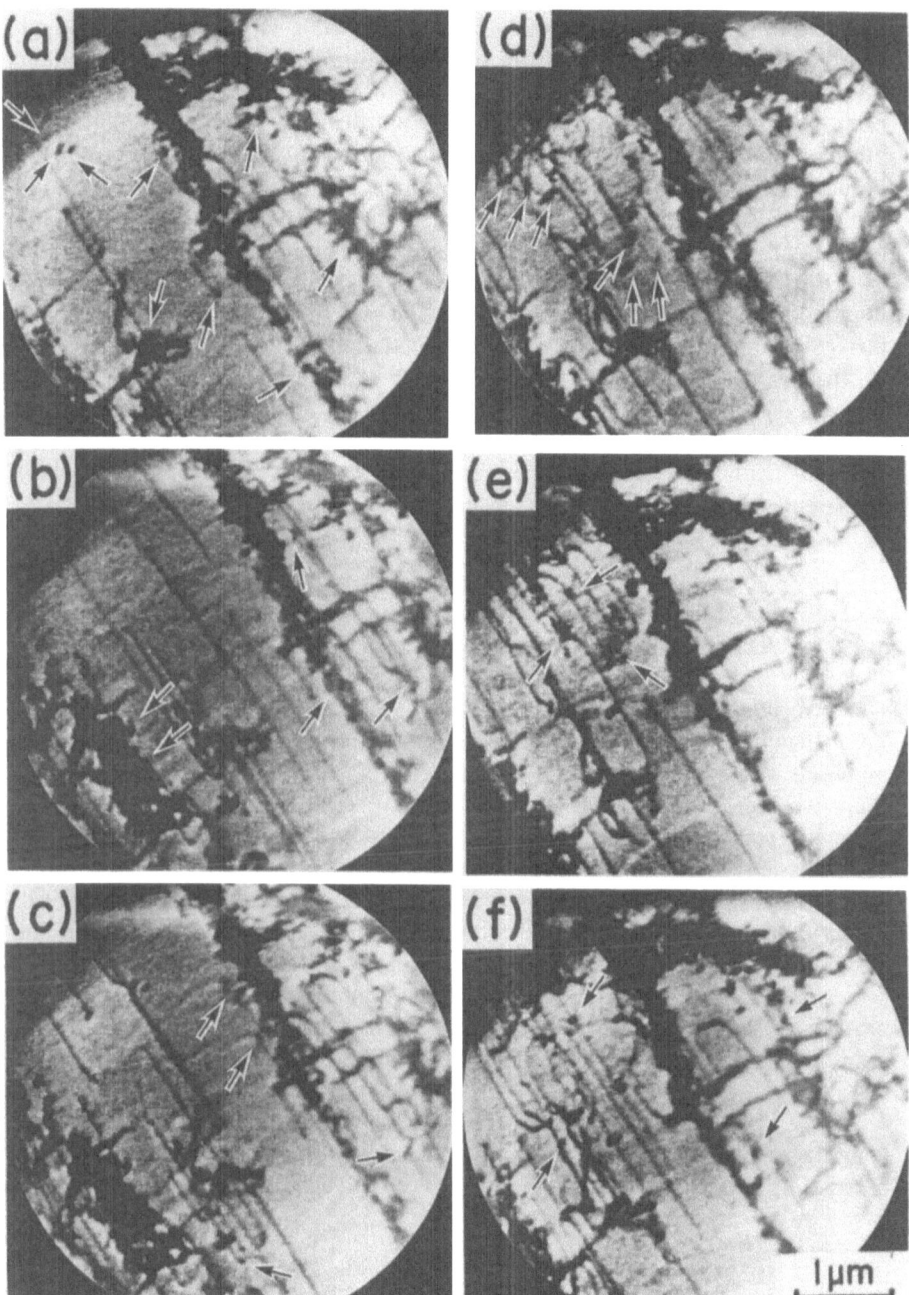

Fig. 8.16. Formation process of dislocation tangles in the $\langle 1\,1\,1 \rangle$ W single crystal at a low homologous temperature, i.e., 0.09 of the melting point (accelerating voltage: 2 MV). Arrows show the prismatic dislocation loops formed by double cross-slip of mobile dislocations

in Fig. 8.16 that the work hardening process of metals with a considerably high stacking fault energy results from the dislocation tangling around prismatic dislocation loops induced by the cross-slip of dislocations, and that not only each individual dislocation in the tangles but also the position of the tangles always move during plastic deformation. Figure 8.17 is an example showing pseudo-elasticity of a specially treated ceramic. Namely, a twinned structure in Fig. 8.17a disappears in b immediately after the stress is applied, then abruptly appears again in c as soon as the applied stress is removed.

Based on these results, it is concluded that both microstructure size and homogeneity of chemical composition in the scale of 0.1 μm are important for increasing the toughness and strength of materials.

Besides the above mechanical behaviour of materials, the effect of homogeneous deformation on the mechanical behaviour, annealing phenomena such as recovery and recrystallization of cold worked materials, chemical behaviour, electron irradiation effects and environment-materials interaction have been investigated in detail on various materials [8.82, 85].

8.4.2 Direct Observation of Co-operative Actions among more than Two Factors in Material Behaviour by the in situ Experiment

Quick response of lattice defects to applied conditions is a remarkable materials' behaviour showing "materials are living". Another important behaviour of materials showing "materials are living" is co-operative action among more than two factors in order to reduce the total free energy of materials. These co-operative actions always occur in various phenomena of materials. The following few examples show the co-operative actions in material behaviour.

Figure 8.18 [8.86, 87] shows the recrystallization process of a cold-worked aluminium crystal. It is recognized in Fig. 8.18 that a misfit angle of the sub-boundary denoted by 11 gradually increases as the subgrain-coalescence is preceded in interior regions, as seen in a–c, and finally sub-boundary 11 changes to a high-angle boundary in d. This fact means that the surface energy (E_S) is gradually changed as the volume free energy difference (ΔE_V) increases [8.87, 88].

Here, it has been shown that a crystal nucleus is formed in a solid at a fixed temperature when the total free energy difference of a block expressed by (8.25) becomes negative:

$$\Delta E_T = \Delta E_V + E_S + E_L + \zeta \Delta V^m. \tag{8.25}$$

Here, E_L: the lattice distortion term due to the coordination number difference between the block and the matrix, which plays an important role at the beginning of nucleation; $\zeta \Delta V^m$: a term corresponding to the specific volume difference between the block and the matrix, in which each of ζ and m is

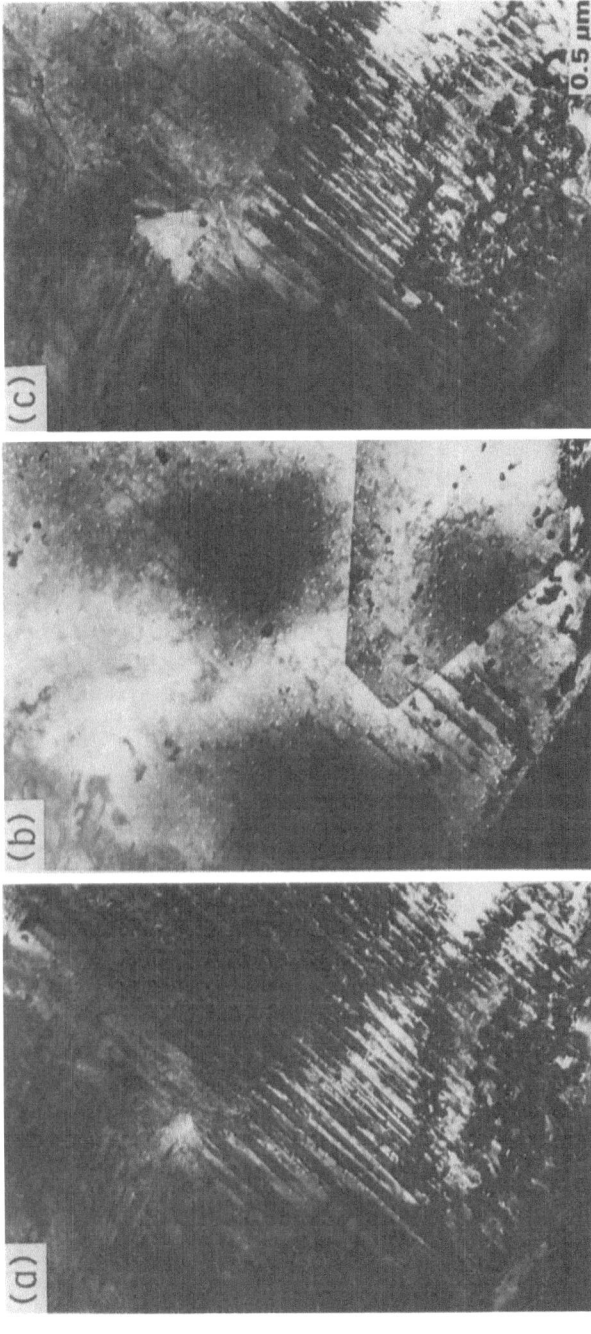

Fig. 8.17. Successive stages of pseudo-elasticity of a monoclinic ZrO_2-3.9 mass%MgO crystal (accelerating voltage: 2 MV)

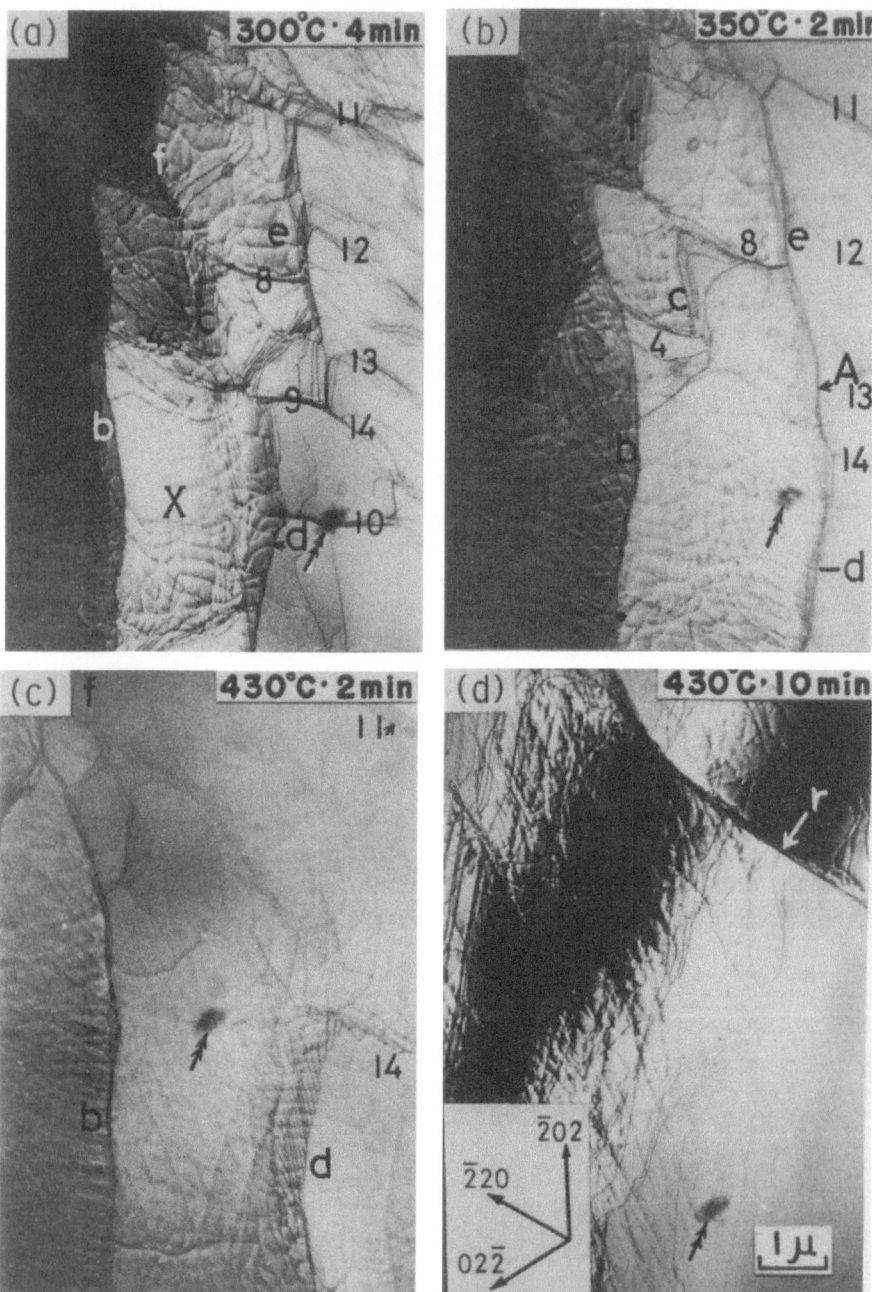

Fig. 8.18. Successive stages of subgrain growth and recrystallization of a stretched aluminium (Accelerating Voltage: 500 kV). In these micrographs, a double arrow and the slip trace show a fixed position and motion of each individual dislocation, respectively. Furthermore, the alphabet characters and arabic figures show the simple-tilt type and twist type sub-boundaries, respectively, and a high-angle boundary of the recrystallized grain is indicated with r (accelerating voltage: 500 kV)

a positive number. Both E_L and $\xi \Delta V^m$ in (8.25) can be neglected when the matrix is a liquid or gas phase.

Suppose the block is approximated to have a spherical shape of radius r in a liquid matrix, ΔE_V and E_S can be expressed [8.83] as

$$\Delta E_V = \int_0^r 4\pi \Delta \varepsilon_V r^2 \, dr \,, \tag{8.26}$$

$$E_S = \int_0^r \frac{4\pi}{a} \gamma_S r^2 \, dr \,, \tag{8.27}$$

where $\Delta \varepsilon_V$ and γ_S show the specific volume free energy difference and the specific surface energy respectively, and a is an interatomic distance. Up to date, both $\Delta \varepsilon_V$ and γ_S have been considered to be constant irrespective of r, and thus E_S becomes larger than the absolute value of ΔE_V until r reaches the critical one. It has been found, however, that both ΔE_V and E_S are functions of r, as shown in Fig. 8.18 [8.87, 88], and that the absolute value of ΔE_V should always be kept larger than E_S for the nucleation. The upper series in Fig. 8.19 [8.89] schematically show the crystal nucleation process in a liquid phase. The linked atoms represent isolated atom-chains such as dimers and trimers of constitutive atoms. The corresponding energy relationship between ΔE_V and E_S in each stage is shown at the lower series in Fig. 8.19. Here the atoms in the liquid phase are assumed to be randomly distributed. Thermal energy of each atom decreases with decrease in the liquid temperature, and then the atom-chains consisting of a few atoms each are formed just like the molecule formation by the local charge transfer among the constitutive atoms. As a result, these molecularized atom-chains interact with each other by a kind of the Coulomb force to form atom clusters, and thus the absolute value of $\Delta \varepsilon_V$ increases at the center of the atom cluster. At the peripheral region of the atom cluster, however, the value of $\Delta \varepsilon_V$ becomes nearly zero since the atomistic structure of the cluster varies continuously from the center to the surrounding matrix in order to reduce the interfacial energy.

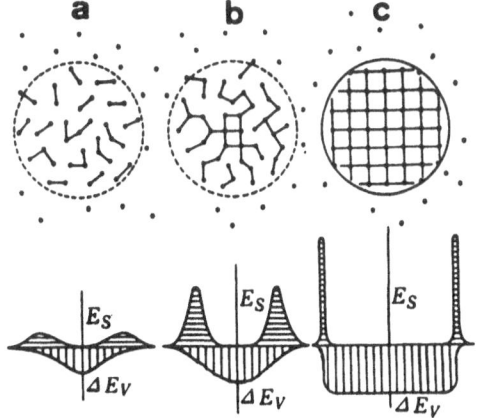

Fig. 8.19. Nucleation process of a crystalline material in a liquid

Based on these results, not only the mechanism of the crystal nucleation but also that of the solid amorphization, which is the reverse process of the crystal nucleation, can be reasonably verified [8.1, 89].

The second example is the formation of complex dislocation structures in deformed crystals, as shown in Fig. 8.20 which is the case of a Cu-10 at.% Al crystal [8.90]. When the crystal is deformed, multiple slip systems are generally activated, so that the dislocation structures formed are frequently affected by the interaction among these multiple slip systems.

Figure 8.21 [8.90] shows the interaction between primary (p) and secondary (s) slips. In Fig. 8.21, slip plane s run across slip bands $p_1, p_2, \ldots$ successively at

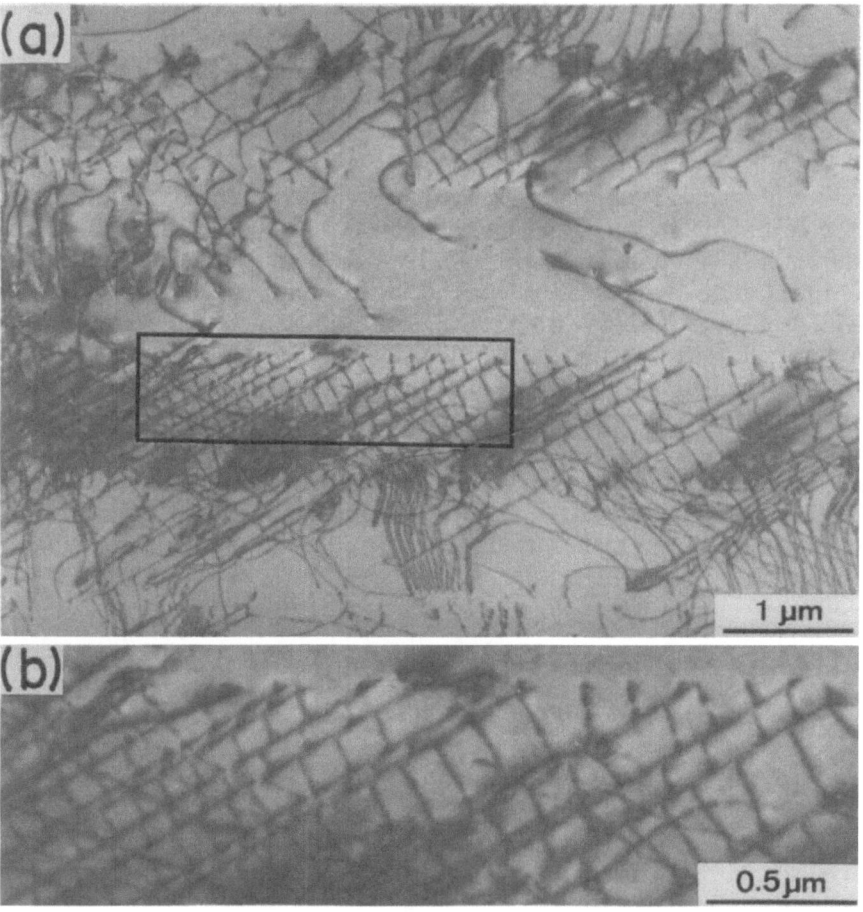

Fig. 8.20. Hexagonal dislocation networks formed on the primary slip planes in a Cu–10at%Al crystal deformed to 10% strain. Micrograph (b) is an enlargement of a framed part in the upper micrograph. Extended and shrunk three-fold nodes of dislocations can be clearly observed alternately in (b) (accelerating voltage: 2 MV)

(a)

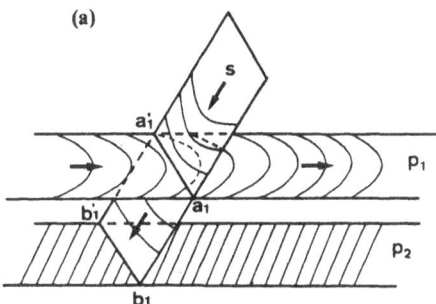

Fig. 8.21. Formation process of hexagonal networks of dislocations in a crystals with a low stacking fault energy. A conjugate slip band(s) in Fig. (a) is shifted to the right-hand side by the primary slip on slip band P_1 in (b), and the conjugate dislocations are dispersed on primary slip band P_2 to form hexagonal dislocation networks, as shown in (b)

(b)

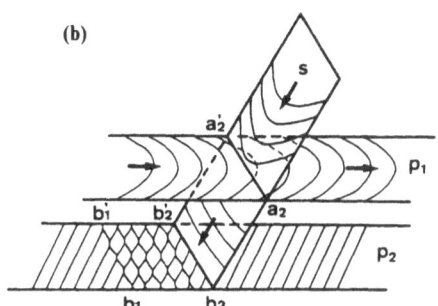

points $a_1 - a'_1$, $b_1 - b'_1$, ... When slip band p_1 is activated in Fig. 8.21, slip band s is shifted from $b_1 - b'_1$ to $b_2 - b'_2$ by the pimary slip on slip band p_2 depending on the slip amount at the very moment, as shown in Fig. 8.21b, and the secondary dislocations are dispersed on primary slip band p_2. This process is one of the most effective process of the work-hardening of the primary slip bands, because many conjugate dislocations hit the primary dislocations in a wide range $(b_1 - b'_1) - (b_2 - b'_2)$ on each individual primary slip band p_2. On the other hand, the propagation of conjugate slip is strongly suppressed in regions where the primary slip markedly occurs, because the stress concentration due to pile-up dislocations does not occur against the primary slip bands. This process is clearly observed in crystals with a low stacking fault energy, as seen in Fig. 8.20. Furthermore, the process shown in Fig. 8.21 is closely related to the latent hardening of the secondary slip as well as the effect of long range stress field of the primary dislocations. Hexagonal networks of dislocations observed in Fig. 8.20 were formed by the process mentioned in Fig. 8.21.

The final example is the interaction between solute atoms and point defects in materials. By using the electron irradiation effect, foreign atoms can easily be implanted into the solid materials [8.91]. Figures 8.22a and b were taken before and after implantation of Pb-atoms into the aluminium matrix at about 175° K and at 2 MV, and c after annealing for 36 ks at 573° K. It is noted in Fig. 8.22 that the diffraction contrast does not appear so clearly in b in spite of a large difference more than 20% in atomic size between Pb- and Al-atoms. These implated Pb-atoms precipitate to the ultrafine particles by annealing in

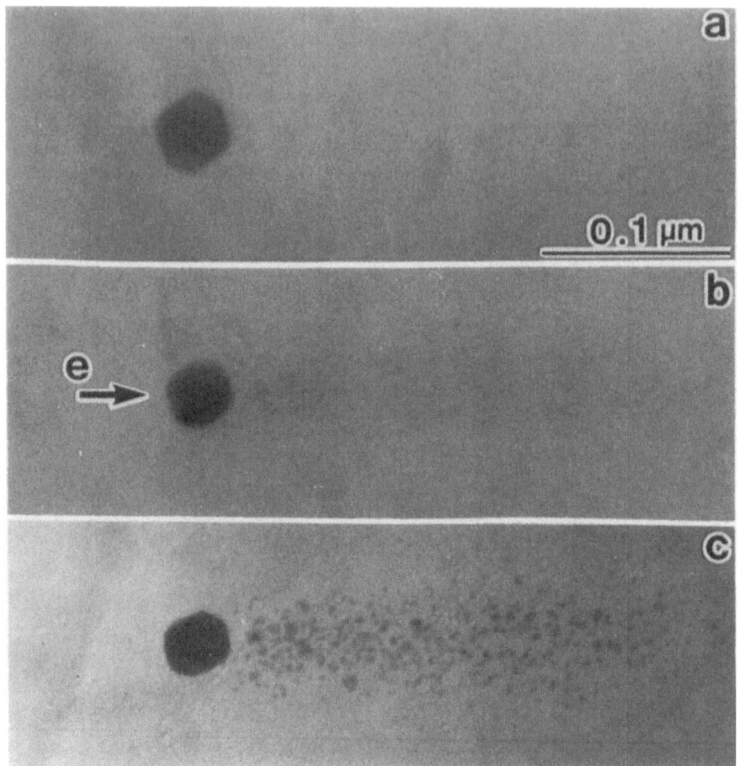

Fig. 8.22. Electron irradiation induced implantation of Pb-atoms into Al-matrix (accelerating voltage: 2 MV). Electron irradiation was made at 2 MV and 175 K, and with a dose rate of $1.2 \times 10^{24}/m^2 s$

Fig. 8.22c. This means that in this case aluminium vacancies induced by the electron irradiation tightly interact with implanted Pb-atoms at low temperatures, as shown in Fig. 8.22b. Namely, the lattice distortion of the Al-matrix, which is caused by the foreign atom implantation, is markedly relieved by the buffer effect of the vacancies [8.1]. Since the irradiation temperature in this method is sufficiently low in Fig. 8.22b, the vacancies do not move to form the voids so that the sharp diffraction contrast does not appear in the Pb-implanted regions. On the contrary, when the implated atoms are small in size, the interstitials induced by the electron irradiation tightly interact with implanted atoms.

The same buffer effect of point defects, such as vacancies and interstitials, also plays an important role to relieve any lattice distortion resulting from fluctuation of both the internal stress and the solute atoms, as in the cases of spinodal decomposition and formation of buffer layers in the vicinity of the interfaces. These facts are also typical examples showing the self-control action in materials being in the living state.

As mentioned above, details of the complex phenomena relating with both point defects and solute atoms can effectively be investigated by HVEMy directly to get indispensable local information on their behaviour.

8.5 New Research Fields by HVEMy "Micro-Laboratory"

Recently, attentions have been paid to the usage of non-equilibrium phases as advanced materials showing new functional behaviour. Non-equilibrium phases are synthesized by two ways: (a) By changing only the atomistic structures, (b) by changing chemical composition as well as atomistic structures. Extreme cases of items (a) and (b) are formation of amorphous solids [[8.1, 83, 89, 92] and extremely super-saturated solid solutions [8.1, 91], respectively. Various novel processes have been applied such as beam annealing, ion implantation and mixing, mechanical mixing and alloying, liquid quenching, vapor deposition, electron irradiation induced method, and others.

The author and his coworkers have succeeded in making amorphous solids of intermetallic compounds by electron beam irradiation [8.91]. Using the high energy electron irradiation method, various conditions for the formation of non-equilibrium solid phases can be easily and precisely controlled. Furthermore, compared with ion implantation and ion mixing methods, this method has such great advantages as small irradiation damage, small temperature rise in the specimens, deep implantation of foreign atoms and very fine-scale fabrication. The formation mechanisms of these non-equilibrium phases can dynamically be investigated in the atomic scale by the in situ experiments using sufficiently thick specimens. Namely, HVEMy has been improved from powerful tools for both characterization and identification of materials to an indispensable "Micro-Laboratory" [8.1, 93] in which various sorts of specimen treatments including formation of non-equilibrium phases can also be carried out precisely in the atomic scale.

By using the HVEMy "Micro-Laboratory", i.e., electron beam science and engineering, formation processes of such non-equilibrium phases as amorphous solids and electron irradiation induced foreign atoms implantation have been verified. Based on the results, necessary conditions [8.1, 83, 89, 93] for the formation of both non-equilibrium phases have also been obtained.

Determination of the atomistic structures and the behaviour of atom clusters [8.94, 95] is one of interesting applications of HVEMy "Micro-Laboratory".

8.6 Conclusions

The progress in materials science and engineering has been assisted by efforts to develop various novel methods for characterization and identification of materials. Although the accuracy and reproducibility of the characterization methods have been greatly improved to a satisfactory level, many of those methods give only average information, both in time and space. And thus, it is difficult to obtain a definite conclusion on the effect of ultrafine microstructures on material behaviour by those methods, as in cases of a) determination of atomistic structures of liquids and amorphous solids from their halo-patterns, and b) estimation of the effect of segregation of solute atoms on the behaviour of semiconductors. Therefore, it is desirable to obtain topographic and dynamic information at the same time in a wide magnification range from the optical microscope scale to the atomic scale. Electron microscopy is the most effective technique to satisfy this demand.

In the present paper, a few examples showing great advantages of EMy for studying materials science were mentioned. By using these advantages, various topographic information in the atomic scale can easily be obtained very accurately, dynamically, and also in a wide magnification range which is very important to increase reliability of information in the observed area.

Acknowledgements. The authors would like to express their thanks to all staff members of the Research Center for Ultra-High Voltage Electron Microscopy, Osaka University for their co-operative work in experiments.

References

8.1   H. Fujita: Materials Trans. JIM **31**, 523–537 (1990)
8.2   R.D. Heidenreich: *Fundamentals of Transmission Electron Microscopy* (Interscience, New York 1964)
8.3   P.B. Hirsch, A. Howie, R.B. Nicholson, D.W. Pashley, M.J. Whelan (eds.): *Electron Microscopy of Thin Crystals*, 2nd edn. (Krieger, New York 1977)
8.4   S. Amelinckx, R. Gevers, G. Remaut, J. Van Landuyt (eds.): *Modern Diffraction and Imaging Techniques in Material Science*, 2nd edn. (North-Holland, Amsterdam 1979)
8.5   U. Valdré (ed.): *Electron Microscopy in Material Science* (Academic, New York 1971)
8.6   U. Valdré, E. Ruedl (eds.): *Electron Microscopy in Materials Science*, Part I–IV (The European Communities, Luxembourg 1975)
8.7   G. Thomas, M.J. Goringe: *Transmission Electron Microscopy of Materials* (Wiley, New York 1979)
8.8   J.M. Cowley: *Diffraction Physics*, 2nd edn. (North-Holland, Amsterdam 1981)
8.9   R. Uyeda: Acta Cryst. **A24**, 175–181 (1968)
8.10 A. Howie: The theory of high energy electron diffraction, in *Modern Diffraction and Imaging Techniques in Material Science*, ed. by Amelinckx, 2nd edn. (North-Holland, Amsterdam 1979) pp. 295–339
8.11 A. Howie, M.J. Whelan: Proc. Roy. Soc. London **A263**, 217–235 (1961)
8.12 N. Kato: Acta Cryst. **16**, 276–281 (1963); ibid 282–290
8.13 H.A. Bethe: Ann. Phys. **87**, 55–129 (1928)
8.14 C.H. MacGillavry: Physica **7**, 329–343 (1940)
8.15 R.D. Heidenreich: J. Appl. Phys. **20**, 993–1010 (1949)
8.16 N. Kato: J. Phys. Soc. Jpn. **7**, 397–406 (1952); ibid 406–414
8.17 H. Hashimoto, A. Howie, M.J. Whelan: Proc. Roy. Soc. London **A269**, 80–103 (1962)
8.18 A. Howie, M.J. Whelan: Proc. Roy. Soc. London **A267**, 206–230 (1962)

8.19 M.F. Ashby, L.M. Brown: Phil. Mag. **8**, 1083–1103 (1963)
8.20 F. Nagata, A. Fukuhara: Jpn. J. Appl. Phys. **6**, 1233–1235 (1967)
8.21 J.S. Lally, C.J. Humphreys, A.J.F. Metherell, R.M. Fisher: Phil. Mag. **25**, 321–343 (1972)
8.22 C.J. Humphreys, L.E. Thomas, J.S. Lally, R.M. Fisher: Phil. Mag. **23**, 87–114 (1971)
8.23 F. Fujimoto, N. Sumida, H. Fujita: J. Phys. Soc. Jpn. **42**, 1274–1281 (1977)
8.24 H. Hashimoto: Jernkont Ann. **155**, 480–490 (1971)
8.25 H. Hashimoto, R. Uyeda: Acta Cryst. **10**, 143 (1957)
8.26 G.A. Bassett, J.W. Menter, D.W. Pashley: Proc. Roy. Soc. London **A246**, 345–368 (1958)
8.27 H. Hashimoto, M. Mannami, T. Naiki: Phil. Trans. Roy. Soc. London **A253**, 490–516 (1961)
8.28 H. Hashimoto, M. Mannami, T. Naiki: Phil. Trans. Roy. Soc. London **A253**, 459–489 (1961)
8.29 S. Miyake, K. Fujiwara, M. Tokonami, F. Fujimoto: Jpn. J. Appl. Phys. **3**, 276–285 (1964)
8.30 J.C.H. Spence: *Experimental High-Resolution Electron Microscopy* (Clarendon, Oxford 1981)
8.31 S. Horiuchi: *High-Resolution Electron Microscopy – Principle and Application* (Kyoritsu, Tokyo 1987) (in Japanese)
8.32 J.W. Cowley, A.F. Moodie: Acta Cryst. **10**, 609–619 (1957)
8.33 K. Ishizuka, N. Uyeda: Acta Cryst. **A33**, 740–749 (1977)
8.34 D. Van Dyck, W. Coene: Ultramicroscopy **15**, 29–40, 41–50, 287–300 (1984)
8.35 D.J.H. Cockayne, I.L.F. Ray, M.J. Whelan: Phil. Mag. **20**, 1265–1270 (1969)
8.36 D.J.H. Cockayne: Z. Naturforsch. **27a**, 452–460 (1972)
8.37 D.J.H. Cockayne: J. Microsc. **98**, 116–134 (1973)
8.38 K. Lonsdale: Proc. Roy. Soc. London **A179**, 315–320 (1942)
8.39 T. Evans, C. Phaal: Proc. Roy. Soc. London **A270**, 538–552 (1962)
8.40 G.S. Woods: Phil. Mag. **34**, 993–1012 (1976)
8.41 A.R. Lang: Phil. Mag. **36**, 495–500 (1977)
8.42 N. Sumida, A.R. Lang: Proc. Roy. Soc. London **A419**, 235–257 (1988)
8.43 N. Sumida, A.R. Lang: J. Appl. Cryst. **15**, 266–274 (1982)
8.44 I.L.F. Ray, D.J.H. Cockayne: Proc. Roy. Soc. London **A325**, 543–554 (1971)
8.45 I.L.F. Ray, R.C. Crawford, D.J.H. Cockayne: Phil. Mag. **21**, 1027–1032 (1970)
8.46 R.C. Crawford, I.L.F. Ray, D.J.H. Cockayne: Phil. Mag. **27**, 1–7 (1973); J. Micros. **98**, 196–199 (1973)
8.47 A.T. Winter, S. Mahajan, D. Brasen: Phil. Mag. **A37**, 315–326 (1978)
8.48 P. Pirouz, D.J.H. Cockayne, N. Sumida, P.B. Hirsch, A.R. Lang: Proc. Roy. Soc. London **A386**, 241–249 (1983)
8.49 D.W. Pashley, J.W. Menter, G.A. Bassett: Nature **179**, 752–755 (1957)
8.50 A.R. Lang: Nature **220**, 652–657 (1968)
8.51 L.A. Bursill, J.L. Hutchison, N. Sumida, A.R. Lang: Nature **292**, 518–520 (1981)
8.52 J.C. Barry, L.A. Bursill, J.L. Hutchison: Phil. Mag. **A51**, 15–49 (1985)
8.53 P. Humble, J.K. Mackenzie, A. Olsen: Phil. Mag. **A52**, 605–621 (1985)
    P. Humble, D.F. Lynch, A. Olsen: Phil. Mag. **A52**, 623–641 (1985)
8.54 R.F. Stephenson: The Partial Dissociation of Nitrogen Aggregates in Diamond by High Temperature-High Pressure Treatments. Ph.D.Thesis, University of Reading (1978)
8.55 J. Maguire: Platelets and Voidites in Diamond. Ph.D.Thesis, University of Reading (1983)
8.56 J.C. Barry, L.A. Bursill, J.L. Hutchison, A.R. Lang, G.M. Rackham, N. Sumida: Phil. Trans. Roy. Soc. London **A321**, 361–401 (1987)
8.57 P.B. Hirsch, P. Pirouz, J.C. Barry: Proc. Roy. Soc. London **A407**, 239–258 (1986)
8.58 P.B. Hirsch, J.L. Hutchison, J. Titchmarsh: Phil. Mag. **A54**, L49–L54 (1986)
8.59 J.L. Hutchison, L.A. Bursill: J. Micros. **131**, 63–66 (1983)
8.60 J.C. Barry: Ultramicroscopy **20**, 169–176 (1986)
8.61 N. Sumida, J.L. Hutchison, P.B. Hirsch: Proc. Diamond Conf. (Oxford 1987) pp. 79–81
8.62 G.S. Woods: Proc. Roy. Soc. London **A407**, 219–238 (1986)
8.63 T. Evans, G.S. Woods: Phil. Mag. **55**, 295–299 (1987)
8.64 T. Mishima, N. Sumida, H. Fujita: Proc. 11th Int'l. Cong. on EM (Kyoto 1986) pp. 995–996
    N. Sumida, T. Mishima, H. Fujita: J. Jpn. Inst. Metals **54**, 1302–1307 (1990) (in Japanese)
8.65 E. Taguchi, N. Sumida, H. Fujita: J. Elect. Microsc. **39**, 164–167 (1990)

8.66 F. Thon: Phase contrast electron microscopy, in *Electron Microscopy in Materials Science*, ed. by U. Valdré (Academic, New York 1971) pp. 571–625

8.67 Y. Hirotsu, R. Akada: Jpn. J. Appl. Phys. **23**, L479–L481 (1984)

8.68 T. Hamada, F. E. Fujita: Jpn. J. Appl. Phys. **25**, 318–327 (1986)

8.69 T. Sakata, H. Fujita, N. Sumida: Proc. 11th Int'l Congr. on EM (Kyoto 1986) pp. 1545–1546

      T. Sakata, N. Sumida, H. Fujita: Inst. Phys. Conf. Ser., No. 90 (IOP, Bristol 1987) pp. 171–174

      T. Sakata, N. Sumida, H. Fujita: J. Jpn. Inst. Metals **54**, 495–501 (1990) (in Japanese)

8.70 G. Thomas, H. Mori, H. Fujita, R. Sinclair: Scr. Metall. **16**, 589–592 (1982)

8.71 S. Iijima, T. Ichihashi: Jpn. J. Appl. Phys. **24**, L125–L128 (1985)

8.72 M. Mitome, Y. Tanishiro, K. Takayanagi: Z. Phys. D **12**, 45–51 (1989)

8.73 T. Sakata, N. Sumida: to be published

8.74 C. Lu, H. Fujita: J. Electr. Microsc. **39**, 412–416 (1990)

8.75 N. Sumida, H. Fujita: to be published

8.76 H. Fujita, T. Taoka and NRIM 500 kV EM Group: J. Electr. Microsc. **14**, 307–308 (1965)

8.77 H. Fujita: Jpn. J. Appl. Phys. **5**, 729 (1969)

8.78 H. Fujita: J. Phys. Soc. Japan **21**, 1605 (1966)

8.79 H. Fujita, Y. Kawasaki, E. Furubayashi, S. Kajiwara: Jpn. J. Appl. Phys. **11**, 214–230 (1972)

8.80 H. Fujita, T. Tabata, K. Yoshida, N. Sumida, K. Katagiri: Jpn. J. Appl. Phys. **11**, 1522–1536 (1972)

8.81 H. Fujita: J. Electr. Microsc. Techn. **3**, 243–304 (1986)

8.82 H. Fujita: J. Electr. Microsc. Techn. **12**, 201–218 (1989)

8.83 T. Tabata, H. Mori, H. Fujita: J. Phys. Soc. Japan **40**, 1103–1111 (1976)

8.84 H. Fujita: Proc. 11th Int'l Congr. on EM (Kyoto 1986) pp. 1025–1030

8.85 H. Fujita: ISIJ, International **30**, 70–81 (1990)

8.86 H. Fujita: J. Phys. Soc. Jpn. **26**, 331–338 (1969)

8.87 H. Fujita: J. Phys. Soc. Jpn. **26**, 1437–1445 (1969)

8.88 H. Fujita: J. Phys. Soc. Jpn. **16**, 396–406 (1961)

8.89 H. Fujita: J. Electr. Microsc. Techn. **3**, 45–56 (1986)

8.90 H. Fujita, S. Kimura: J. Phys. Soc. Jpn. **52**, 157–167 (1983)

8.91 H. Fujita, H. Mori: Trans. JIM. **29** (suppl.) 37–40 (1988)

8.92 H. Mori, H. Fujita: Jpn. J. Appl. Phys. **21**, L494–L496 (1982)

8.93 H. Fujita: *Advanced Materials-II* (The Joint Committee for Advanced Materials Research under the Auspices of the Jpn. Soc. of Appl. Phys., the Ceramics Soc. Jpn. and Jpn. Inst. of Metals, 1990) pp. 113–122

8.94 H. Fujita: Ultramicroscopy **39**, 369–381 (1991)

8.95 H. Fujita: Trans. Materials Research Soc. Jpn. **2**, 38–50 (1992)

8.96 L. Reimer (ed.): *Energy-filtering Transmission Electron Microscopy*, Springer Ser. Opt. Sci., Vol. 71 (Springer, Berlin Heidelberg 1995)

8.97 S. J. Pennycook: Annu. Rev. Mater. Sci. **22**, 171–195 (1992)

8.98 H. Mori, M. Komatsu, H. Fujita: Ultramicroscopy **51**, 31–40 (1993)

8.99 H. Mori, H. Yasuda, N. Sumida, H. Fujita: *New Functional Materials*, Vol. C, ed. by T. Isuruta, M. Doyama, M. Seno (Elsevier, Amsterdam 1993) pp. 221–226

8.100 H. Yasuda, H. Mori, K. Takeda, H. Fujita: Proc. of Int'l Symp. on Diffusion in Materials (DIMAT-92, 1993) pp. 697–702

8.101 H. Fujita: Materials Trans. Jpn. Inst. Met. **35**, 563–575 (1994)

8.102 H. Fujita: Science and Technology No. 7 (Kinki University, 1995) pp. 9–20

9 Mössbauer Spectroscopy in Materials Science

U. Gonser

As far back as we can look into the past, mankind has been constantly in the process of envisioning, developing and sophisticating tools and methods. The question we might ask is, how, in general, are methods conceived and created? In most cases they are the result of a deeper understanding of nature, for one part, and technical ingenuity, for the other. Also, not just one, but usually a number of scientists are involved in the development and improvement of a method, and it is often difficult to trace the individual contribution which has led to a method's ensuing success.

The various scientific methods can be divided into three general categories. In the first group are the descriptive methods. These are the old and classical methods such as electrical and thermal conductivity, calorimetry and specific heat. Their very name describes them. In the second group are what we might call name-tag methods. They have been named after their discoverers, as for instance, the Raman effect, Rutherford backscattering, the Laue and Bragg methods and, of course, the Mössbauer effect. Finally we have as a third group the abbreviated methods. Everywhere, nowadays, abbreviations constitute a new form of communication, and methods, too, are becoming increasingly known by some short form or other, sometimes just a sequence of letters which may or may not be pronounceable. Examples of such methods, with their characteristic probe particles, are listed in Table 9.1. It should be noted that this table is far from being complete. A list of all the existing acronyms would be an order of magnitude larger and we would need a code book to decipher all the meanings. Recently, a book was published dealing with selected modern microscopic methods in metals [9.1].

Most of the abbreviated methods have been discovered in the second half of this century and they work according to the scheme shown in Fig. 9.1. Emitted from a source, a beam strikes the sample (photon, electron, neutron, proton, phonon, positron, muon, etc.) and an interaction takes place. In backscattering or in transmission geometry detectors made it possible to analyze the radiation in terms of energy, spread in energy, mass, intensity, polarization, lifetime, etc. In Fig. 9.1 the symbols γ, e and x (γ-radiation, conversion electron and X-ray) designate the situation for Mössbauer spectroscopy.

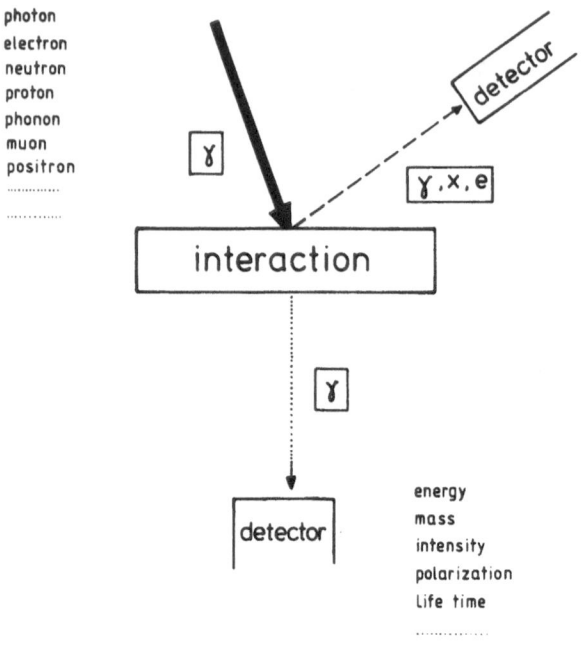

photon
electron
neutron
proton
phonon
muon
positron
....
....

Fig. 9.1. Schematic representation of interactions of probe particle or radiation with a sample

Table 9.1. Techniques, their acronyms and "probe particles"

Acronym	Technique	Probe particle
AES	Auger electron spectroscopy	Electron
APS	Appearance potential spectroscopy	Electron
ARPES	Angle-resolved photo-emission spectroscopy	Photon
BIS	Bremsstrahlung isochromate spectroscopy	Electron
EAPFS	Extended appearance potential fine structure	Electron
EDXD	Energy dispersive X-ray diffraction	Photon
EELS	Electron energy-loss spectroscopy	Electron
ENDOR	Electron nuclear double resonance	Photon
EPMA	Electron probe micro-analyzer	Electron
EPR	Electron paramagnetic resonance	Photon
ESCA	Electron spectroscopy for chemical analysis	Photon
ESR	Electron spin resonance	Photon
EXAFS	Extended X-ray absorption fine structure	Photon
FEM	Field electron microscopy	Electron
FES	Field emission spectroscopy	Electron
FIM	Field ion microscopy	Atom
FMR	Ferro-magnetic resonance	Photon
FQHE	Fractional quantum Hall effect	Electron
HEED	High-energy electron diffraction	Electron
HREELS	High-resolution electron-energy-loss spectroscopy	Electron

Table 9.1. Contd.

Acronym	Technique	Probe particle
HREM	High-resolution electron microscopy	Electron
ILEED	Inelastic low-energy electron diffraction	Electron
IMMA	Ion-microprobe mass analysis	Atom
LEED	Low-energy electron diffraction	Electron
LEERM	Low-energy electron reflection microscopy	Electron
LEIS	Low-energy ion scattering	Atom
LEPD	Low-energy positron diffraction	Positron
MAS-NMR	Magic angle spinning-NMR	Photon
μSR	Muon spin rotation	Muon
NMR	Nuclear magnetic resonance	Photon
NQR	Nuclear quadrupole resonance	Photon
PAC	Perturbed angular correlation	Photon
PAS	Positron annihilation spectroscopy	Positron
PES	Photo-electron spectroscopy	Photon
PIXE	Proton-induced X-ray emission	Proton
PSD	Photon-stimulated desorption	Photon
RHEED	Reflection high-energy electron diffraction	Electron
SAM	Scanning Auger electron microscopy	Electron
SAM	Scanning acoustic microscopy	Phonon
SANS	Small-angle neutron scattering	Neutron
SAXS	Small-angle X-ray scattering	Photon
SHEED	Scanning high-energy electron diffraction	Electron
SLAM	Scanning laser acoustic microscopy	Photon
SNMS	Secondary neutral mass-spectroscopy	Atom
SPLEED	Spin-polarization low-energy electron diffraction	Electron
SRXAS	Synchrotron radiation X-ray absorption spectroscopy	Photon
STEM	Scanning transmission electron microscopy	Electron
STM	Scanning tunneling microscopy	Electron
SXAPS	Soft X-ray appearance potential spectroscopy	Electron
TDPAC	Time differential perturbed angular correlation	Photon
TEM	Transmission electron microscopy	Electron
UPS	Ultraviolet photo-electron spectroscopy	Photon
XANES	X-ray absorption near-edge structure	Photon
XAS	X-ray absorption spectroscopy	Photon
XFS	X-ray fluorescence spectroscopy	Photon
XPS	X-ray photo-electron spectroscopy	Photon

9.1 Historical Remarks

It might be claimed that every discovery has some kind of precursor. In the case of the Mössbauer effect, four stages of development might be distinguished, as shown schematically in Fig. 9.2. First, there was the phenomenon of resonance playing a significant role in natural science. Resonance can take place between two bodies – a source and an absorber – on a macroscopic or microscopic scale. No one knows who discovered this phenomenon. Atomic resonance was found

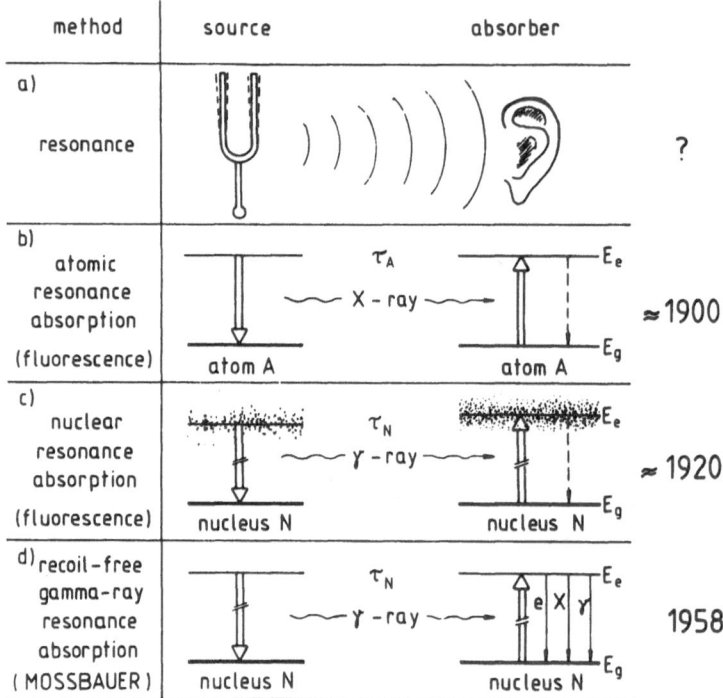

Fig. 9.2. The four stages in the development of the Mössbauer effect

at the turn of this century; nuclear resonance was carried out in the twenties and recoil-free nuclear resonance absorption was discovered by R. Mössbauer in 1958. *Mössbauer* [9.2] realized that γ-ray resonance was relatively simple to achieve, and an important new method was born [9.2].

In retrospect, we should ask ourselves why the Mössbauer effect was not found earlier. Theoretically, the general concept was already provided in the twenties. It seems surprising that we are genuinely so wrapped up in our macroscopic environment, where we experience the phenomenon of recoil everywhere, that it took the analysis of an accidental observation – by a student working on his Ph.D. degree – to open our eyes to this microscopic recoil-free quantum effect.

The keyword in Mössbauer's finding is "**recoil-free**". If an isotope is fixed in a solid it might emit γ-radiation of relatively low energy without creating or annihilating any phonons, or in other words, exchanging any energy in the γ-transition process. The γ-ray carries the total energy of the transition and might be absorbed in the reverse process, again, of course in a recoil-free fashion. Thus, resonance may occur between two nuclei, as shown schematically in their excited and ground states in Fig. 9.3. The separation represents the nuclear transition energy E_0. The two bold arrows connecting the excited and ground

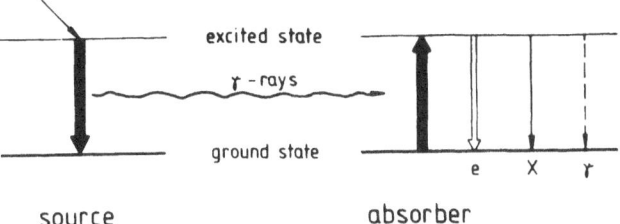

Fig. 9.3. Schematic representation of the nuclear resonance transitions between excited and ground states in two nuclei (bold arrows). Also shown are the non-resonance de-excitation modes in the absorber

Table 9.2. Abbreviated methods involving Mössbauer spectroscopy

CEMS	Conversion electron Mössbauer spectroscopy
CRIME	Coulomb-recoil implantation Mössbauer effect
CXMS	Conversion X-ray Mössbauer spectroscopy
DCEMS	Depth selection conversion electron Mössbauer spectroscopy
GEMS	Gamma emission Mössbauer spectroscopy
RFMS	Radio-frequency Mössbauer spectroscopy
RSMR	Rayleigh scattering Mössbauer radiation
SRMS	Synchrotron radiation Mössbauer spectroscopy
STRMS	Simultaneous triple radiation Mössbauer spectroscopy
TDMS	Time-differential Mössbauer spectroscopy

nuclear states represent the actual resonance effect. The de-excitation radiations – also shown in Fig. 9.3 – bear important information, as will be demonstrated later.

The discovery of the Mössbauer effect is, on the one hand, unique in its fantastic combination of ingenuity and simple good luck and, on the other hand, one might say that this new method was typical in as far as it gained entrance into science with a wide range of applications, advancing instrumentation and data evaluation and, above all, providing a deeper understanding of nature.

Numerous books and articles have been written describing the principles and application of the Mössbauer effect [9.3–9]. Also, specific modifications have been developed and they have come to be known by their abbreviations (Table 9.2) ("M" stands for "Mössbauer"). Without Mössbauer's personal achievement we might have called this method "recoil-free gamma resonance absorption" and abbreviated it as "RGRA".

9.2 Principles

In Mössbauer spectroscopy resonance occurs between the same type of nuclei in the source and in the absorber by virtue of recoil-free γ-radiation. The probability of recoil-free events can be expressed quantitatively by the Debye–Waller factor,

$$f = \exp(-k^2\langle x^2 \rangle),$$

where k is the magnitude of the wave vector of the γ-ray and $\langle x^2 \rangle$ is the mean-square vibrational amplitude of the resonating atom in the direction of observation. Thus the recoil-free fraction f increases with decreasing γ-ray energy k. If no exchange of energy occurs with the surroundings in these nuclear transition processes, the resulting line is extremely sharp. The narrow width of the Mössbauer line Γ, or the spread in energy, is governed basically by the Heisenberg uncertainty principle,

$$\Gamma = \hbar/\tau.$$

The product of the conjugate variables of energy and time – in this case the mean lifetime of the excited state – is related to Planck's constant. The sharp resonance line with natural line width and Lorentzian shape made it possible for the first time to resolve the various hyperfine interactions.

The first 14.4 keV excited state of ^{57}Fe has a half lifetime of about 10^{-7} s; thus the natural line width is absolute $\Gamma \sim 5 \times 10^{-9}$ eV, and relative in regard to the γ-energy one reaches a resolution of $\Gamma/E_0 \sim 3 \times 10^{-13}$. The energy of the resonance lines E_0 can be easily and accurately modulated by a first order linear Doppler effect,

$$E_D = \frac{v}{c} E_0.$$

v represents the velocity between the source and the absorber, and c the velocity of light. In a Mössbauer spectrum the γ-ray intensity arriving at the detector is plotted as a function of v.

Table 9.3. Relevant properties of the most popular Mössbauer transitions [a: isotopic abundance, E_0: nuclear transition energy, $t_{1/2}$: half lifetime, I_e and I_g: nuclear spin quantum number of excited (e) or ground (g) state, respectively, σ_0: maximum resonance cross-section in barn, 2Γ: natural line width, theoretical full width at half maximum, E_R: recoil energy of the free atom] (courtesy John and Virginia Stevens)

Isotope	a [%]	E_0 [keV]	$t_{1/2}$ [ns]	I_e	I_g	σ_0 [10^{-20}cm^2]	2Γ [mm/s]	E_R [10^{-3}eV]	U [%]
1. ^{57}Fe	2.19	14.4125	97.81	3/2	1/2	256.6	0.1940	1.957	66.2
	2.19	136.46	8.7	5/2	1/2	4.300	0.2304	175.4	
2. ^{119}Sn	8.58	23.871	17.75	3/2	1/2	140.3	0.6456	2.571	14.6
3. ^{151}Eu	47.82	21.64	9.7	7/2	5/2	11.42	1.303	1.665	2.0
4. ^{125}Te	6.99	35.46	1.48	3/2	1/2	26.56	5.212	5.401	1.7
5. ^{121}Sb	57.25	37.15	3.5	7/2	5/2	19.70	2.104	6.124	1.6
6. ^{129}I	(radioactive)	27.77	16.8	5/2	7/2	40.32	0.5863	3.210	1.5
7 ^{197}Au	100	77.35	1.90	1/2	3/2	3.857	1.861	16.31	1.2
8. ^{161}Dy	18.88	25.65	28.1	5/2	5/2	95.34	0.3795	2.194	
	18.88	43.84	920	7/2	5/2	28.29	0.006782	6.410	1.1
	18.88	74.57	3.35	3/2	5/2	6.755	1.095	18.55	
9. ^{237}Np	(radioactive)	59.537	68.3	5/2	5/2	32.55	0.06727	8.031	0.8
10. ^{170}Yb	3.03	84.262	1.60	2	0	23.93	2.029	22.43	0.7

Relevant properties of the 10 most popular resonance transitions are listed in Table 9.3. The right-hand column U shows the percentage of publications using this specific isotope, with 66.2% ^{57}Fe having the lion's share. This is mainly because of the favourable nuclear parameters and the importance of the element iron in technology, science and everyday life.

9.3 Hyperfine Interaction

The greatness and wide applicability of Mössbauer spectroscopy came with the realization and resolution of the hyperfine interaction. The sharpness of the line made it possible to measure small shifts and splittings. The hyperfine coupling mechanisms consisting of interaction between a nuclear property (nuclear charge distribution, nuclear magnetic dipole moment and nuclear quadrupole moment) and an electronic property (electronic density in source and absorber, magnetic field and electric field gradient (EFG) tensor at the nucleus), respectively. These three main hyperfine interactions correspond to the nuclear moments determining the nuclear levels:

(i) Electric monopole interaction leads to the isomer shift.
(ii) Magnetic dipole interaction leads to the nuclear Zeeman effect.
(iii) Electric quadrupole interaction leads to the quadrupole splitting.

The nuclear level scheme of the isotope, ^{57}Fe, the allowed transitions, the spectral observations and their angular dependence are schematically represented in Table 9.4.

9.4 Polarization and Thickness Effects

Up to the present, α-Fe has been the most widely used object in Mössbauer spectroscopy. It has served as a standard for the relative velocity between source and absorber and also as the center of the isomer shift. Computer simulated spectra of α-Fe are shown in Fig. 9.4. In the upper set an external longitudinal magnetic field ($\delta_m = 0°$) has been assumed. Fig. 9.4a. In the lower set an external transverse field ($\delta_m = 90°$) was assumed Fig. 9.4b. δ_m (δ_q) represents the angle between the propagation direction of the γ-radiation and the direction of the magnetic field (or principal axis of the EFG, respectively). The upper spectrum of Fig. 9.4a and the one at the top of Fig. 9.4b represent curves with small effective absorber thicknesses

$$T_A = \sigma_0 f_A d_A na.$$

The absorption lines become large – in width and intensity – with increasing T_A, as seen in Fig. 9.4. σ_0 is the maximum resonance absorption cross section, f_A is

Table 9.4. Mössbauer parameters and effects

Hyperfine interactions		
Isomer shift	Magnetic hyperfine splitting	Quadrupole splitting

Interaction of the nuclear charge distribution with the electron density at the nucleus in source and absorber (electric monopole interaction)	Interaction of the nuclear magnetic dipole moment μ with a magnetic field H at the nucleus. (magnetic dipole interaction): Nuclear Zeeman effect	Interaction of the nuclear quadrupole moment eQ with EFG V_{zz} at the nucleus (electric quadrupole interaction)
$\delta = C \dfrac{\delta R}{R}\,[\,|\psi_A(0)|^2 - |\psi_S(0)|^2\,]$	$E_m = -\dfrac{\mu H m_I}{I}$	$E_Q = \pm 1/4\,eQV_{zz}\,(1+1/3\,\eta^2)^{1/2}$

the recoil-free fraction and d_A the physical thickness in cm. n is the number of atoms/cm^3 of the particular element and a the isotopic abundance of the resonance isotope. The angular dependence of the magnetic hyperfine pattern of ^{57}Fe with its six allowed transitions is shown in Table 9.5 and the relative line intensities of one half of the spectra are represented at the bottom of Table 9.4 in the thin absorber approximation. For $\delta_m = 90°$ the relative line intensities are $3:4:1:1:4:3$ and for $\delta_m = 0°$, $3:0:1:1:0:3$. For the unique orientation $\delta_m = 54°44'$ and also for a random orientation of the magnetic moments the relative line intensities become $3:2:1:1:2:3$. In all cases the sum of the line intensities is isotropic, as indicated by the circle (cut of the sphere) in Table 9.4.

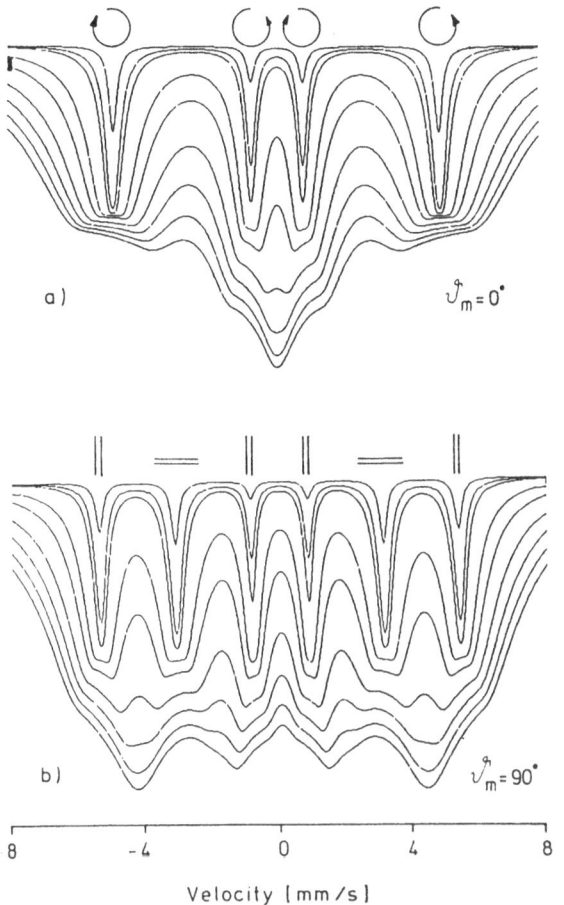

Fig. 9.4. Computer simulated α-Fe spectra with increasing thickness: $T_A = 1, 5, 10, 50, 100, 200, 300, 400, 500$. (a) in an external longitudinal magnetic field ($H_{ext}\|\gamma$), $H_{int} = 300$ kOe, (b) in an external transverse magnetic field ($H_{ext}\perp\gamma$), $H_{int} = 330$ kOe

Table 9.5. Angular dependence of the magnetic hyperfine six-line pattern of ^{57}Fe. $\varDelta m$ represents the angle between the propagation direction of the γ-radiation and the direction of the magnetic field. This relationship is also shown in the lower part of Table 9.4

Transition	$\varDelta m$	Angular dependence
$\pm 3/2 \rightarrow \pm 1/2$	± 1	$3/4(1 + \cos^2 \vartheta_m)$
$\pm 1/2 \rightarrow \pm 1/2$	0	$\sin^2 \vartheta_m$
$\mp 1/2 \rightarrow \pm 1/2$	∓ 1	$1/4(1 + \cos^2 \vartheta_m)$

The situation of the radiation pattern produced at the ^{57}Fe nucleus by an electric field gradient (EFG) with axial symmetry (asymmetry parameter $\eta = 0$) is similar. As for the relative line intensities of the quadrupole split lines (again in the thin absorber approximation) they are $3:1$ for $\delta_q = 0°$ and $3:5$ for $\delta_q = 90°$. Random orientations and an angle of $54°44'$ between the propagation direction of the γ-radiation and principal axis of the EFG produce doublets with equal line intensities. Again the sum of the line intensities is isotropic, as indicated by the circle (or cut of the sphere).

However, if the mean-square displacements of the vibrating resonating atoms are not isotropic and are, let us say, larger in the direction of the principal axis of the EFG, the Debye–Waller factor becomes smaller and along with it the line intensities (see figure at bottom of Table 9.4). In such a situation the circle (or sphere) in Table 9.4 also becomes flattened (not shown here) and thus the recoil-free radiation becomes anisotropic.

The line intensities reflect the polarization of the emitting and absorbing γ-radiation. With a longitudinal external magnetic field the four resonance lines are circularly polarized with left- and right-hand polarization, as indicated at the top of the Mössbauer spectra of Fig. 9.4a. With a transverse external magnetic field the six lines are linearly polarized as shown at the top of the lower part of Fig. 9.4b. From an unpolarized γ-ray source each polarized absorber line (circular or linear) can absorb by itself only up to one half of the resonant γ-rays. For the other half, with the orthogonal helicity or linearity, the material is transparent. This phenomenon where one polarized component (circular or linear) is absorbed, while the other component with orthogonal helicity or linearity is transmitted, is known in optics as dichroism and is used – for instance – in sun glasses. When two lines of orthogonal polarity are in close proximity a certain overlap – and that is absorption of the orthogonal polarity – occurs. This is schematically shown in Fig. 9.5. In very thick samples with large T_A an apparent line may develop *in between* the actual resonance lines, as is seen in Fig. 9.5 [9.10]. Such a phenomenon is also observed in Fig. 9.4a with circular

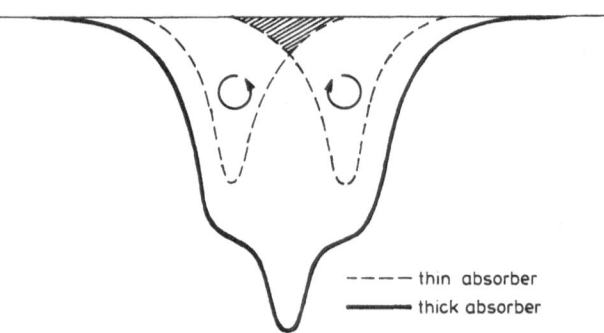

Fig. 9.5. Schematic representation of thin and thick absorber lines when polarized components of opposite sign are in close proximity

polarization in the center ($v = 0$) and in Fig. 9.4b with linear polarization between the outer two lines at the velocities $v \sim \pm 4.2$ mm/s.

9.5 Phase Analysis

The Mössbauer effect entered with ease all disciplines of natural science: physics, chemistry, biology, metallurgy, engineering, geology and even such fields as archaeology, art, medicine and etc. Considering the applicability of Mössbauer spectroscopy in general, the most beneficial aspects were those of *phase analysis*.

In general, every phase containing Mössbauer isotopes exhibits a spectrum with characteristic hyperfine interaction parameters which can be described as "fingerprints" of the phase. By means of these typical spectra, certain phases can be identified and quantitatively determined. Mössbauer spectroscopy will hardly ever be able to compete with chemical analysis, but is often superior when interest is focused on the analysis of phases as distinct from composition.

The following approach has proved to be very successful: appropriate amounts of any number of reference spectra are subtracted from the unknown total with the amount being varied iteratively until the residuals become satisfactorily small. This approach has been particularly useful in analyzing multi-phase components occurring in minerals, alloys, meteorites, etc. [9.11].

In physical metallurgy the constitution diagrams showing the phases existing at various temperatures normally serve as the starting point and guide line. The term "phase" is not restricted to crystallographic phases. Order/disorder, amorphous, ferro-, ferri- and antiferro-magnetic ordering, spin-glass, super- and semi-conducting phases, etc. should also be included. In addition to the thermodynamically stable phases appearing in the phase diagram a number of metastable and intermediate phases are often formed under certain conditions. In some cases the final stable phases – such as various ordered phases – can be produced only with great difficulty because at relatively low temperatures diffusion practically ceases.

Microscopic methods have given us detailed knowledge of the local atomic structure and consequently the word "phase" has acquired a new meaning or the definition has become blurred. The question might be asked: how large must a cluster be in order to constitute a new phase? In the following section we would like to present some typical examples of phase analysis in metals.

9.6 Cu–Fe System

Fe exists in two modifications: at low temperatures (α-phase) and at high temperatures (δ-phase). In both it has the body-centered-cubic (bcc) structure. In the temperature range of 1183–1663 K, the face-centered-cubic (fcc) structure

– the γ-phase – is stable. At low temperatures the α-phase exhibits ferromagnetic ordering with a Curie temperature of 1046 K (β-phase). The internal magnetic field of the α-phase is 330 kOe at room temperature.

The magnetic properties of the bcc α-phase are well known. However, in the case of γ-Fe the properties are still controversial with respect to the magnetic ordering of the fcc γ-phase. Three methods are used to stabilize the γ-Fe at low temperatures:

(i) By alloying with certain elements, the γ-phase region can be extended.

(ii) In supersaturated C̲u̲–Fe alloys iron precipitates coherently with the lattice as fcc γ-Fe.

(iii) By epitaxial growth Fe vapour condenses on cold copper substrate and forms fcc γ-Fe.

In a laboratory course the usefulness of Mössbauer spectroscopy might well be demonstrated appropriately by the following sequence of specimen preparation and spectral analysis:

(a) A copper alloy is prepared containing a small amount (< 1 at. %) of ^{57}Fe. This amount is soluble above $\cong 800°C$. By the choice of an appropriate quenching rate a large fraction of the Fe is in solution and Fe dimers (Fe pairs) are also formed during quenching or due to the mobility of the quenched-in vacancies. The Mössbauer spectrum of Fig. 9.6 can be decomposed into the components: a single line representing the monomers, that is, isolated Fe atoms with 12 nearest-neighbour Cu atoms, and a quadrupole split component representing the dimers. The relative integral intensities of the resonance lines are roughly proportional to the fraction of monomers and dimers. The existence of the quadrupole splitting indicates that an electric field gradient of the nucleus is present and that the cubic symmetry is locally removed.

(b) When this specimen is annealed at $\sim 650°C$ coherent precipitation of fcc γ-Fe occurs. γ-Fe is paramagnetic at room temperature and consequently the

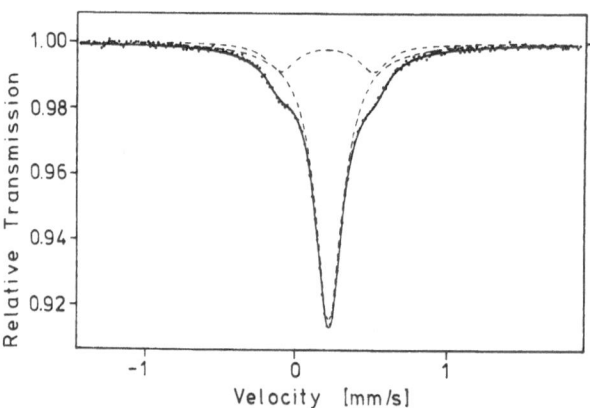

Fig. 9.6. Mössbauer spectrum of Cu-0.2 at. % Fe alloy at room temperature

Mössbauer spectrum consists of a single line, as seen in Fig. 9.7a. While in the foregoing state Fe was mostly surrounded by Cu atoms, now most of the Fe atoms have 12 nearest-neighbour Fe atoms. By comparison this difference caused an isomer shift.

(c) At low temperatures antiferromagnetic ordering of γ-Fe becomes evident by a broadening of the resonance lines. The magnetic hyperfine splitting corresponds to 18 kOe, which is insufficient to resolve the six-line hyperfine pattern. The close positions of the six-line magnetic patterns are indicated by the stick diagram in Fig. 9.8. The Néel temperatures were determined by the thermal scan technique. Instead of taking a series of full spectra, only the ratio of the resonant count rate at one special velocity to the off-resonance count rate as a function of temperature is measured, as shown in Fig. 9.9 (thermal scan). The change in absorption indicates the Néel temperature. In this way the Néel temperature was first observed for γ-Fe coherent precipitates in Cu and found to be 67 K [9.12].

(d) Coherent γ-Fe precipitates in Cu are metastable. The transformation to the stable state ($\gamma \rightarrow \alpha$) can be accomplished by plastic deformation at room temperature provided the precipitates are larger than about 40 Å. After deformation the single-line spectrum of paramagnetic γ-Fe has changed to a strong six-line pattern – typical of α-Fe – as seen in Fig. 9.7b. The remaining line in the center represents the retained paramagnetic fcc γ-Fe and possibly superparamagnetic contributions.

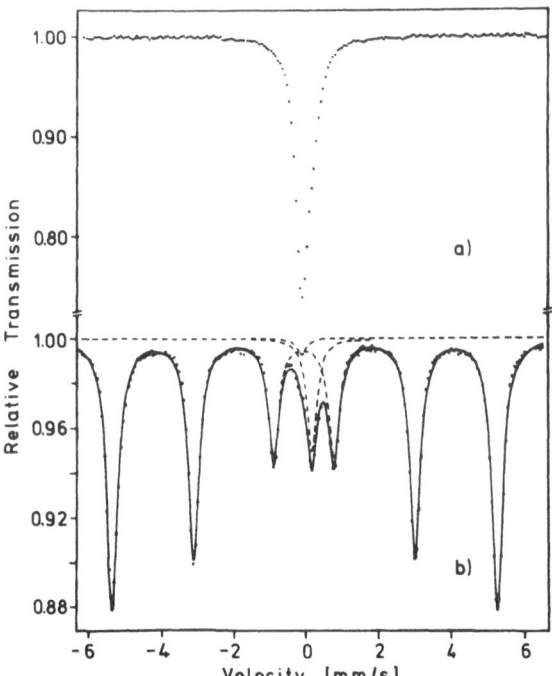

Fig. 9.7. Mössbauer spectra at room temperature (a) before and (b) after cold rolling of a Cu sample originally containing only coherent γ-Fe precipitates

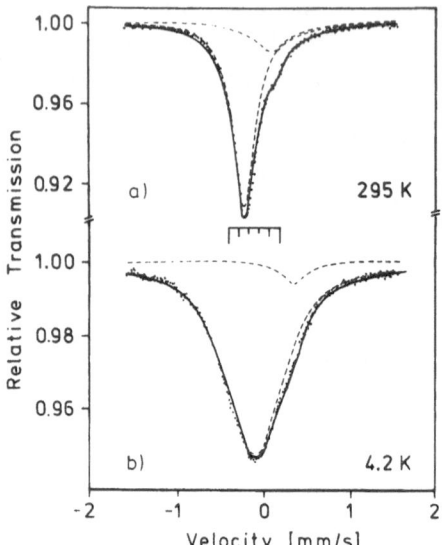

Fig. 9.8. Mössbauer spectra of epitaxial γ-Fe thin films (a) at 295 K in the paramagnetic state, (b) at 4.2 K in the antiferromagnetic state. The stick diagram indicates the line positions of the hyperfine pattern. The additional small lines in the two spectra result from Fe atoms at the interface

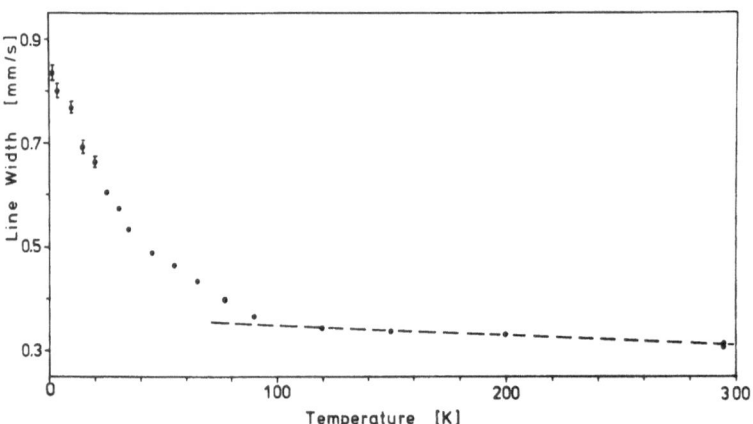

Fig. 9.9. Ratio of the resonant count rate at zero velocity to the off-resonance count rate normalized to line width as a function of temperature (thermal scan)

9.7 Precision Phase Analysis

In certain cases multi-phase alloys can be analyzed quantitatively with great precision [9.13]. For this purpose it is required that the phases containing resonance isotopes be distinguishable by differences in at least one of the hyperfine interactions. On the other hand, one assumes that the Debye–Waller factor for the phases involved is the same. In the analysis the distribution of the resonance atoms is of interest and one correlates the fraction of a spectral component with the fraction of a particular phase and applies the lever relationship to locate the phase boundaries. For instance, let us consider the solubility of Fe in Ti. This is an interesting problem because it can be solved with a precision unmatched by any other method. Specimens with a starting composition C_0 of about 0.1 at.% Fe – within the two-phase region – were annealed in the temperature range of 850–1080 K and then quenched. Owing to the differences in the isomer shift, the spectra can be decomposed into a component representing iron dissolved in α-Ti (hexagonal closed-packed – hcp) and one representing β-Ti (body-centered cubic – bcc). The fraction of the components depends on the temperature. If the spectral intensities of the two components are roughly equal, the total amount of Fe is rather evenly shared by the two phases. This near equality was achieved deliberately by choosing C_0 very close to the equilibrium line. As a result.

$$|C_0 - C_\alpha| \ll |C_\beta - C_0|.$$

Applying the lever relationship,

$$\frac{C_0 - C_\alpha}{C_\beta - C_0} = \frac{M_\beta}{M_\alpha},$$

it follows that the amount of the two phases, $M_\alpha \gg M_\beta$. From measurements on specimens with various values of C_0, annealed at various temperatures and then quenched, and with the use of the lever relationship, the curve of Fig. 9.10 was obtained and then thermodynamical data were determined. The data indicate that precision phase analysis is possible in the range down to 0.001 at.%.

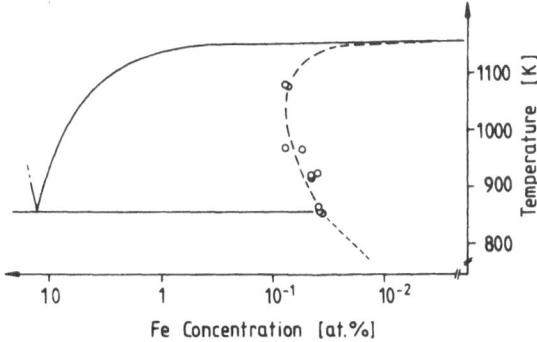

Fig. 9.10. α-phase boundary (solubility) of Ti–Fe alloys

9.8 Amorphous Metals, General

The word "amorphous" is not well defined. The answer to the question, what do we understand by "amorphous", depends on the person asked and is normally given in negative terms, as the Greek origin of the word amorphous itself means: lacking structure or form. Thus, we may say that the use of the word amorphous expresses our ignorance or inability to find a generally accepted definition concerning the structure. Even more confusion was introduced by those who started using the term "metallic glass".

Up to the middle of this century it was believed that solid metals never exist in the amorphous state and that ferromagnetic order cannot exist in the amorphous state. In the second half of the century two important kinds of materials were discovered: amorphous metals and high-temperature superconductors. These exciting materials have two things in common. First, they represent typical defect structures, and second, they have received considerable attention in scientific research and also for their potential technological importance.

Our particular interest here is concentrated on amorphous metals, for which, in the meantime, seven different methods of production have been developed [9.14–21]. These are shown in Fig. 9.11 which gives the year of discovery, describes the method and names its founders.

It was realized that a relatively high stability exists in alloys with an approximate composition of $T_{80}M_{20}$ (T = transition elements such as Fe, Co, Ni,... and M = metalloid atoms such as B, C, N, P, ...). This composition coincides roughly with very pronounced eutectics. The commercially available amorphous metals (Metglas®, Amomet® and Vitrovac®) are produced as ribbons by the melt-spinning technique, where liquid metals impinge with high

1950	electro-deposition	Brenner, Couch, Williams	14
1952	vapour deposition	Buckel, Hilsch	15
.958	irradiation (d)	Gonser, Okkerse, Fujita	16,17
1960	quenching from the melt	Duwez, Willens, Klement	18
1969	melt-spinning	Pond, Maddin	19
1981	nano-crystal-deposition	Gleiter	20
1983	solid state reaction	Yeh, Samwer, Johnson	21

Fig. 9.11. Methods of producing amorphous metals

velocity on a rapidly rotating metal drum and large quantities can be produced in continuous operation. A large variety of materials can be "tailor-made" by changing composition, quenching rate, annealing, etc. Especially attractive is the combination of high mechanical strength with extreme hardness and magnetic softness.

One question is of particular interest: are amorphous metals undercooled liquids or do regions exist with atomic arrangements similar to those of the crystalline state? The variety of models suggested in this field might be reduced to three extreme concepts shown schematically in Fig. 9.12. In the upper corner one assumes nano- or microcrystalline regions separated by "grain boundaries". These regions must be smaller than 20 Å. Otherwise they would be revealed by diffraction methods.

As the microcrystals become increasingly smaller – proceeding down the left-hand side of the triangle – we finally arrive at some kind of unit cells or "molecules". The difficulty lies in putting such building elements together and understanding the structure as a whole. At the lower right-hand side of Fig. 9.12 the Bernal model (random dense packing of hard spheres) is located. As Bernal demonstrated, random dense packing can easily be realized by hard spheres. The structure contains a certain proportion of characteristic polyhedral holes and a certain proportion of characteristic coordination numbers also found in grain boundaries.

The large variety of amorphous alloys cannot be represented by one unique structure. The real structures might fall somewhere within the triangle, perhaps close to one of the corners. A number of theoretical publications have considered certain influences and principles which are thought to play a major role in the amorphization of metals: metastable phases, stoichiometry, packing effects, ratio of atomic radii (Laves), coordination and packing densities

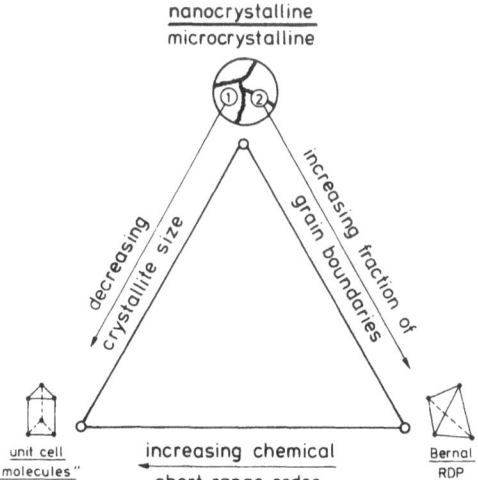

Fig. 9.12. Schematic representation of the three alternative models used in the interpretation of the structure of amorphous metals

(Kasper), valence electron concentration (Hume–Rothery, Jones), electron-egativity (Zintl), covalency (Grimm, Sommerfeld). The names in parentheses designate the crystalline intermetallic phases which are related to these influ-ences. In addition, a "confusion principle" has been suggested, according to which the tendency toward amorphization is strong if many states (phases) are available for the system.

A large amount of literature has been published covering the field of amorphous metals. For further information one might consult the proceedings of the three conference series "Rapidly Quenched Metals" (RQM), "Liquid and Amorphous Metals" (LAM) and "Non-Crystalline Metals" (NCM). Schemati-cally, one might illustrate the contrast between the crystalline and the amorph-ous state as follows:

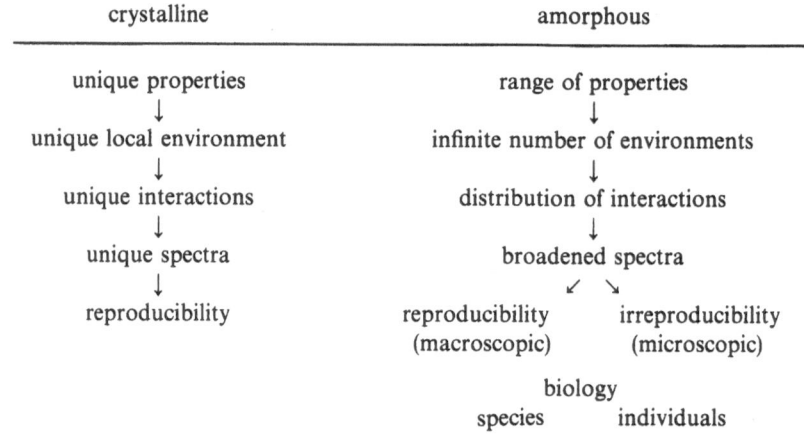

Concerning the lower part one might state that from a microscopic point of view the amorphous state should be regarded as irreproducible, while the macro-scopic properties can be reproduced as demonstrated in the industrial appli-cations. One might compare the situation with biology and nature, where the various species (macroscopically) are reproducible, but individuals (micro-scopically) are totally different from each other and thus irreproducible.

9.9 Amorphous Metals, Experimental

Mössbauer spectroscopy is an "ideal" tool for probing the local surroundings of resonance atoms. Consequently, it was thought that the atomistic structure of amorphous metals would be revealed by the hyperfine spectra. However, the effectively infinite number of atomistic environments in the amorphous state produces an infinite number of spectral components. Thus, only distributions are observed – in terms of isomer shift, magnitude of the magnetic hyperfine field, magnitude, of the sign and asymmetry parameters of the quadrupole interaction, orientation (texture) of the hyperfine field, EFG principal axes, polarization, dichroism, thickness effects, etc.

As an example, the Mössbauer spectrum of amorphous $Fe_{80}B_{20}$ is shown in Fig. 9.13. Contrary to crystalline materials one finds very broad Mössbauer lines which indicate distributions of hyperfine interactions, reflecting various neighbouring field contributions. Thus, the interpretation of Mössbauer spectra becomes dubious and controversial. In such situations scientists adopt or assume models which allow in some way the data to be fitted, but that is not enough. Actually, the data should lead by a unique interpretation to the correct model.

In all fairness – maybe with the exception of nanocrystals – we must admit that Mössbauer data have not yet provided clear evidence that any of the models of amorphous metals really exist. Various models have simply been assumed, and for each of them it has been found that some kind of fitting of the broad lines could be accomplished. However, the actual realization could not be proved. It can be shown that, in general, it is impossible to deduce the correct distribution function from the experimental spectra [9.22] and there is a real danger of artefacts being observed.

Similar attempts to deduce the atomistic structure by NMR, EXAFS, etc. have shared the fate of Mössbauer spectroscopy. The optimism of earlier years that atomistic structures could be determined has given way to a more realistic and cautious view of the situation.

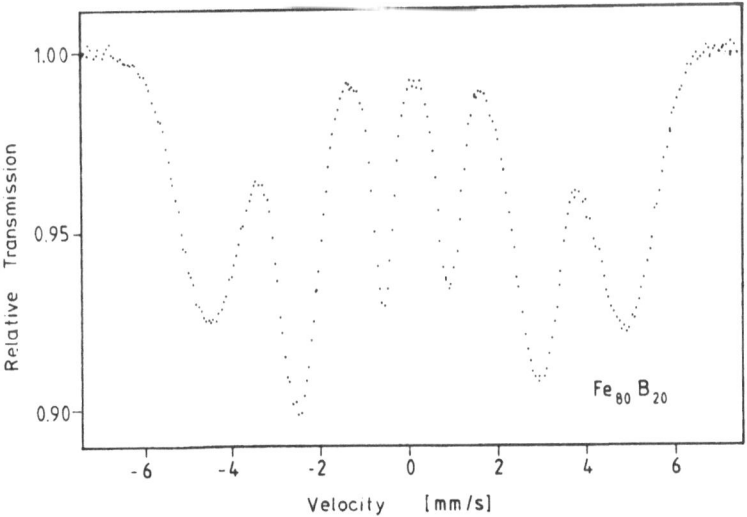

Fig. 9.13. Typical Mössbauer transmission spectrum of a ferromagnetic amorphous alloy of the type $T_{80}M_{20}$

9.10 Nanocrystalline Materials

Considering the schematic representation of the three alternative models used in the interpretation of amorphous metals (Fig. 9.12) one finds in the upper corner the term "nanocrystalline". The understanding is that these materials are basically polycrystals in which the size of the individual crystallites is in the order of 1–10 nm. The nanocrystalline materials can be visualized as consisting of two components: on the one hand, the small crystallites and, on the other, atoms forming the interfacial component comprising all the atoms in the grain boundaries between the grains. The structure of the interface depends on the orientation relationship between adjacent crystals. Thus, boundary inclination and interatomic spacings are important parameters in the characterization of the nanocrystalline arrangements. The situation might also be described in the following way: each atom with crystallographically "ideal" close surroundings belongs to the crystals, while the atoms in the grain boundaries are dislocated or have different coordinations. A schematic model of an idealized nanocrystalline material is shown in Fig. 9.14.

Up to the present two methods have been introduced to produce nanocrystalline materials. In the one case, a metal is evaporated into a noble-gas atmosphere. The mean free path of the atoms is rather small, thus the kinetic energy of the metal vapor is reduced quickly and homogeneous nucleation and growth occur. In addition, crystallization of the agglomerates takes place. The size of the nanocrystals formed can be controlled by the experimental conditions. After cold-trapping and compacting the density of the material might be of the order of 90%, as compared with the common bulk material.

The other method of production uses a precursor, an amorphous metal quenched from the melt. By appropriate annealing, crystals are formed which are embedded in an amorphous matrix. The density of the resulting nanocrystalline material is usually higher than that of the nanocrystalline material prepared by the evaporation technique, but the two components (nanocrystals and interfacials) are not likely to be so easily distinguished.

It is an advantage for the experiments if about one half of the atoms is settled in the nanocrystals and the other is located in the interfacial region. In this case, the two different atomic configurations should be detectable by the two corresponding subspectral components with rather similar intensities.

Representative Mössbauer spectra of a nanocrystalline Fe sample are shown in Fig. 9.15 at 295 K (a) and 10 K (b). Indeed, two subspectra can be distinguished, in particular, by using standard fit methods. The sharp components can be identified and represent well crystallized α-Fe, while the broad components can be associated with the Fe atoms in the interface or grain boundaries, comprising a large variety of close configurations giving rise to a wide distribution of internal fields [9.23]. This assignment is supported by the observation that the broad component – due to interfacial atoms – has at lower temperatures a larger mean internal magnetic field as compared to the internal field of the

Fig. 9.14. Model of a nanocrystalline material

sharp component resulting from the α-Fe crystals (Fig. 9.15b). This enhanced hyperfine field in the interfaces can be explained by the Bethe–Slater curve which relates the exchange energy to the atomic spacings. Especially in the case of iron an expansion of the nearest neighbours results in an increase of the magnetic moment and also of the hyperfine field. By plotting the hyperfine fields of the two components as a function of temperature, see Fig. 9.16, it can be observed that the open circles – representing the interfacial component – indeed exhibit larger fields at low temperatures, as compared to the nanocrystalline component – represented by the closed circles. At higher temperatures, however, a cross-over is observed. The more rapid decrease indicates for the more open structures of the interfacial component a lower Curie temperature than that of crystalline α-Fe.

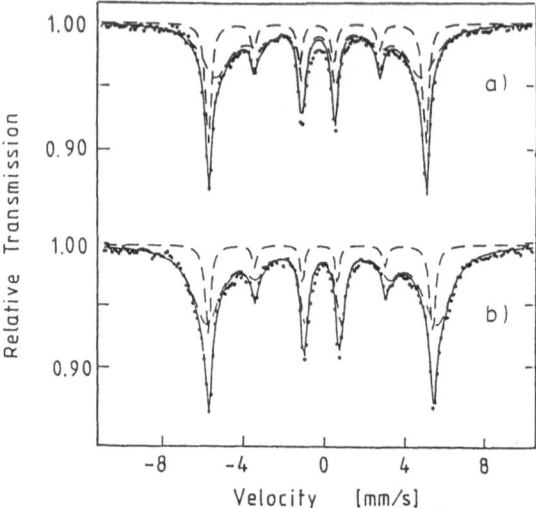

Fig. 9.15. Mössbauer spectra of a nanocrystalline iron sample at (a) 295 K and (b) 10 K

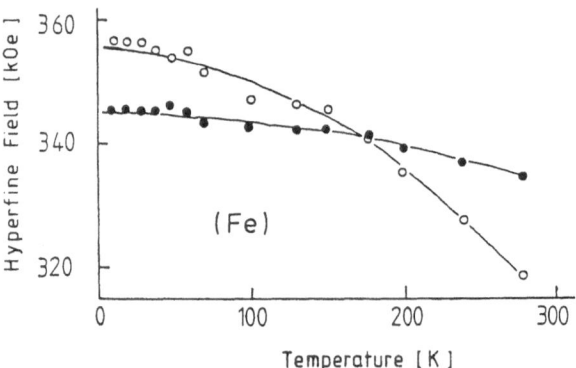

Fig. 9.16. Mean hyperfine magnetic fields of the two Mössbauer subspectra representing nanocrystals (closed circles) and interfaces of Fe (open circles)

Finally, another rather important observation has been made concerning the dynamical behavior. It turns out that the Debye–Waller factor – effectively a way of evaluating the mean square displacement of the atoms involved – is lower for the interfacial component than for the nanocrystalline component. These findings clearly indicate that the atoms in the interface have more room and therefore an enhanced mean square displacement.

In general, another comparison is appropriate. In Fig. 9.17 typical distributions of the hyperfine fields are shown for (a) nanocrystalline materials and (b) amorphous metals quenched from the melt. In the former case, two different components can easily be distinguished, whereas amorphous metals (or metallic glasses) have broad distributions but otherwise rather uniform behavior.

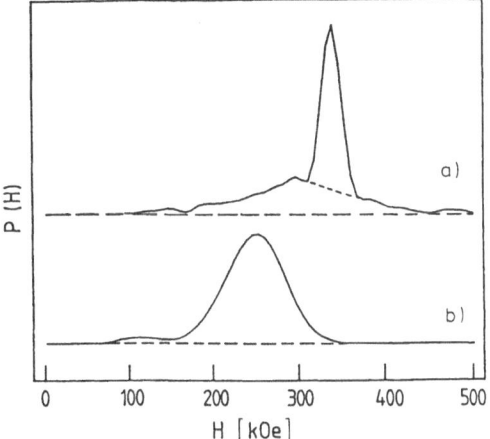

Fig. 9.17. Schematic representation of the distributions of hyperfine fields in (a) nanocrystalline Fe and (b) amorphous $Fe_{84}B_{16}$ at 295 K

9.11 Crystallization

So far, success in the elucidation of the structure of nanocrystalline material has been achieved only to a certain degree. Our methods have failed us completely in our efforts to obtain conclusive information on the structure of the large amount of amorphous metals which have been produced. On the other hand, however, Mössbauer spectroscopy has become the most powerful tool for investigating or tracing any type of ordering, and also where and at what temperature crystallization commences and in what sequence crystalline phases precipitate.

Amorphization by the melt-spinning technique will be influenced by certain parameters. Of course, the most significant factor is the speed of the roller surface V_u because it determines the quenching rate. The amorphous alloy $Fe_{80}B_{20}$ is a typical example. When the surface velocity of the wheel drops below 25 m/s new lines appear indicating the start of precipitation and crystallization, (Fig. 18b). The new lines can be used to identify the phases. In our case, α-Fe and Fe_2B are the first crystalline phases to appear. A further reduction of the velocity to 18.7 m/s leads to the occurrence of the Fe_3B phase (Fig. 18c). The crystalline phases were identified from the well-known parameters of their Mössbauer spectra. It is interesting to note that the sequence of crystallization of the various phases is different when the amorphous sample is subjected to an annealing heat treatment (starting from lower temperatures) or produced by reducing the quenching rate (coming from the melt at higher temperatures). When amorphous $Fe_{80}B_{20}$ alloys are annealed the first phase to crystallize is Fe_3B followed by α-Fe. The crystallization of Fe_2B requires prolonged annealing at elevated temperatures. Contrary to this behaviour, the lowering of the quenching rate immediately produces Fe_2B and α-Fe before the Fe_3B phase is formed [9.24].

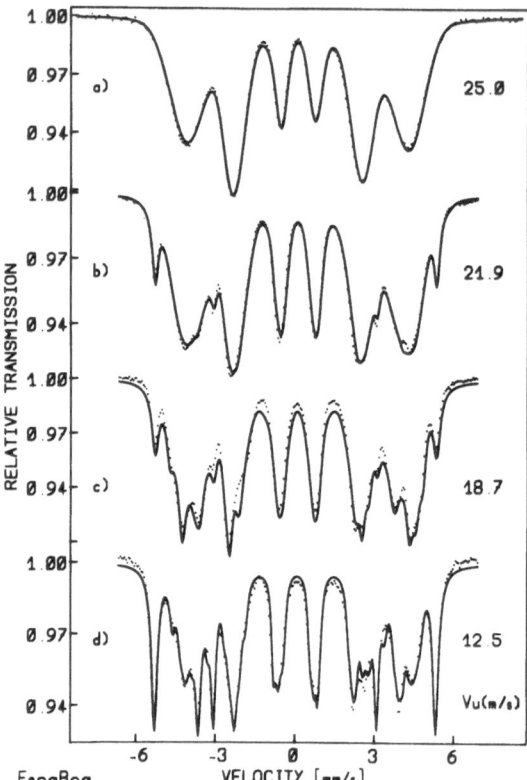

Fig. 9.18. Mössbauer spectra of $Fe_{80}B_{20}$ produced by the melt-spinning technique at decreasing surface velocities V_u and thus decreasing cooling rates

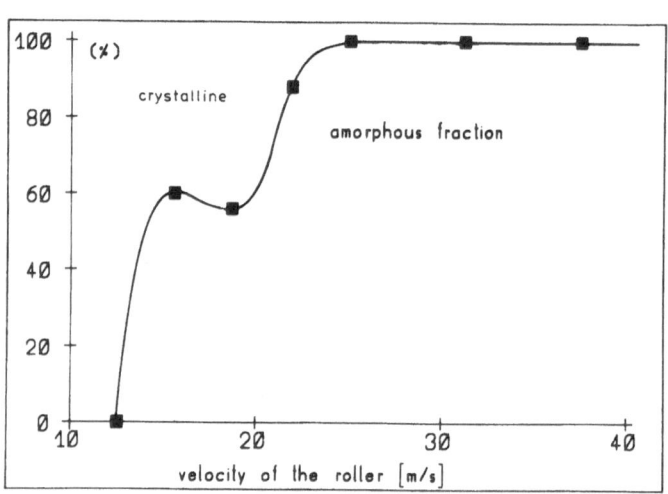

Fig. 9.19. Amorphous and crystalline fraction versus surface velocity V_u of the roller in the melt-spinning process

The data allow us to plot a phase diagram of the amorphous fraction versus the surface velocity of the rollers (Fig. 9.19). The crystalline area is extremely well known with regard to the identity of the phases, the sequence of their appearance and their relative abundance. Here is where the success lies and where Mössbauer spectroscopy has almost no competitors.

Figure 9.20 shows three spectra from an amorphous $Fe_{40}Ni_{40}B_{20}$ metal in the as-quenched state. The upper spectrum was obtained with a wheel surface velocity of $V_u = 13$ m/s and the two lower spectra were obtained with velocities of $V_u = 10$ and 8 m/s, respectively. The spectra can be decomposed into two phases. Both have in common a subspectrum corresponding to a face centered cubic (fcc) (Fe, Ni, B) phase. However, the additional phases have similar

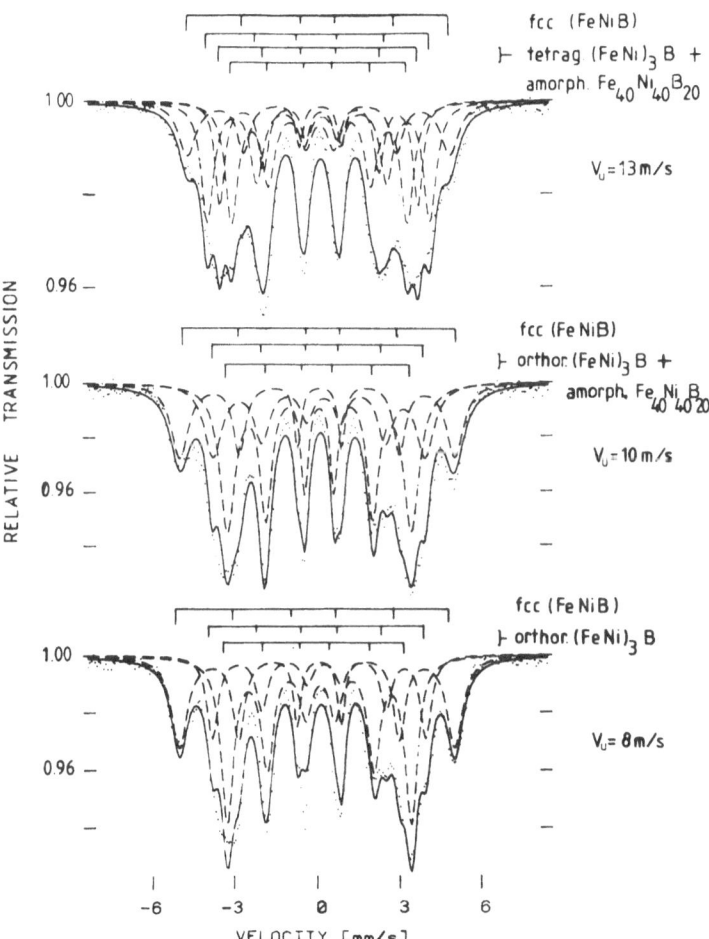

Fig. 9.20. Mössbauer spectra obtained from $Ni_{40}Fe_{40}B_{20}$ alloy quenched at different surface velocities of the rollers (V_u)

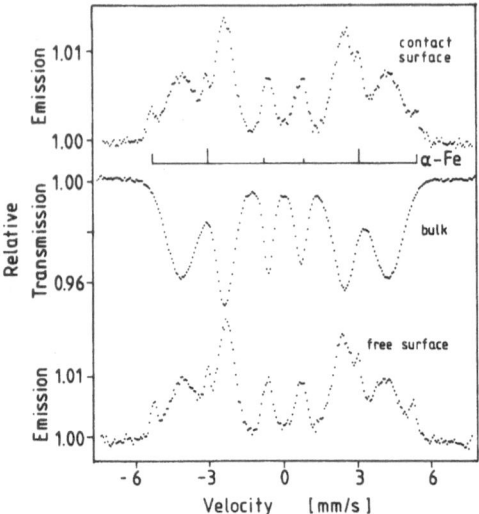

compositions $(Fe, Ni)_3B$, but different crystallographic structures: a tetragonal phase resulting in the quenching ($V_u = 13$ m/s) and an orthorhombic phase resulting in the slow quenching ($V_u = 10$ or 8 m/s). In these structures iron is in two or three distinguishable environments, respectively, as indicated in Fig. 9.20 by the stick diagram. It is of interest that controlled quenching rates generate the metastable tetragonal crystalline phase which is otherwise inaccessible by heat-treatment.

A very useful combination is that of conversion electron Mössbauer spectroscopy (CEMS) and γ-ray transmission Mössbauer spectroscopy, which makes simultaneous scanning of the surface and the bulk possible. By annealing one can produce a situation where the γ-ray transmission spectrum (scanning the bulk) exhibits no traces of crystallization, while the electron emission spectrum clearly shows – from both the contact and the free surface – new lines, indicating the formation of crystalline α-Fe. This is demonstrated in Fig. 9.21.

9.12 Simultaneous Triple-Radiation Mössbauer Spectroscopy (STRMS)

In the final section of this chapter a new type of Mössbauer methodology is presented. It consists basically of a combination of conversion electron Mössbauer spectroscopy (CEMS), conversion X-ray Mössbauer spectroscopy (CXMS) and transmission Mössbauer spectroscopy [9.25].

In the "classical" Mössbauer setup a source emits recoil-free γ-rays resulting from the nuclear transition from the excited state to the ground state. In the absorber the reverse transition process leads to the excited state in the same isotope. This Mössbauer resonance absorption is indicated by the bold arrows in Fig. 9.3 and is mostly measured as transmission Mössbauer spectroscopy (TMS). The de-excitation in the absorber occurs by means of conversion electrons and conversion X-rays (Here we omit the re-emitted γ-rays). Because of their common precursor, the Mössbauer resonance, both the electron radiation and the X-radiation carry effectively the information of a conversion

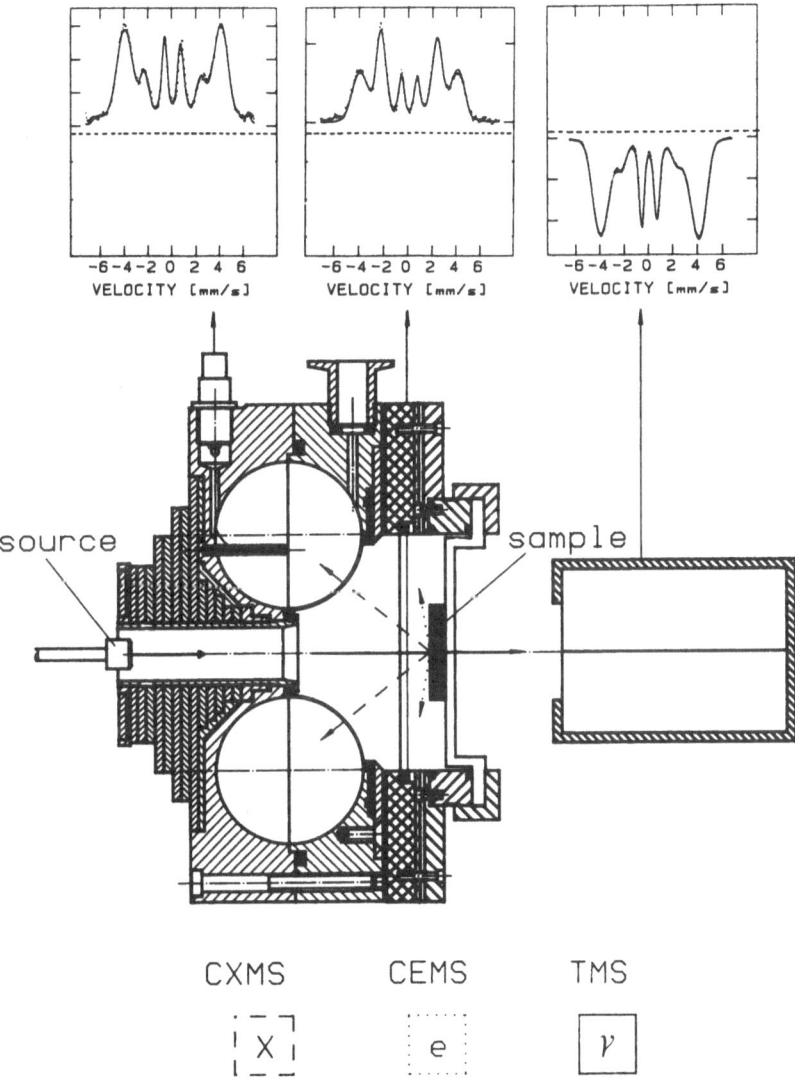

Fig. 9.22. Setup for simultaneous triple-radiation Mössbauer spectroscopy (STRMS) using back-scattering of conversion electron, conversion X-ray and transmission γ-rays. The counts from the three detectors are registered in separate multichannel analyzers

electron Mössbauer spectrum (CEMS) and of a conversion X-ray Mössbauer spectrum (CXMS). Both can be measured in the backscattering geometry.

An apparatus has been constructed to measure simultaneously three spectra corresponding to the three characteristic radiations: γ-rays, conversion electrons and conversion X-rays. The arrangement of the source, the absorber and the three counters for these radiations is shown in Fig. 9.22. The γ-rays from the source pass through the windows of the electron counter to reach the absorber (sample). The transmitted γ-rays are registered in the γ-ray counter. The re-emitted conversion electrons are backscattered into the electron counter. The conversion X-rays are backscattered through a window into a toroidal ("doughnut"-like) proportional detector which offers a large solid angle. Effective shielding minimizes direct source radiation.

The usefulness of the STRMS setup can be demonstrated with an amorphous ribbon of $Fe_{78}B_{13}Si_9$. In the as-quenched state the three spectra taken simultaneously by electrons, X-rays and γ-rays have a very similar appearance, as seen in Fig. 9.23. Thus $\Delta m = 0$ lines are strong in all three spectra, indicating that the spins are preferentially oriented in the plane of the ribbon.

After treatment by two subsequent laser pulses of an excimer laser with a pulse duration of 40 ns and an energy density of 15 mJ/mm^2 the three spectra

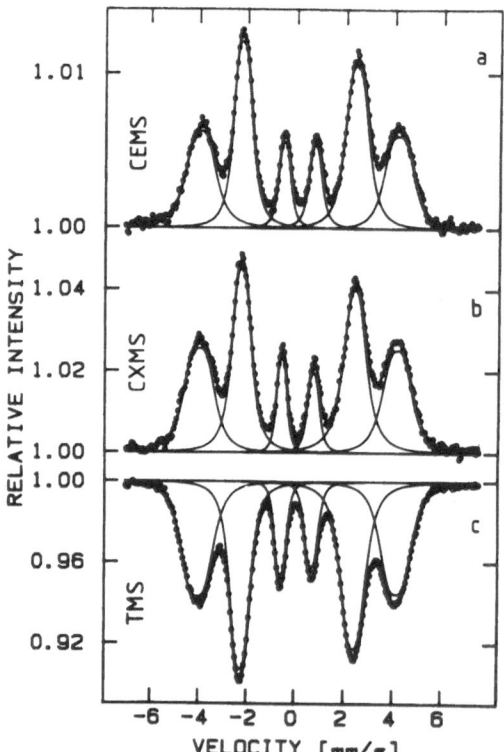

Fig. 9.23. STRMS of amorphous $Fe_{78}B_{13}Si_9$ at room temperature obtained by simultaneous CEMS, CXMS and TMS

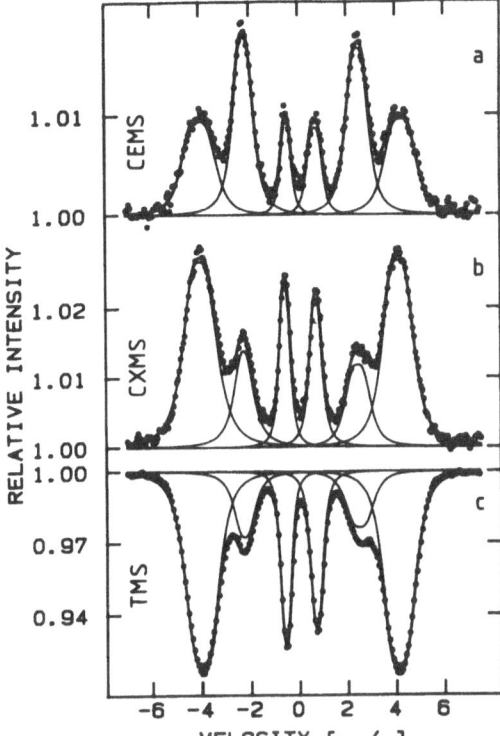

Fig. 9.24. STRMS of amorphous $Fe_{78}B_{13}Si_9$ after laser irradiation at room temperature, again obtained by simultaneous CEMS, CXMS and TMS. Note the intensity changes of the $\Delta m = 0$ lines by CXMS and TMS

change considerably by comparison, as seen in Fig. 9.24. In particular, the relative intensity of the $\Delta m = 0$ lines is still strong in the electron spectrum. On the other hand, the $\Delta m = 0$ lines become weak in the X-ray and γ-ray spectra. This indicates a spin reorientation from an in-plane magnetization to an orientation along the γ-ray propagation direction, that is, perpendicular to the amorphous ribbon. In this interpretation of the data one must realize that the electrons originating from the ^{57}Fe transition can escape from the sample only in the surface region within a range of about 1000 Å. This is illustrated on the left-hand side of Fig. 9.25. The X-rays, on the other hand, have an escape range of nearly two orders of magnitude higher ($\cong 10$ μm), so that a deeper part of the sample is observed. Finally, the γ-rays scan the whole sample, effectively measuring the bulk. Thus, the three radiations enable a discrete depth profile analysis.

The following interpretation can be given: by means of the laser irradiation a very thin surface layer hardens, causing by magnetoelastic effects a compressive stress. Because the material exhibits positive magnetostriction the spins tend to align perpendicular to the compressive stress, with the exception of the surface area. This is accomplished by the formation of closure domains, see right-hand side of Fig. 9.25.

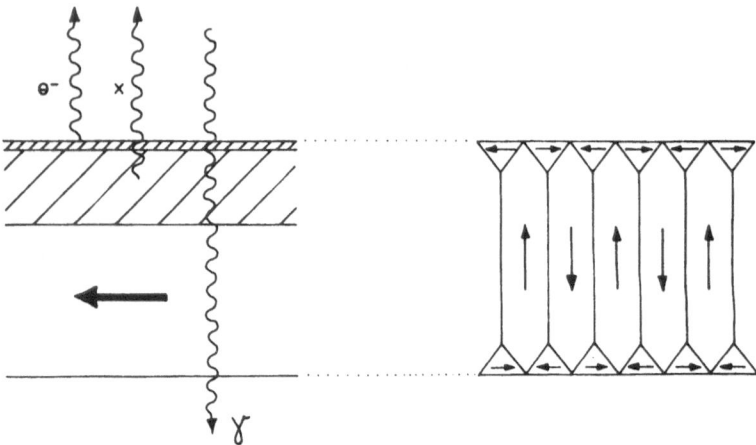

Fig. 9.25. Schematic cross-section of the ribbon sample. On the left, the ranges of the conversion electrons and the conversion X-rays which are able to escape from the sample are shown. γ-rays are transmitted. Spins are oriented in the plane of the ribbon before irradiation, indicated by the bold arrow. On the right, the suggested closure domains originating from the laser radiation are shown

The special features of this experiment should be summarized. From *one single nuclear event* – that is the transition from the nuclear first excited state of ^{57}Fe – three spectra can be obtained simultaneously:

(1) A recoil-free γ-ray is absorbed by exciting a nucleus. This transition is observed by transmission Mössbauer spectroscopy (TMS).

(2) The transition from the excited state to the ground state is accompanied by the emission of conversion electrons with small penetration depths ($\cong 1000$ Å). Observation in backscattered geometry leads to conversion electron Mössbauer spectroscopy (CEMS).

(3) Similarly, the nuclear transition also leads to conversion X-rays with larger penetration depths of about 10 μm. Again, in backscattering one measures conversion X-ray Mössbauer spectroscopy (CXMS).

In this STRMS experiment it is essential that the emission of conversion electrons and conversion X-rays occurs with isotopes having just experienced a foregoing γ-Mössbauer resonance. Needless to say, this is Mössbauer spectroscopy at its highest efficiency.

9.13 Quo Vadis?

The new techniques which have been developed during the past decades will play an ever increasing role in our laboratories and their range of applications will continue to grow, with team work and collaboration becoming more and more necessary for the comparison of results obtained with other tools of investigation.

In this chapter we have focused our attention on Mössbauer spectroscopy, a method which has gained easy entrance into all fields of science. One of the main reasons for its attractiveness is the existence and use of that special gift of nature, the isotope ^{57}Fe, with its favourable nuclear properties. However, the more frequent use of other isotopes might also be envisioned, for instance the rare earth isotopes.

In summary, the following three points should be emphasized:

First, Mössbauer spectroscopy has progressed from a simple tool – in fact, it was first demonstrated with a children's toy – to a sophisticated and advanced technique in terms of new counters, detectors, drive systems, and electronics.

Second, Mössbauer spectroscopy has proved to be highly suited for the disclosure of important facets of nature, particularly where new fields of investigation have opened up or new materials have been discovered. Quite naturally, Mössbauer spectroscopists are always among the first to get on any new bandwagon.

The third point is that not only did Mössbauer make a great discovery, he also foresaw the inherent potential of his effect, intuitively putting it into words at the conclusion of his speech in acceptance of the Nobel Prize on 10 December 1961: "We may therefore hope that this young branch of physics stands only at its threshold, and that it will be developed in the future, not only to extend the application of existing knowledge but to make possible new advances in the exciting world of unknown phenomena and effects" [9.26].

References

9.1   U. Gonser (ed.): *Microscopic Methods in Metals*, Topics Curr. Phys., Vol. 40 (Springer, Berlin, Heidelberg 1986)
9.2   R.L. Mössbauer: Naturwissensch. **45**, 538 (1958)
9.3   A.H. Muir, Jr., K.J. Ando, H.M. Coogan: *Mössbauer Effect Data Index* (Interscience, New York 1958–1965); J.G. Stevens, V.E. Stevens: *Mössbauer Effect Reference and Data Journal*, Vol. 1–13 (Mössbauer Effect Data Center, Asheville, NC 1977–1990)
9.4   H. Frauenfelder: *The Mössbauer Effect* (Benjamin, New York 1962)
9.5   I.J. Gruverman (ed.): *Mössbauer Effect Methodology*, Vol. 1–10 (Plenum, New York 1965–1976)
9.6   V.I. Goldanskii, R.H. Herber (eds.): *Chemical Applications of Mössbauer Spectroscopy* (Academic, New York 1968)
9.7   N.N. Greenwood, T.C. Gibb: *Mössbauer Spectroscopy* (Chapman & Hall, London 1971)
9.8   U. Gonser (ed.): *Mössbauer Spectroscopy I* and *II*, Topics Appl. Phys., Vol. 5 and Topics Curr. Phys., Vol. 25 (Springer, Berlin, Heidelberg 1975, 1981)
9.9   R.L. Cohen (ed.): *Application of Mössbauer Spectroscopy I* and *II* (Academic, New York 1976, 1980)
9.10 U. Gonser, H. Fischer: in *Mössbauer Spectroscopy*, ed. by U. Gonser, Topics Curr. Phys., Vol. 25 (Springer, Berlin, Heidelberg 1981) p. 99

9.11 A.H. Muir, Jr.: *Analysis of Complex Mössbauer Spectra by Stripping Techniques in Mössbauer Effect Methodology*, Vol. 4 (Plenum, New York 1968)

9.12 U. Gonser, C.J. Meechan, A.H. Muir, H. Wiedersich: J. Appl. Phys. **34**, 2373 (1963)

9.13 A. Bläsius, U. Gonser: J. Physique **C-6**, 397 (1976)

9.14 A. Brenner, D.E. Couch, E.K. Williams: J. Res. Natl. Bur. Stand. **44**, 109 (1950)

9.15 W. Buckel, R. Hilsch: Z. Phys. **132**, 420 (1952)

9.16 U. Gonser, B. Okkerse: J. Phys. Chem. Solids **7**, 55 (1958)

9.17 F.E. Fujita, U. Gonser: J. Phys. Soc. Jpn. **13**, 1068 (1958)

9.18 P. Duwez, R.H. Willens, W. Klement: J. Appl. Phys. **31**, 1136 (1960)

9.19 R. Pond, R. Maddin: Trans. Metall. Soc. AIME **245**, 2475 (1969)

9.20 H. Gleiter: Proc. 2nd Riso Int'l. Symp. on Metallurgy and Materials Science (1981) p. 15

9.21 X.L. Yeh, K. Samwer, W.L. Johnson: Appl. Phys. Lett. **42**, 242 (1983)

9.22 G. Le Caer, J.M. Dubois, H. Fischer, U. Gonser, H.-G. Wagner: Nucl. Instrum. Methods **B5**, 25 (1984)

9.23 U. Herr, J. Jing, R. Birringer, U. Gonser, H. Gleiter: Appl. Phys. Lett. **50**, 472 (1987)

9.24 U. Gonser, C.T. Limbach, F. Aubertin: J. Non-Cryst. Solids **106**, 395 (1988)

9.25 U. Gonser, P. Schaaf, F. Aubertin: Hyperfine Interactions **66**, 95 (1991)

9.26 Les Prix Nobel (Nobel Foundation, Stockholm 1961–1962) p. 136

9.27 H. Wiedemann: *Particle Accelerator Physics* (Springer, Berlin Heidelberg 1993)

9.28 E. Gerdau, R. Ruffer, H. Winkler, W. Torksdorf, C.P. Klages, J.P. Hannon: Phys. Rev. Lett. **54**, 835 (1985)

9.29 A. Jayaraman: Rev. Mod. Phys. **55**, 65 (1983)

9.30 M.P. Pasternak, S. Nasu, K. Wada, S. Endo: Phys. Rev. B **50**, 6446 (1994)

9.31 S. Nasu: Hyp. Int. **90**, 59 (1994)

9.32 S. Yamamoto, X. Zhang, H. Kitamura, T. Shioya, T. Mochizuki, H. Sugiyama, M. Ando, Y. Yoda, S. Kikuta, H. Takai: J. Appl. Phys. **74**, 500 (1993)

9.33 A. Heiming, K.H. Steinmetz, G. Vogl, Y. Yoshida: J. Phys. F. **18**, 149 (1988)

9.34 Y. Yoshida, W. Miekeley, W. Petry, R. Stehr, K.H. Steinmetz, G. Vogl: Material Science Forum **15–18**, 301 (1987)

9.35 F.E. Fujita, Y. Yoshida: Proc. Int'l Conf. Perspectives in Materials Science (Bangalore, India 1995) to be published in J. Mat. Sci.

10 Further Progress

Progress in sciences and technologies in our world is extraordinarily fast, probably incomparably fast in the history of human kind. Among others, remarkable developments have been done within these thirty years especially in the fields of electronics and computer, space science and astronomy, and medicine and biology. In addition, progress in the field of materials science and technology, producing various advanced materials to support today's human life and, at the same time, the developments of the above-mentioned other fields, must be noticed.

After having published the first edition of this book, it was already recognized that some additional descriptions to introduce further developments in all of the fields described and discussed were necessary because of the unexpectedly fast progress mentioned above. In the second edition, therefore, it was decided to add Chap. 10 as an addendum to introduce briefly the most recent developments in each subject. Each section in this chapter corresponds to each preceding chapter with the same number. All the contributors were requested to make their description as short and dense as possible, since this chapter was planned to be a simple addendum in order not to deteriorate the original style of the book. For a further understanding to keep up the newest knowledge and information, readers are recommended to search the additional references cited in the original chapter. It must be emphasized again that, as mentioned in the conclusion of Chap. 6, the purpose of this book is to introduce the underlying physical ideas, i.e., the ground stream, of a big flow in the development of advanced materials. This implies that the above-mentioned newest knowledge and information do not imply technological applications, but still the physics of the newest materials. Readers who wish to know about the practical or engineering applications of newer materials are recommended to study more specialized articles on technology.

10.1 Electronic Structure and Magnetism of Transition Metal Systems

Various attempts to improve the Local Density (LD) functional approach to *ab initio* calculations of electronic structure have been made with notable success in extending its capability of quantitative prediction. One is to correct the self-interaction of electrons, which has not been properly taken into account in

LD. Readers are referred to [2.70] for recent developments following those in [2.7, 38]. In addition, [2.71] describes another hopeful approach which may enable us to extend the calculation to transition-element oxides (Sects. 2.3.4 and 2.5).

In relation to Cu and Ni compounds such as oxides and sulfides an additional reference [2.72] is given; it is quoted in Fig. 2.8 of Sect. 2.5 as an unpublished work. It suggests the existence of a metallic phase of Ni compounds in which electrons are highly correlated to keep a well-developed magnetic moment of Ni atoms.

One of the remarkable developments in the calculation of impurity electronic structure of ferromagnetic transition-metal systems (Sect. 2.3.2) has been discussed in [2.73]. It is the optimization of the atomic arrangements which surrounds an impurity.

Calculations on ferromagnetic disordered alloys (Sect. 2.3.3) have extensively been carried out recently [2.74, 75]. Fe alloys with interstitial light elements (Sect. 2.3.5) provided a subject which has attracted some recent attention. In particular, iron-based new magnetic materials containing nitrogen as an interstitial component opened a field called nitromagnetics [2.76], although the enhanced magnetism in $Fe_{16}N_2$ reported in [2.40, 77] is not conclusive yet, both experimentally and theoretically. A calculation on iron with light interstitials such as B, C and N has been carried out within the framework of the density-functional approach [2.78]. While it supports the general discussion of the enhancement of the magnetic moment of iron atoms distant from the interstitials in Sect. 2.3.5, it concludes that in the case of a Fe–N system Fe atoms neighboring N can also have an enhanced magnetic moment due to partial ionization in contrast with the cases of Fe–B and Fe–C. Here, Fe atoms close to interstitials have a magnetic moment smaller than in the case of pure iron. Though implications of the results need to be examined more carefully, the calculation may hint the possibility of a new phase in the Fe–N system.

Artificially made materials opened a new field in material science. The giant magnetoresistance found in Fe–Cr multilayers on a substrate has attracted much attention [2.79]. Also monolayers of ferromagnetic metals on a substrate such as noble metals have been the subject of intensive recent studies. An extensive review of theoretical efforts for the electronic structure and magnetism of monolayer systems has been given in [2.80].

The above-mentioned topics have rather arbitrarily been selected examples of recent developments.

10.2 Structure Characterization of Solid-State Amorphized Materials by X-Ray and Neutron Diffraction

The study of amorphous solid structures has rapidly been progressing towards characterization of not only metallic alloys but also of nonmetallic systems such as semiconducting elements, oxides and so on.

The structural change from crystalline Se to amorphous Se induced by mechanical milling at room temperature was observed by *Fukunaga* et al. [3.45] using pulsed neutron diffraction. This conversion is closely related to a medium-range conformation change between ring and chain. The mechanical milling of elemental C results in the partial evolution of the valence electron configuration sp^2 to sp^3 [3.46]. *El-Eskandarany* et al. [3.47] have demonstrated that the multilayer structure of an agglomerated powder prepared in the early stage of mechanical alloying is converted to an amorphous structure by heating it at a temperature far below the crystallization temperature. This exothermic reaction called "thermally assisted solid-state amorphization" has been found for a series of pairs of transition metals (Ti, Zr, Nb, Ta) and aluminum metal.

The low-energy excitation, often called the Boson peak, was found at an energy transfer range between 2 to 3 meV for metal-metalloid amorphous alloys, such as Pd–Si [3.48], Pd–Ge [3.49] and Pd–Ni–P [3.50], by inelastic pulsed neutron scattering. Since the Boson peak is observed in a low momentum-transfer range far below the main diffraction peak, the low-energy excitation has to be correlated to a collective motion of atoms included within the medium-ranged nanometer scale spatial area. The Boson peak drastically disappears when the structure relaxation of an amorphous alloy occurs below the crystallization temperature.

The structure of Si–C–Ti–O fibers prepared by pyrolysing polytitanocarbosilane fibers as an organic precursor consists of two contributions; one is the anisotropic fiber structure in the range of 10 to 100 nm in length, while the other is the isotropic distribution of β-SiC nanoclusters embedded in an amorphous matrix [3.51]. The latest measurement of small-angle X-ray scattering by *Wang* et al. [3.52] shows that the very high mechanical strength of the Si–C–Ti–O fibers essentially originates from the formation of β-SiC nanoclusters which have a sharp interface against the amorphous matrix.

10.3 Recent Progress in Nanophase Materials

The area of nanophase materials has expanded explosively since the writing of the original Chap. 4. This explosion is manifested in the number of papers published in Nanostructured Materials, a new international journal dedicated to materials with nanometer-scale structures, and in numerous other papers in more traditional journals. Three international conferences on nanostructured materials have already been held, including the last one in July 1996, and their Proceedings have appeared as separate volumes of Nanostructured Materials. A NATO advanced Study Institute dedicated to nanophase materials was held in 1993 [4.109] and books on the subject are forthcoming [4.110].

Our knowledge of the synthesis, structure, properties, as well as the utility of nanophase materials has expanded considerably in the past few years owing to the increased activity in this field worldwide. It may be useful here to highlight some areas in which particular progress has been made and let the reader search the cited literature for further examples.

A wide range of new methods for the synthesis of nanophase materials, and especially chemical methods for the formation of nanoscale powders that can be consolidated to make them, are now becoming readily available [4.109–111]. Many of these methods should yield materials that will find useful application in research and eventually in practice. The gas condensation method has been scaled up successfully to produce tonnage quantities of nanophase materials for commercial use [4.112]. Activities in understanding and developing the processing of nanophase materials to form useful parts still lag behind the synthesis capabilities, although some progress is now being made. For example, it has been demonstrated that net-shape forming of nanophase ceramics is indeed a feasible process yielding fully dense parts with good shape definition and mechanical behaviour.

The atomic scale structures of nanophase metals and ceramics have come into much clearer focus during the past few years through a wide range of careful studies reviewed elsewhere [4.113]. The essential features are (1) equiaxed nanoscale grains with random crystallographic orientations that result from grain-boundary sliding mechanisms that are operative during consolidation, (2) grain boundaries that are fundamentally similar in atomic structure to those observed in coarse-grained polycrystals, and (3) porosity on a variety of length scales that can be removed or retained, if high-surface-area materials are desired.

The unique properties of nanophase materials are also becoming far better understood. It is now rather evident that completely new areas of property space can indeed be accessed by the nanostructuring of matter, and that this extends to all material properties (electrical, magnetic, mechanical, optical, etc.) of interest. A few useful examples can be given here. The high nonlinear voltage-current response of ZnO, used worldwide in voltage-stabilizing varistors, can be engineered over much wider ranges by using nanophase ZnO in pure [4.114] and doped [4.115] forms. The area of giant magnetoresistance, with so much potential for a variety of important applications in magnetic recording, is developing rapidly through the clever use of nanostructuring [4.116, 117]. The fundamental reasons underlying the unique mechanical behaviour of a wide range of nanophase materials, including metals, intermetallics and ceramics, is now becoming more clear through a rather extensive body of recent experiments reviewed elsewhere [4.118]. These show clearly that nanophase metals are strengthened and nanophase ceramics are rendered more ductile as the grain size is decreased, while nanophase intermetallics exhibit a more complex transitional behaviour. The ability to alter the optical response of nanoscale powders by size control has been known for some time [4.119], but quite recently the commercial application of this concept to ultraviolet scattering and absorption in nanophase TiO_2-loaded sun screens [4.112] is making a significant impact. While there is only room for a few such examples in this brief update, it is now abundantly clear that nanophase materials will continue to be an area of strong research interest in the coming years and that these unique new materials will make a significant impact on the world in which we live.

10.4 Further Progress in the Theory of Intercalation Compounds

After completing the manuscript for the first edition, important progress has been made in a couple of fields of intercalation compounds of transition-metal dichalcogenide. One is the microscopic theory of the influence of intercalation on superconductivity of 2H–NbS$_2$ and on the phonon anomaly (lattice instability) of 2H–NbSe$_2$ and 2H–NbS$_2$ [5.73–75]. Another one is the experimental study of the electronic structure of 1T–TiS$_2$ and its intercalation compounds M$_x$TiS$_2$ (M: transition metal) by photo emission and X-ray absorption spectroscopy [5.76–78].

2H–NbSe$_2$ shows a CDW transition that accompanies the structural transition (lattice instability), which is driven by complete softening of the Σ_1 phonon at $q = (2/3)\Gamma M$. The frequency softening (phonon anomaly) of the Σ_1 mode around $q = (2/3)\Gamma M$ is caused by the strong electron-phonon interaction and the frequency renormalization of the Σ_1 phonon is not large enough to cause the lattice instability. Both NbS$_2$ and NbSe$_2$ become superconductors at $T_C = 6$ K and $T_C = 7.3$ K, respectively.

By using the electron-phonon interaction and the renormalized phonon frequencies of NbS$_2$, *Nishio* et al. [5.73, 74] have calculated the superconducting spectral function and evaluated the transition temperature T_C by solving the linearized Eliashberg equation. The calculated value T_C of NbS$_2$ agrees in the order of magnitude with the observed one. Then, they have studied the effects of intercalation on the phonon anomaly of NbSe$_2$ and NbS$_2$, and on the superconductivity of NbS$_2$ in the framework of the rigid-band approximation [5.73, 75]. Their results are briefly summarized as follows:

In the case of doping acceptor-type intercalants into NbS$_2$, the point of phonon anomaly in q-space is moved to $q = \Gamma M$ from $q = (2/3)\Gamma M$. Further, reflecting an increase of the electronic density of states at the Fermi level, the frequency renormalization is enhanced and the CDW transition temperature is expected to increase. The same phenomena are seen for the phonon anomaly in NbS$_2$, and hence doping of acceptors into NbS$_2$ may cause a lattice instability which corresponds to $q = \Gamma M$. In case of doping donor-type intercalant, on the other hand, the frequency renormalization is weakened in both NbSe$_2$ and NbS$_2$ and the CDW transition in NbSe$_2$ may disappear by donor-type intercalation. As for the superconducting transition temperature T_C of NbS$_2$, the donor-type intercalant lowers T_C as observed in the case of alkali-metal doping, while the acceptor-type intercalant raises T_C as long as a lattice instability does not take place.

10.5 Various New-type Carbon Materials

As a recent development in the study of carbon materials, discovery of various types of fullerenes C_{60}, C_{70} etc. [6.25], and carbon nanotubes [6.26] must be introduced, in addition to the content of Chap. 6. In the study of the former, the formation of thin films, single crystals and C_{60}–C_{70} alloys have been realized and their electronic properties [6.27], phase-transition, mechanical behaviour, intercalation and lattice defect structures [6.28] were observed and measured. For instance, by changing the conditions of intercalation, C_{60} thin films change their electrical conductivity from insulator to semiconductor, and to superconductor [6.29].

Theoretical investigations on the structures of single and multi-layered nanotubes predict that their conductivity changes periodically from metallic to semiconducting, and *vice versa*, with the change of the chiral vector, which determines the amount of twist of the tube [6.30]. In addition to their interesting mechanical behaviour, carbon nanotubes exhibit various odd properties and will be a useful advanced material in parallel with fullerenes.

In the study of amorphous and nanocrystalline materials, the triangular relationship between the crystalline, nanocrystalline and amorphous states is often considered to discuss their structural transitions [6.31]. In the case of carbon, fullerene and carbon nanotube must be added to the already discussed diamond, graphite and amorphous structures, making a pentagonal relationship, for better understanding and further development of carbon new materials.

It is worthy of note that most recently a substantial improvement, or almost a different new method, appeared in the technique to make diamond by chemical vapour deposition (CVD) which was introduced in Sect. 6.5.3. In the usual CVD method, mixing of reactant gases, including hydrogen, the gas pressure and temperature, etc. must be carefully controlled to make suitable hydrogen radicals for diamond formation. In the new method invented by *Hirose* [6.32], however, no hydrogen is needed but only alcohol or ether is evaporated and thermally decomposed. The resultant is deposited on a substrate under simple conditions. Even drinking alcohol, like whisky, beer and sake, can be changed to high quality diamond particles at a very fast growth rate. Another surprisingly simple method to produce diamond particles found by *Hirose* is the deposition of carbon from the reducing part of an oxy-acetylene combustion flame [6.33]. These new techniques open not only a new way for the production of artificial diamond, but also seem to suggest a hidden mechanism of the formation of natural diamond without any sophisticated conditions in the universe; the simple conditions in *Horose*'s method must be found commonly in the space. Structure changes and phase changes to create matters in the space are not always irrelevant to the physics and the technologies for the creation of advanced materials in our world.

More recently, *Banhart* [6.34] obtained the surprising result when studying an onion-type fullerene under an electron microscope that the fullerene structure starts to be transformed into diamond. A similar transformation can be achieved with strong irradiation with an ion beam.

10.6 New Findings in Ordered Structures

Recently, there has been renewed interest in the behaviour of lattice vacancies in certain ordered phases. The classical problem here is that of constitutional vacancies, first discovered in B2-NiAl by *Bradley* and *Taylor* in 1937 [7.159]. NiAl has a broad homogeneity range, and *Bradley* and *Taylor* discovered, by a critical comparison of densities and lattice parameters, that the Al-rich phase, $Ni_{50-x}Al_{50+x}$, accommodates its stoichiometry by introducing 'constitutional' vacancies on the Ni sublattice; the adjective denotes that the vacancy concentration is determined by the composition, not by temperature, and that the concentration may reach several percent. Recent experimental research has confirmed the old findings in good detail [7.160]. Early attempts to explain constitutional vacancies in terms of the electron-to-atom ratio failed, and their origin (confirmed in a number of other phases, such as CuGa) has remained a mystery. Older attempts of interpretation were surveyed by *Cahn* [7.161]. Now *Cottrell* [7.162] has provided what appears to be an entirely satisfactory explanation, in terms of the modern theory of cohesion in metals. He compared the total cohesive energy (in terms of the numbers and energies of different kinds of nearest-neighbour bonds) for a non-stoichiometric alloy with and without constitutional vacancies, and found that the vacancies lower the total free energy. A little later [7.163], he extended his theory to Al–Ni–Cu alloys, the constitutional vacancy content of which was first studied experimentally as a function of composition in 1938, and was able to make sense of these old findings.

Another aspect of vacancies in ordered alloys has arisen from the greatly enhanced interest, for technological reasons, in ordered iron-aluminum alloys, especially those near the FeAl (B2-type) composition. In an important paper, *Chang* et al. [7.164] established by lattice parameter measurements that the Al-rich FeAl contains constitutional vacancies proportional to the deviation from stoichiometry. In addition, they found that the microhardness increases linearly with the square root of the vacancy concentration (this is a large effect). In fact, these observations apply to both Fe-rich and Al-rich alloys. In Fe-rich Fe–Al, a range of research outlined by *Munroe* [7.165], going back to the 1970s, demonstrates that exceptionally high concentrations of thermal vacancies are present. They can be modified by heat-treatment following quenching, with a concomitant large reduction in the hardness. As *Munroe* showed, the effect of heat-treatment on hardness can be moderated by the addition of ternary additives, notably nickel. Research on vacancies in FeAl is burgeoning.

Another field which has greatly expanded in the last few years is the study of mechanical disordering of intermetallics, notably by milling in ball-mills. A series of specialized conferences [7.166, 167] has been devoted to this recondite topic and, very recently, an excellent survey article by *Bekker* et al. has appeared [7.168]. This survey describes some unexpected findings, notably the creation of magnetic spin-glasses in certain milled intermetallics such as near-stoichiometric AuFe and CuMn. Recent research by *Le Caër* in Nancy [7.169] on ferromagnetic (L2₁) Heusler phases, Ni_2MnSn and Co_2MnSn, reveals that quenching from high temperatures partly disorders them without destroying the ferromagnetism. However, ball-milling disorders them in a different way, apparently, with a replacement of ferromagnetic order by a spin-glass structure. This is one of a growing number of cases where thermal and mechanical disordering appears to lead to different structural changes.

Finally, it is important here to record the publication of an outstanding 2-volume overview of the entire field of ordered intermetallic compounds; the first, larger, volume is devoted to basic science, the second to engineering applications. Between the two volumes, 101 researchers have covered the whole field in an exemplary fashion. These volumes, edited by *Westbrook* and *Fleischer* [7.170], will be the standard reference for many years to come.

10.7 Recent Developments in High-Resolution and High-Voltage Electron Microscopy

The field of high-resolution electron microscopy has seen remarkable progress in recent years: (1) decreasing the electron wavelength by increasing the accelerating voltage, (2) increasing the coherence of the electron beam, and (3) excluding the inelastic scattering of electrons from the image formation are major routes towards higher resolution.

The first mentioned attempt aims at a high point resolution with the advantage of a straightforward interpretation of structure images taken under the Scherzer optimum focus conditions. 1250 kV high-resolution electron microscopes (HREM) installed in Tsukuba and Stuttgart proved to reach a point resolution of 0.1 nm and enabled us to observe oxygen atoms in high-T_C superconducting materials.

A second improvement has been achieved by an electron microscope equipped with a field-emission gun. In the improved microscope, the coherence of the electron beam is excellent, and extremely-high-contrast images are formed. So, we can detect the weak contrast produced by light elements in the structure image.

Third, the inelastic-scattering electrons involved in the scattering process can be eliminated from the image formation with an energy filtering Transmission Electron Microscope (TEM) [8.96]. The resolution limit in the zero-loss image becomes high, because the chromatic aberration reduces in the contrast transfer function shown in Fig. 8.7 b.

Finally, a new trend in high-resolution electron microscopy must be pointed out. A scanning transmission electron microscope forms a focused electron probe of atomic dimension, say 0.2 nm. This capability allows us to form an incoherent z-contrast image [8.97]. The intensity of the coherent electrons produced by Rutherford scattering depends strongly on the composition of the material through the z^2 dependence of the scattering cross section, z representing the atomic number. The incoherent electrons are collected by an annular detector and used to form the z-contrast image.

If the electron probe, whose size is sufficiently fine compared to the separations of the atomic columns, scans across thin crystalline material, it will map out the location of each atom column. Furthermore, the intensity of each column in the image will directly reflect its composition. Therefore, the z-contrast electron microscope provides a new view of materials on the atomic scale. That is, the image of atomic-structure composition can be interpreted without the need for any preconceived model structure.

After the first edition, in which the research field of high-voltage electron microscopy (HVEMy) was mentioned in Sect. 8.5, further applications of Ultra-HVEM have extensively been carried out on the following subjects:

(1) Amorphization induced by high-energy electron irradiation has been examined on various materials of more than 60 intermetallic compounds and 30 ceramics [8.98]. Based on the experimental results, a general rule for the crystalline-amorphous transition was obtained. The rule is applicable to other amorphization processes, such as liquid quenching, vapour deposition, mechanical mixing and alloying and others.

(2) Chemical amorphization of various alloys by absorbing hydrogen and oxygen gases was investigated by *in situ* experiments using an environmental cell. The above-mentioned general rule on the amorphization process was successfully applied.

(3) Foreign-atom implantation by electron irradiation has been carried on various materials, for instance, Au into Si crystals, Si and Ge into Al crystals, etc. The atoms can be implanted as a mass into the matrix along a desired direction when the cross-section of the former element is sufficiently large. The mechanism of this phenomenon seems to be closely related to electron channeling [8.99].

(4) Structure and behaviour of various atom clusters have been investigated, and anomalous phenomena such as spontaneous alloying (or mixing) [8.100], sudden shape changes and others have been discussed [8.101, 102].

(5) Three-dimensional observation of microstructures in IC devices has been carried out, and the break-down mechanism of aluminum lead-wires was clarified [8.98].

Most of the above-mentioned investigations were done by 3 MV HVEM of the Osaka University. This 3MV HVEM was replaced in June 1996 by a new 3 MV Ultra-HVEM to be operated at 3.5 MV at maximum. Further new applications are expected not only in material science but also in natural science and industries.

10.8 New Directions in Mössbauer Spectroscopy

Undoubtedly, one of the most remarkable progresses in the Mössbauer spectroscopy of today is the use of synchrotron-radiation source instead of radioactive isotope sources of specific nuclides. When an electron beam is warped by a magnetic field, it radiates electromagnetic waves, the wavelength of which depends upon the acceleration energy and the strength of the magnetic field (or the curvature of the orbit) [9.27]. Therefore, the synchrotrons and electron storage rings inevitably radiate electromagnetic waves from the bent part of their orbits. For instance, a storage ring for 2.5 GeV electrons with a radius of 8.8 m produces strong white rays ranging from 0.5 Å to the infrared region. Especially, from a suitably designed undulator inserted in a linear part of the orbit, a well collimated high-brilliant electromagnetic beam can be taken out and monochromated to, for instance, 14.4 keV to be used as a Mössbauer source to study ^{57}Fe.

In the case of ^{57}Fe measurement using synchrotron radiation, the absorption spectra as resolvable as those obtained from an ordinary ^{57}Co source were already observed by *Gerdau* et al. [9.28]. Many sophisticated experiments have been done by this group in Hamburg. They show a large possibility to use synchrotron radiation for Mössbauer spectroscopy of any desired appropriate nuclides, which are not always available from the radioactive isotope sources.

Development of the Mössbauer measurements under high pressure is also noticeable. A small handy Diamond Anvil Cell (DAC) can produce high pressures of the order of 100 GPa, which are almost hydrostatic [9.29]. A thin Mössbauer specimen of about 1 mm^2 in area is sandwiched between two diamond anvils with ruby fine particles for pressure calibration. The assembly is compressed by a screw-and-lever pinch mechanism and penetrated by γ-rays from a radioactive source placed very close to the assembly. With this DAC-type apparatus, many Mössbauer measurements have been carried on iron, magnetite [9.30, 31] and other materials. New phases and structures were found. A new attempt to combine the DAC high-pressure apparatus with synchrotron radiation to observe the nuclear resonant scattering under high pressure was successfully done [9.32].

Another important extension in the world of Mössbauer spectroscopy is the measurement at high temperatures, which sounds somewhat contradictory because of the general rule that the recoil-free fraction in γ-ray absorption will easily be deteriorated by thermal vibrations of atoms in the specimen when the temperature is raised. In fact, people had scarcely paid attention to high-temperature Mössbauer spectroscopy. By careful considerations and experiments, however, *Vogl* et al. [9.33, 34] obtained clear Mössbauer spectra of iron and its alloys at temperatures elevated to above 1000 °C, even in the range of δ-Fe which is around 1500 °C. Likewise, Ti–Fe, NiAl–Fe, Ti$_3$Al–Fe and other intermetallic compounds have been studied and their structural changes, magnetic changes and diffusion processes were precisely analyzed [9.35]. According to the capability of high-temperature measurements, Mössbauer spectroscopy has become more useful in the study of metallic, intermetallic and ceramic new materials.

Subject Index

Springer Series in *Materials Science*

Advisors: M. S. Dresselhaus · H. Kamimura · K. A. Müller
Editors: U. Gonser · R. M. Osgood Jr. · H. Sakaki
Managing Editor: H. K. V. Lotsch

* The 2nd edition is available as a textbook with the title: *Laser Processing and Chemistry*

Springer
and the
environment

At Springer we firmly believe that an international science publisher has a special obligation to the environment, and our corporate policies consistently reflect this conviction.
We also expect our business partners – paper mills, printers, packaging manufacturers, etc. – to commit themselves to using materials and production processes that do not harm the environment. The paper in this book is made from low- or no-chlorine pulp and is acid free, in conformance with international standards for paper permanency.

 Springer